NEW AND EMERGING PLANT VIRUSES

The Threat to Food Security

NEW AND EMERGING PLANT VIRUSES

The Threat to Food Security

Edited by

Alireza Golnaraghi, PhD

Rajarshi Kumar Gaur, PhD

AAP | APPLE ACADEMIC PRESS

First edition published 2025

Apple Academic Press Inc.
1265 Goldenrod Circle, NE,
Palm Bay, FL 32905 USA

760 Laurentian Drive, Unit 19,
Burlington, ON L7N 0A4, CANADA

CRC Press
2385 NW Executive Center Drive,
Suite 320, Boca Raton FL 33431

4 Park Square, Milton Park,
Abingdon, Oxon, OX14 4RN UK

Library and Archives Canada Cataloguing in Publication

CIP data on file with Canada Library and Archives

Library of Congress Cataloging-in-Publication Data

CIP data on file with US Library of Congress

ISBN: 978-177-491-658-2 (hbk)
ISBN: 978-1-77491-659-9 (pbk)
ISBN: 978-1-00349-748-6 (ebk)

About the Editors

Alireza Golnaraghi, PhD
Assistant Professor, Department of Plant Protection, Science and Research Branch, Islamic Azad University, Tehran, Iran; Researcher, Department of Earth, Ocean & Atmospheric Sciences, University of British Columbia, Vancouver, British Columbia, Canada; Researcher, Department of Biodiversity, BoomZista Institute, Vancouver, British Columbia, Canada

Alireza Golnaraghi, PhD, is currently working as Assistant Professor in the Department of Plant Protection, Science and Research Branch, Islamic Azad University. He is also a researcher in the Department of Earth, Ocean and Atmosphere Sciences, University of British Columbia, Vancouver, British Columbia, Canada, and in the Department of Biodiversity, BoomZista Institute, Vancouver, British Columbia, Canada. His PhD focused on biological, serological, and molecular characteristics of a Tospovirus isolated from potato and its detection through modern techniques, i.e., antibody microarrays. Dr. Golnaraghi has made significant efforts and contributions on viruses and endophytic microbes, especially in wild plants in natural ecosystems. He has published over 50 national/international papers, authored several books, contributed in writing two book chapters, and presented more than 65 papers in national and international conferences. In 2017, he joined a famous environmental microbiology/virology group at the Department of Earth, Ocean and Atmosphere Sciences (University of British Columbia, Canada) and, since then, has been working on algal viruses together with a well-known scientist (Prof. Curtis Suttle). With the help of international researchers and talented students, he established a not-for-profit corporation in Canada in 2020 (BoomZista Institute; boomzista.org) that focuses on various issues of the planet in an effort to disseminate nature-based sciences to people all over the world, especially children and young adults.

Rajarshi Kumar Gaur, PhD
Professor, Department of Biotechnology, Deen Dayal Upadhyaya Gorakhpur University, Gorakhpur, Uttar Pradesh, India

Rajarshi Kumar Gaur, PhD, is currently serving as a Professor and Head of the Department of Biotechnology at Deen Dayal Upadhyaya Gorakhpur University, Gorakhpur, Uttar Pradesh, India. He earned his PhD in 2005, focusing on the molecular characterization of sugarcane viruses such as mosaic, streak mosaic, and yellow luteovirus. He was awarded a MASHAV fellowship by the Government of Israel for his postdoctoral studies and conducted research at The Volcani Centre, Israel, and Ben-Gurion University, Negev, Israel. In 2007, he received a Visiting Scientist Fellowship from the Swedish Institute Fellowship, Sweden, and worked at Umea° University, Umea°, Sweden. Additionally, he was granted a postdoctoral fellowship from International Center for Genetic Engineering and Biotechnology (ICGEB), Italy, in 2008. Dr. Gaur has made significant contributions to the field of sugarcane viruses and has published 130 national and international papers, authored 22 edited books, and presented approximately 50 papers at national and international conferences. He has been honored as a Fellow of the Linnaean Society, Fellow of the Royal Society of Biology, Fellow of the Society of Plant Research, Fellow of the Society of Applied Biology (FSAB), Fellow of the International Society of Biotechnology (FISBT), and Fellow of the Indian Virological Society. He has received numerous other awards, including the Prof. B.M. Johri Memorial Award from the Society of Plant Research (SPR), the Excellent Teaching Award from Astha Foundation, Meerut; the UGC Research Teacher Award and the Young Scientist Award 2012 in Biotechnology from the Society of Plant Research (SPR); and the Scientific and Applied Research Centre Gold Medal Award 2011 for Outstanding Contribution in the Field of Biotechnology. Dr. Gaur is also a member of the National Academy of Science India (NASI). He has visited several laboratories in the United States, Canada, New Zealand, United Kingdom, Thailand, Sweden, and Italy. Currently, he is involved in managing various national and international grants and international collaborative projects related to plant viruses and disease management.

Contents

Contributors

Víctor Albores-Flores
Institute of Biosciences, Autonomous University of Chiapas, Tapachula, Chiapas, Mexico

Alangar Ishwara Bhat
ICAR–Indian Institute of Spices Research, Kozhikode, Kerala, India

Bhavin Bhatt
Shree Ramkrishna Institute of Computer Education and Applied Sciences, Faculty of Science, Sarvajanik University, Surat, Gujarat, India

Milena Georgieva Bozhilova-Sakova
Institute of Animal Science, Department of "Genetics, Breeding, Selection, Reproduction, and Biotechnologies of Farm Animals," Kostinbrod, Bulgaria

Sayyad Ali Raza Bukhari
Department of Biotechnology, University of Sargodha, Sargodha, Pakistan

Ma. Guadalupe Bustos-Vázquez
Mante-Center Multidisciplinary Academic Unit, Autonomous University of Tamaulipas, Cd. Mante, Tamaulipas, Mexico

Shailja Chauhan
Department of Plant Biotechnology, School of Biotechnology, Madurai Kamaraj University, Madurai, Tamil Nadu, India

Aparna Chodon
Department of Plant Biotechnology, School of Biotechnology, Madurai Kamaraj University, Madurai, Tamil Nadu, India

V. Devappa
Department of Plant Pathology, College of Horticulture, Bangalore; University of Horticultural Sciences, Bagalkot, Karnataka, India

Ivona Vassileva Dimitrova
University of Forestry, Agronomy Faculty, Department of Plant Protection, Sofia, Bulgaria

Rajarshi Kumar Gaur
Department of Biotechnology, Deen Dayal Upadhyaya Gorakhpur University, Gorakhpur, Uttar Pradesh, India

Alireza Golnaraghi
Assistant Professor, Department of Plant Protection, Science and Research Branch, Islamic Azad University, Tehran, Iran; Researcher, Department of Earth, Ocean & Atmospheric Sciences, University of British Columbia, Vancouver, British Columbia, Canada; Researcher, Department of Biodiversity, BoomZista Institute, Vancouver, British Columbia, Canada

Shridhar Hiremath
Department of Plant Pathology, College of Agriculture, University of Agricultural Sciences, GKVK, Bangalore, Karnataka, India

Khushboo Jain
Department of Biotechnology, Vigyan Bhawan, New Campus, Mohanlal Sukhadia University, Udaipur, Rajasthan, India

Arjunan Jeevalatha
ICAR–Indian Institute of Spices Research, Kozhikode, Kerala, India

P. Lekshmipriya
Department of Plant Biotechnology, School of Biotechnology, Madurai Kamaraj University, Madurai, Tamil Nadu, India

Hermilo Lucio-Castillo
Mante-Center Multidisciplinary Academic Unit, Autonomous University of Tamaulipas, Cd. Mante, Tamaulipas, Mexico

Lalit Mahatma
Department of Plant Pathology, N.M. College of Agriculture, Navsari Agricultural University, Navsari, Gujarat, India

Mahsa Mansourpour
Department of Plant Protection, College of Agriculture and Natural Resources, University of Tehran, Karaj, Iran

M. Mantesh
Department of Plant Pathology, College of Agriculture, University of Agricultural Sciences, GKVK, Bangalore, Karnataka, India

Humberto Martínez-Montoya
Mante-Center Multidisciplinary Academic Unit, Autonomous University of Tamaulipas, Cd. Mante, Tamaulipas, Mexico

Avinash Marwal
Department of Biotechnology, Vigyan Bhawan, New Campus, Mohanlal Sukhadia University, Udaipur, Rajasthan, India

Mukesh Meena
Laboratory of Phytopathology and Microbial Biotechnology, Department of Botany, Mohanlal Sukhadia University, Udaipur, Rajasthan, India

M. Mohanraj
Department of Plant Pathology, College of Horticulture, Bangalore; University of Horticultural Sciences, Bagalkot, Karnataka, India

Chitra Nehra
Department of Biotechnology, School of Life Sciences, Mody University, Lakshmangarh, Sikar, Rajasthan, India

Gopal Pandi
Department of Plant Biotechnology, School of Biotechnology, Madurai Kamaraj University, Madurai, Tamil Nadu, India

Nikolay Manchev Petrov
New Bulgarian University, Sofia, Bulgaria

U. Premchand
Department of Plant Pathology, College of Horticulture, Bangalore; University of Horticultural Sciences, Bagalkot, Karnataka, India

Punam Ranjan
Department of Botany, Patna Science College, Patna University, Bihar, India

Neha Rauniyar
Department of Botany, Deen Dayal Upadhyaya Gorakhpur University, Gorakhpur, Uttar Pradesh, India

C. N. Lakshminarayana Reddy
Department of Plant Pathology, College of Agriculture, University of Agricultural Sciences, GKVK, Bangalore, Karnataka, India

Muhammad Saeed
Technical University Kaiserslautern-Landau, Phytopathology, Department of Biology, Paul-Ehrlich-Straße 22, Kaiserslautern 67653, Germany

Rodolfo Torres-de Los Santos
Mante-Center Multidisciplinary Academic Unit, Autonomous University of Tamaulipas, Cd. Mante, Tamaulipas, Mexico

Niayesh Shahmohammadi
Department of Plant Protection, College of Agriculture and Natural Resources, University of Tehran, Karaj, Iran

K. S. Shankarappa
Department of Plant Pathology, College of Horticulture, Bangalore; University of Horticultural Sciences, Bagalkot, Karnataka, India

Abhishek Shukla
Department of Entomology, N.M. College of Agriculture, Navsari Agricultural University, Navsari, Gujarat, India

Achuit Kumar Singh
Crop Improvement Division, ICAR–Indian Institute of Vegetable Research, Jakhini (Shahanshahpur), Varanasi, Uttar Pradesh, India

Deepa Srivastava
Department of Botany, Deen Dayal Upadhyaya Gorakhpur University, Gorakhpur, Uttar Pradesh, India

Mariya Ivanova Stoyanova
Institute of Soil Science, Agrotechnologies and Plant Protection (ISSAPP) "Nikola Pushkarov," Agricultural Academy, Sofia, Bulgaria

Shalini Tailor
Department of Biotechnology, Vigyan Bhawan, New Campus, Mohanlal Sukhadia University, Udaipur, Rajasthan, India

Jebasingh Tennyson
Department of Plant Sciences, School of Biological Sciences, Madurai Kamaraj University, Madurai, Tamil Nadu, India

R. Ivonne Torres-Acosta
Mante-Center Multidisciplinary Academic Unit, Autonomous University of Tamaulipas, Cd. Mante, Tamaulipas, Mexico

Daniel Trujillo-Ramírez
Mante-Center Multidisciplinary Academic Unit, Autonomous University of Tamaulipas, Cd. Mante, Tamaulipas, Mexico

V. Venkataravanappa
Division of Crop Protection, ICAR–Indian Institute of Horticultural Research, Bangalore, Karnataka, India

Preface

The threat posed by newly emerging or re-emerging viruses is constantly increasing in frequency and posing potential threats even in new regions that were previously free from such diseases, despite intensive efforts to manage and prevent plant viruses and their potential vectors in crop production processes. In all of the damages caused by these viruses, we can find virus-diversifying events, appearance, and dissemination of efficient vectors, weather events, and, most importantly, human activities. On the other hand, the conventional method of surveying individual symptomatic agricultural plants has hampered the detection of plant viruses.

Recent research has unequivocally shown that plant viruses may be found in a wide range of settings, including natural vegetation, vectors, water, ice, and soil, as well as on the faces of animals, people, and organisms that eat plant virus vectors. However, due in part to the absence of effective diagnostic tools to detect such viruses in challenging matrices and low titers, the ecological role of plant viruses outside of their host plants and the potential implications to food security has not been thoroughly examined. It appears that thousands of new viruses will still be found in the future. These hitherto unknown viruses may serve as crucial building blocks for the recombination and pseudo-recombination of known harmful plant viruses, resulting in the evolution of more virulent diseases, and possibly increasing the complexity of the plant's microbial population. The development of management techniques to reduce crop losses will greatly benefit from an understanding of the complexity and heterogeneity of viral quasi-species present in natural habitats, as well as the causes of the observed genetic variation.

Plant viruses are widespread and can be found in a variety of places. They are regarded as a serious threat to sustainable and productive agriculture globally, as they jeopardize food security. Due to changes in land use, significant exchange of plant materials, introduction of new plants to native ecosystems, and decrease in biodiversity, human activities have increasingly enhanced the possibility of emergencies and outbreaks of new viruses worldwide. This, in turn, highlights the requirement for the application of multiplex detecting systems for the accurate detection of a greater number of viruses, especially in quarantine programs. Alternatively, a broad-spectrum or universal detection of viruses may be required when no prior knowledge of the target pathogen exists. Early detection of plant viruses has significant consequences for disease forecasting programs and is crucial for predicting and monitoring their epidemics. The potential emergence of novel plant virus species may, in turn, highlight the need for the development of novel plant virus management methods in the future.

To address the unknown world of plant viruses and emphasize their impact on agricultural production and food security, the editors of the multi-authored book undertook the very important task of reviewing the progress made and the challenges ahead in this field. *New and Emerging Plant Virus Sources: The Threat to Food Security* covers various aspects of plant viruses in two parts: (I) Different Sources of Plant Viruses; and (II) Plant Virus-Host Interaction/Evolution. Many distinguished scientists from all around the world have shared their experiences and opinions in the area of science. This book sheds light on undiscovered plant viruses and provides information on the detection, host interaction, evolution, and management of these tiny creatures, which will undoubtedly attract the general public's attention to the area of plant viruses.

We hope readers of the book find it useful in balancing their thoughts about the mysterious world of plant viruses and their potential threats to plant production.

—Alireza Golnaraghi, PhD
Department of Plant Protection, Science and Research Branch, Islamic Azad University, Tehran, Iran; Department of Earth, Ocean & Atmospheric Sciences, University of British Columbia, Vancouver, British Columbia, Canada; Department of Biodiversity, BoomZista Institute, Vancouver, British Columbia, Canada

—Rajarshi Kumar Gaur, PhD
Deen Dayal Upadhyaya Gorakhpur University, Gorakhpur, Uttar Pradesh, India

PART I

Different Sources of Plant Viruses

CHAPTER 1

Tracking of Plant Viruses in Different Sources: A Huge Gap Between Estimated and Known Diversity

MAHSA MANSOURPOUR,[1] NIAYESH SHAHMOHAMMADI,[1] and ALIREZA GOLNARAGHI[2–4]

[1]*Department of Plant Protection, College of Agriculture and Natural Resources, University of Tehran, Karaj, Iran*

[2]*Department of Plant Protection, Faculty of Agricultural Sciences and Food Industries, Science and Research Branch, Islamic Azad University, Tehran, Iran*

[3]*Department of Earth, Ocean & Atmospheric Sciences, University of British Columbia, Vancouver, British Columbia, Canada*

[4]*Department of Biodiversity, BoomZista Institute, Vancouver, British Columbia, Canada*

ABSTRACT

Plant viruses have been studied for more than a century; however, agricultural and horticultural plants are conventionally the most studied sources of plant viruses. Nevertheless, modern detection techniques have revolutionized plant virology by revealing the abundance of plant viruses in a wider variety of sources. Wild plants and vectors are proven to be unlimited sources of plant virus diversity, serving as platforms for the emergence and re-emergence of plant diseases worldwide. Aside from that, plant viruses are found in diverse environmental samples to a great extent. In this chapter, we explore the ever-growing list of plant viruses found in less-studied/less-known sources.

New and Emerging Plant Viruses: The Threat to Food Security.
Alireza Golnaraghi and Rajarshi Kumar Gaur (Eds.)

Moreover, the enormous previously undiscovered genetic diversity of newly discovered plant viruses in sources that have recently come to light will be discussed. We will conclude by considering the potential ecological impact of plant viruses from unknown sources on agricultural ecosystems and global food security.

1.1 DIFFERENT SOURCES FOR DETECTION OF PLANT VIRUSES

1.1.1 HOST PLANTS

1.1.1.1 CROP AND HORTICULTURAL PLANTS: CONVENTIONAL SOURCES OF PLANT VIRUSES

Plant viruses are obligate intracellular entities that depend on the host's protein-synthesizing machinery. They consist of one or more nucleic acid template molecules, either RNA or DNA (deoxyribonucleic acid), enclosed in a protective coat protein (CP) (Hull, 2014). The tobacco mosaic virus (TMV) was the first virus discovered in 1898, marking the beginning of virology (Beijerinck, 1898). Viral infections can reduce photosynthesis, alter the metabolism, and affect resource allocation in host plants, leading to reduced growth (Prendeville, 2010). The loss of crops, horticultural varieties, and pasture plant production caused by plant viruses poses a significant threat to global food security (Prendeville, 2010). Consequently, extensive efforts have been dedicated to identifying and managing viruses in agricultural systems worldwide, with approximately 77% of plant viruses initially isolated from cultivated hosts (Lacroix, Renner, Cole, Seabloom, Borer, & Malmstrom, 2016; Susi, Filloux, Frilander, Roumagnac, & Laine, 2019; Farzadfar, Golnaraghi, & Pourrahim, 2002; Yazdani-Khameneh & Golnaraghi, 2017). It is worth noting that almost all plant viruses described in the past 120 years have been pathogens of important agronomic species (Roossinck, 2010). The world's food supply heavily relies on 13 major crops and horticultural species, including rice, wheat, maize, beet, cane, tomato, potato, sweet potato, cassava, bean, soybean, coconut, and banana (Hull, 2014). On the other hand, agriculture is one of the most important sources of income in many parts of the world, particularly in tropical and subtropical regions. Plant viruses are estimated to be responsible for over $30 billion in economic losses each year (Jones & Naidu, 2019). To ensure that humanity can feed the expected 10 billion people by 2050, it is essential to minimize the risk of crop diseases (Jones & Naidu, 2019). Thus, part of this focus is understandable due to the role of agriculture in human history (Stobbe & Roossinck, 2016). However, a large number of cultivated plant viral studies does not seem to be enough to give a complete picture of plant virus biodiversity (Roossinck, 2011). Nevertheless, viral metagenomics approaches have been used to explore and recover viral sequences in different contexts, including wild plants, viral vectors, and various environmental samples such as water, soil, etc. (Roossinck, 2012). Interestingly, these findings reveal the stability of plant viruses and explain the possible reason for their widespread distribution throughout the world (Roossinck, 2012).

1.1.1.2 WILD PLANTS: SOURCES OF EMERGING PLANT VIRUSES

Although most research historically has focused on viruses causing diseases in crop plants, recent investigations have revealed the abundance of viruses in wild plants, with an incidence rate of 60% (Lacroix, Renner, Cole, Seabloom, Borer, & Malmstrom, 2016; Roossinck, 2015). In a study conducted to evaluate the incidence of five different viruses in wild *Cucurbita pepo* plants, the prevalence of viral infection ranged from 4% to 74% (Prendeville, 2010). As wild populations typically exhibit greater genetic variability than agricultural systems, virus prevalence may also appear to be more variable in wild populations. Virus infection in wild plants affects different life traits of the host, such as plant growth, mortality, or seed production, and ultimately the fitness of different species in wild plant communities (Prendeville, 2010). Thus, the consequences of virus infection are mostly seen at the population level (Prendeville, 2010). Until recently, studies on viruses of wild plants were mostly limited to finding alternative hosts to the crop viruses, and they didn't go beyond the weeds around the crops (Cooper & Jones, 2006), mostly due to the difficulties involved. Viruses in wild plants are usually symptomless or only result in minor changes in the appearance of the plant (Cooper & Jones, 2006; García-Arenal, & Zerbini, 2019). Up to 80% of viral infections in wild plants can be visually unapparent (Prendeville, 2010), and the virus titer is usually low (García-Arenal, & Zerbini, 2019). They may play a role in their hosts' life cycle or even benefit them in special circumstances, perhaps because of the relatively long period that they have been in association with their host plants (Susi, Filloux, Frilander, Roumagnac, & Laine, 2019; Roossinck, 2012). For instance, symptomless infection with viruses can even increase the fitness of wild plants in comparison with the uninfected plants (Prendeville, 2010). In addition, systematic sampling of as many wild species as possible is necessary to gather enough data for analysis (García-Arenal, & Zerbini, 2019). However, metagenomic studies using deep sequencing approaches have been used to find previously undefined and unknown viruses in wild plants (Roossinck & García-Arenal, 2015). These types of reports are on the rise worldwide and will bridge the knowledge gap about viruses in wild plants. Today, it has become clear that most wild plants harbor viruses, specifically unknown viruses (Roossinck, 2011; Kyrychenko, Shcherbatenko, & Kovalenko, 2021). The majority of viruses found in natural systems are novel viruses, which are being uncovered by metagenomic approaches (Hasiów-Jaroszewska, Boezen, & Zwart, 2021). The challenge of viral metagenomics studies, however, is the lack of any universally conserved sequence, such as ribosomal RNA, to use for the discovery of new viruses in environmental samples (Roossinck, 2011, 2012). To overcome this problem, approaches like small interfering RNA (siRNA) sequencing or enriched virus-like particles (VLPs), and dsRNAs are used for the enrichment of samples for viruses (Roossinck, 2011, 2012). Studies of viruses in wild plants (virome) and vectors (vectorome) in natural ecosystems are crucial as they can influence the dynamics of disease in crops by acting as viral reservoirs or vectors, respectively (Susi, Filloux, Frilander, Roumagnac, & Laine, 2019; Hasiów-Jaroszewska, Boezen, & Zwart, 2021).

Viruses are the largest group of emerging entities (García-Arenal, & Zerbini, 2019). There are many ecological and genetic factors involved in the process of emergence, including virus infection of wild plants (García-Arenal, & Zerbini, 2019; Roossinck & García-Arenal, 2015). Emergent viruses are naturally multi-host viruses that have been maintained in the population of a reservoir host permanently. They encounter new hosts, adapt to them, and then spread increasingly in the population of the new hosts with efficient transmission ways (García-Arenal, & Zerbini, 2019; Roossinck & García-Arenal, 2015). Plant viruses can travel both ways between wild and cultivated plants, with potential consequences for both types of ecosystems (Hasiów-Jaroszewska, Boezen, & Zwart, 2021). However, wild plants have been demonstrated to serve as the reservoir host for emergent viruses in agricultural systems, since they often grow side by side with cultivated plants (Cooper & Jones, 2006; Roossinck & García-Arenal, 2015). Putative novel viruses/strains found in wild plants generally belong to the viral taxa with a wide host range and are transmitted by generalist vectors (Susi, Filloux, Frilander, Roumagnac, & Laine, 2019). Thus, by understanding the diversity of viruses in wild plants, we can better understand mechanisms that regulate pathogen populations, allowing us to predict when and where diseases will emerge (Susi, Filloux, Frilander, Roumagnac, & Laine, 2019).

The role of wild plants as reservoirs for viral emergence has been investigated for potyviruses (Gibbs, Trueman, & Gibbs, 2008; Nguyen et al., 2013) and geminiviruses (García-Arenal, & Zerbini, 2019). Phylogeographic studies have suggested that turnip mosaic virus (TuMV) and bean common mosaic virus (BCMV) first spread to crops and ornamentals from wild plant species (Gibbs, Trueman, & Gibbs, 2008; Nguyen et al., 2013), in the case of potyviruses. A member of the *Tymoviridae* family, frequently found in asymptomatic green milkweed plants, has been shown to cause symptoms in *Nicotiana benthamiana.* Over the years, this virus has been spreading in nature (Min et al., 2012), indicating potential danger to cultivated host plants. Interestingly, wild plants infected by geminiviruses, the most emergent plant viruses, are more prevalent in simplified anthropic habitats compared to highly biodiverse ecosystems (García-Arenal & Zerbini, 2019). Pollen-associated viruses, mostly belonging to the *Bromoviridae*, *Partitiviridae*, and *Secoviridae* families, are also more prevalent in regions surrounded by less natural and more anthropogenic manipulated environments (Fetters et al., 2022).

1.1.1.3 VEGETATIVE AND REPRODUCTIVE MATERIALS: THE POORLY-STUDIED PLANT VIRUS SOURCES

To meet the demands of international trade, vegetative propagation is widely used to produce large quantities of plants. However, this can lead to the spread of viruses across national borders (Chauhan, Hamon, Rajakaruna, Webb, Payton, & Verchot, 2015). In many economically important plants such as cassava, potato, sugarcane, banana, sweet potato, onion, as well as most fruit and ornamental plants, viruses can be transmitted through plant vegetative materials such as tubers, rhizomes, stolons, corms, and bulbs (Sastry, 2013; Sastry & Zitter, 2014). Testing of plant materials seems to be crucial for bulbous plants like garlic, onion, and narcissus, which are propagated from bulbs (Sastry, 2013). Infected bulbs and bulbils can result in garlic plants with significantly lower yields due to the infection of multiple viruses, mostly from the families *Potyviridae* and *Flexiviridae* (Sastry, 2013; Sastry & Zitter, 2014). In onion, two major groups of viruses belonging to the *Potyvirus* and *Allexivirus* genera can infect bulbs and reduce the yield of infected plants (Lee, Lee, Wang, & Jan, 2021). Additionally, propagation of ornamental plants mainly through vegetative tissues provides easy and rapid pathways for viral transmission (Lee, Lee, Wang, & Jan, 2021). For example, the gladiolus plant propagates vegetatively, and viral infections can be passed down from generation to generation through corms (Katoch, Abdin, Ram, & Zaidi, 2003). Bean yellow mosaic virus (BYMV) and cucumber mosaic virus (CMV), among the most common gladiolus viruses, have been reported in different parts of the world (Katoch, Abdin, Ram, & Zaidi, 2003), making reliable detection of these viruses in corms of major importance (Stein, Levy, & Loebenstein, 1988). Similarly, saffron, a precious crop propagated vegetatively by corms (Parizad et al., 2019), can host various plant viruses such as nepoviruses, tospoviruses, and especially potyviruses, which can be easily dispersed through plant propagation (Parizad et al., 2018). Canna is grown from rhizomes, and viruses present in the rhizomes, such as potyviruses and badnaviruses, can emerge in the progeny plant (Chauhan, Hamon, Rajakaruna, Webb, Payton, & Verchot, 2015).

Vegetative propagation has been suggested as a possible reason not only for mixed virus infections but also for the low genetic diversity of some plant viruses, such as narcissus potyviruses (Ohshima et al., 2016; Valouzi, Shahmohammadi, Golnaraghi, Moosavi, & Ohshima, 2021). In potato cultivation, single or mixed virus infections are frequently reported and are responsible for major economic losses (Klerks, Leone, Verbeek, Van, & Schoen, 2001). Vegetatively propagated tubers transmit the majority of potato viruses (Kaiser, 1980). Therefore, planting virus-free tuber seeds can effectively reduce the damage (Agindotan, Shiel, & Berger, 2007; Zhang, Gao, He, Wang, Qian, Zhou, & Wu, 2020). Potato mop-top virus (PMTV) and tobacco necrosis virus (TNV), the causal agents of spraying in potato tubers, are spread by planting infected seed tubers (Davey, Carnegie, Saddler, & Mitchell, 2014; Arif, Torrance, & Reavy, 1994; Beuch, Persson, Edin, & Kvarnheden, 2014), with an 18%–54% rate of PMTV transmission from infected seed tubers to the daughter tubers (Davey, Carnegie, Saddler, & Mitchell, 2014), and a high occurrence of symptomless tuber infections by either PMTV or TRV (Beuch, Persson, Edin, & Kvarnheden, 2014; Latvala-Kilby, Aura, Pupola, Hannukkala, & Valkonen, 2009). Therefore, reliable and sensitive indexing techniques must be employed to assess the virus status of potato seeds (Agindotan, Shiel, & Berger, 2007; Zhang, Gao, He, Wang, Qian, Zhou, & Wu, 2020). Cassava mosaic disease (CMD), which is currently the most important factor limiting cassava production in many countries, is spread by stem cuttings used for plant propagation (Chikoti, Mulenga, Tembo, & Sseruwagi, 2019). Virus transmission *via* vegetative propagation is particularly important when dealing with viruses of fruit trees since their propagation usually involves asexual reproduction (Sastry, 2013). Many important viruses, including citrus tristeza virus (CTV) and cacao swollen shoot virus (CSSV), which cause devastating diseases throughout the world, are proven to spread through the host vegetative materials (Sastry, 2013). Vegetative transmission is the simplest way of virus transmission and could be one of the first modes of transmission to evolve in virus-plant interactions when plants did not produce any flowers, seeds, or fruits.

In addition to vegetative propagules, reproductive plant materials such as pollens and seeds can also serve as sources of plant viruses (Bhat & Rao, 2020). A virus can be transmitted *via* pollen in two ways: horizontally through infected flowers or vertically *via* infected seeds (Bhat & Rao, 2020; Card, Pearson, & Clover, 2007); the latter being more common (Card, Pearson, & Clover, 2007). Compared to other sources of plant material, pollen has fewer associated pathogens (Card, Pearson, & Clover, 2007). However, approximately 50 plant viruses, mostly members of the genera *Ilarvirus, Nepovirus, Sobemovirus, Idaeovirus,* and *Potyvirus,* are considered to be pollen-transmitted (Bhat & Rao, 2020). Seeds can carry virus infection through external contamination or infection of the embryo. About 18% of plant viruses are seed transmissible in one or more hosts (Lal, Pant, & Rani, 2015; Johansen, Edwards, & Hampton, 1994). For some virus-host combinations, seed transmission can be extremely high. For example, barley stripe mosaic virus (BSMV), which has no biological vector, relies on seed transmission as the major mechanism for virus dispersal. Pea seed-borne mosaic virus (PSbMV) has a seed transmission rate of up to 100% in peas, and seed transmission of BCMV in French beans can reach 83% (Sastry & Zitter, 2014). Four different viruses, soybean mosaic virus (SMV), CMV, tobacco ring spot virus (TRSV), and tobacco streak virus (TSV), are seed transmitted up to 100% in soybeans (Sastry & Zitter, 2014); frequent single or mixed virus infections have been reported in this crop (Golnaraghi et al., 2004). When seeds or vegetative plants are contaminated with viruses, the virus infections occur at an earlier stage in the seedling, causing more severe disease in the field. Furthermore, each infected seedling can serve as a reservoir for subsequent vector transmission (Sastry & Zitter, 2014). For example, even a 0.001 incidence of seed transmission of lettuce mosaic virus (LMV) in lettuce can threaten lettuce production due to subsequent spread by aphid vectors (Johansen, Edwards, & Hampton, 1994). Therefore, eliminating the virus and producing virus-free plant materials is a major management strategy for viral disease control. Methods such as meristem-tip culture, thermotherapy, electrotherapy, chemotherapy, and cryotherapy are used for the production of virus-free materials (Sastry & Zitter, 2014). Studies on the transmission of plant viruses through vegetative/reproductive means have primarily focused on crop plants. However, not all (crop) hosts have been evaluated for a specific virus. The process and rate of such transmissions depend on various factors, including the host, virus species/strain, and environmental conditions (Hull, 2014). While it remains a topic of debate, a few cases of seed transmission of non-seed transmissible viruses, such as tospoviruses and geminiviruses, have been reported (Groves, German, Dasgupta, Mueller, & Smith, 2016; Kil, Park, Choi, Kim, & Lee, 2017). The modes of transmission serve as efficient ways for overwintering/oversummering and survival of plant viruses in different seasons, as well as for their dispersal. It appears that the vegetative/reproductive transmission of hundreds of other plant viruses in different hosts, particularly in natural ecosystems, is yet to be identified. In contrast to the significance of seeds in plant virology, little is known about the pool of viral nucleic acids present in plant seeds and their potential role in virus evolution. To differentiate between viruses existing inside plants, referred to as 'plant virome' (Maclot et al., 2020), and those present in seeds, we propose the terms 'seedvirome' and 'seedviromics.' These terms describe viruses and viral-like nucleic acids carried by different parts of plant seeds, some of which are transferred to daughter plants.

1.1.1.4 GERMPLASMS: IMPORTANT SOURCES FOR WORLDWIDE DISSEMINATION OF PLANT VIRUSES

The term "germplasm" is applied to the seeds or other plant parts (vegetative propagules) that are maintained in collections to be used in crop breeding, research, and conservation efforts. As described in Section 1.1.1.3, seeds and plant vegetative propagules can harbor devastating viral diseases. Therefore, safe conservation and global movement of germplasm are crucial to minimize virus spread through germplasm export and import (Gillaspie, Hopkins, & Pinnow, n.d.; Kumar et al., 2021). Several studies have reported the prevalence of germplasm infection with important plant viruses worldwide. For example, virus infections of cowpea collections have been reported in different studies (Gillaspie, Hopkins, & Pinnow, n.d.; Kumar et al., 2021; Salem, Ehlers, Roberts, & Ng, 2010; Kateřina, Helena, & Karel, 2017). A survey conducted to analyze cowpea germplasm seed samples for seed-borne viruses revealed that 67% of samples were infected by at least one cowpea virus (Salem, Ehlers, Roberts, & Ng, 2010). Similarly, the Czech collection of shallots (*Allium cepa* var. *ascalonicum*) genetic resources has been shown to be infected with various viruses, with an incidence ranging from 53% to 93% for different genotypes (Kateřina, Helena, & Karel, 2017). In another survey, 75.3% of cherry and sour cherry germplasm in the country was found to be infected with different viruses (Přibylová, Lenz, Fránová, Koloniuk, & Špak, 2020). In different germplasm collections in Brazil, numerous viruses have been found to infect diverse host plants in several studies (Kitajima, 2020). Recently, the occurrence of various kinds of viruses has been confirmed in 54.6% of samples in a *Prunus* germplasm collection in the United States (Diaz-Lara, Stevens, Klaassen, Golino, & Al Rwahnih, 2020). Moreover, endornaviruses, persistent RNA viruses which are present in every cell and transmitted exclusively vertically with high efficiency to the progeny, have been found in germplasm used for cultivar development in the breeding of crops, such as bell pepper, common bean, and melon (Khankhum, 2016). The movement of germplasms around the world generally involves a risk of introducing pathogens, particularly those with no obvious symptoms, to a new environment (Thomas, 2015). Thus, research into finding viruses in different plant germplasms is relevant and informative from both phytosanitary and breeding perspectives in the worldwide collections that collect, maintain, study, and exchange germplasm (Salem, Ehlers, Roberts, & Ng, 2010). Also, such studies can minimize the risk of virus dissemination and future epidemics (Thomas, 2015).

1.1.2 VECTORS AS THE MOVING SOURCES OF PLANT VIRUSES

1.1.2.1 FUNGI AND FUNGI-LIKE ORGANISMS: THE OLD FRIENDS OF PLANTS

Plants are sessile organisms and, except for some instances of vegetative or reproductive transmission (*see* Sections 1.1.1.2 to 1.1.1.4), cannot transmit viruses. This means that plant viruses spread mainly through vectors of different types of organisms (Singh, Verma, & Varma, 2008). Among them, zoosporic fungi of the genus *Olpidium* (phylum *Olpidiomycota*) have been proven to act as the vector of plant viruses, mostly belonging to *Virgaviridae, Aspiviridae,* and *Tombusviridae* (Singh, Verma, & Varma, 2008; Sutela, Poimala, & Vainio, 2019), in two different ways: *in vitro* and *in vivo* transmission (Singh, Verma, & Varma, 2008; Navarro, Sanchez-Navarro, & Pallas, 2019). *In vitro* transmission is reported for viruses with icosahedral virions (*Tombusviridae*) and involves the acquisition of the virus by the virus-free zoospore from the surrounding environment, since the resting spore or sporangia are not infected. It is believed that through the interaction between the receptors in the zoospore membrane and the coat protein (CP) of the virions, virions are tightly and specifically adsorbed to the membranes (Singh, Verma, & Varma, 2008; Navarro, Sanchez-Navarro, & Pallas, 2019). Conversely, *in vivo* transmission occurs through the direct absorption of the virus by resting spores, sporangia, and zoospores produced as the fungus grows in the root (Navarro, Sanchez-Navarro, & Pallas, 2019). Other fungal-like virus vectors include plasmodiophorids in the genera *Polymyxa* and *Spongospora* (Sarwar, Aslam, Sarwar, & Iftikhar, 2020). There are 20 species of plasmodiophorid-transmitted plant viruses belonging to five genera: *Furovirus, Pecluvirus, Pomovirus* (family Virgaviridae), *Benyvirus* (family Benyviridae), and *Bymovirus* (family Potyviridae) (Sarwar, Aslam, Sarwar, & Iftikhar, 2020). Although these fungi-like organisms survive in soil, they can move and disseminate the soil-borne viruses through water or wind, resulting in the widespread occurrence and prevalence of these viruses in various regions (Sarwar, Aslam, Sarwar, & Iftikhar, 2020; Farzadfar, Pourrahim, Golnaraghi, & Ahoonmanesh, 2007).

Moreover, there are plant viruses that have recently been reported to naturally infect fungi, and *vice versa,* in a phenomenon known as cross-kingdom viral infection. An exceptional study has reported the natural infection of a phytopathogenic fungus, *Rhizoctonia solani*, with CMV. Most importantly, the transmission of CMV from infected to uninfected plants by *R. solani* has been demonstrated under laboratory conditions (Andika, Wei, Cao, Salaipeth, Kondo, & Sun, 2017). These findings suggest that plant viruses can spread *via* fungal transmission to a much greater extent than previously thought. Additionally, it has been demonstrated that a wild-type isolate of TMV can replicate in the phytopathogenic fungi *Colletotrichum acutatum* and persist for up to subcultures (Mascia et al., 2014). Similarly, a mutant of TMV is able to replicate and trigger the RNA interference (RNAi) mechanism in oomycete *Phytophthora infestans* (Mascia, Labarile, Doohan, & Gallitelli, 2019). Recently, it has been shown that taxonomically different plant viruses, including artichoke mottled crinkle virus (AMCV), potato virus X (PVX), potato virus Y (PVY), and CMV plus its satellite RNA, can replicate and persist in *C. acutatum*, at least in the first subculture. Artichoke Italian latent virus (AILN), AMCV, PVX, PVY, TMV, and CMV plus its satellite RNA can replicate and persist in *P. infestans*, at least in the first subculture (Mascia et al., 2019). On the other hand, mitoviruses – mycoviruses with small RNA (+) genomes that replicate in mitochondria – have been shown to infect several land plant species from both ferns and flowering plants (Nibert, Vong, Fugate, & Debat, 2018).

Phylogenetic studies support the idea of virus exchange between plants and fungi, and there are some virus families, such as *Partitiviridae* and *Endornaviridae*, with genera shared between the two kingdoms (Roossinck, 2015). The genera *Alphapartitivirus* and *Betapartitivirus* contain a mixture of fungal and plant viruses (Dolja & Koonin, 2018). The replication of *Penicillium aurantiogriseum partitivirus 1*, a mycovirus of a marine plant endophytic fungus, is demonstrated in the *N. tabacum* BY2 cell line (Nerva, Varese, Falk, & Turina, 2017). More interestingly, there are some newly-identified plant viruses suggested to be members of fungal virus families, such as *Totiviridae* and *Chrysoviridae*. More than half of the new viruses identified in metagenomics studies of wild plants belong to families that share fungal and plant viruses or those that are traditionally classified as fungal virus families (Roossinck, 2015). Likewise, fungal viruses have been identified with genomes related to plant viruses from the *Benyviridae, Betaflexiviridae, Virgaviridae, Tombusviridae,* and *Tymoviridae* families (Dolja & Koonin, 2018).

Plants are almost always associated with different types of pathogenic, endophytic, or mycorrhizal fungi. The world of plant-colonizing fungi is more diverse in wild plants grown in natural environments; many of them are unknown fungal species (Yazdani-Khameneh, Golnaraghi, & Wylie, 2019; Teimoori-Boghsani et al., 2020; Garcés-Ruiz, Senés-Guerrero, Declerck, & Cranenbrouck, 2019). So, there has been an ongoing complex interaction between plants and fungi for millions of years of evolution (Roossinck, 2018). Consequently, opportunities for virus transmission naturally exist between the two kingdoms and may have played a role in the evolution of several virus families (Roossinck, 2018). Therefore, it would not be surprising to find many other plant virus vectors if we deeply investigate the virome and vectorome (*see* Section 1.1.2.2) of plant-colonizing fungi in the future.

1.1.2.2 ARTHROPODS: THE DIVERSE ANIMAL WORLD ASSOCIATED WITH PLANTS

Viruses can efficiently move from cell to cell and systematically infect all organs and tissues of a single host (Blanc, Uzest, & Drucker, 2011). In the next step, they need effective ways to jump between hosts. Approximately 70% of viruses infecting plants use vectors, including organisms as diverse as plant-parasitic fungi and fungi-like organisms (*see* Section 1.1.2.1), plant root nematodes, and arthropods (Blanc, Uzest, & Drucker, 2011; Van, Hogenhout,

& Van, 1999), with arthropods playing the most important role in plant virus transmission (Dáder, Then, Berthelot, Ducousso, Ng, & Drucker, 2017). The most dominant arthropod vectors of plant viruses are from Thysanoptera (thrips), Coleoptera (chrysomelid beetles), Acarina (eriophyid mites), and four families of Homoptera (Aphididae, Aleyrpdidae, Cicadellidae, and Delphacidae) (Nault, 1997). The prevalence of arthropod-borne plant viruses in various hosts and regions is well documented (Golnaraghi, Shahraeen, Pourrahim, Farzadfar, & Ghasemi, 2002; Farzadfar, Pourrahim, Golnaraghi, & Ahoonmanesh, 2005, 2006, 2007). Interestingly, numerous arthropods feeding on plants are not listed as plant virus vectors (Hull, 2014). Plant viruses that use arthropods as their vectors have been detected within the vector's body or in the vector's mouthparts using diverse methods (Khelifa, 2019). The use of vectors as targets for virus discovery is an effective way to provide significant information about the diversity of plant viruses without the limitation of the collection of plant tissues. Such investigations allow us to find potential, even asymptomatic viral diseases present in a particular region and monitor the emerging viruses which may lead to future epidemics (Shahmohammadi, Mansourpour, & Golnaraghi, 2022). They have been proven particularly useful in the case of geminiviruses, which are difficult to detect in wild plants but may be readily identified by examining their whitefly vectors (Roossinck, 2012).

Conventional PCR and real-time PCR techniques were used to sensitively detect different plant viruses in their vectors (Khelifa, 2019; Kim, Cha, Kwon, & Maharjan, 2016; Olmos, Bertolini, Gil, & Cambra, 2005; Fabre et al., 2003; Nagata, Inoue-Nagata, Ávila, & Giordano, 2004; Barragan & Guzmán-Barney, 2014; Hasanvand et al., 2018). More recently, vector-enabled metagenomics (VEM), an innovative approach to discovering viruses, enables the discovery of viral particles directly from the insect vectors and provides insight into how viruses are circulating in various ecosystems (Stobbe & Roossinck, 2016; Claverie et al., 2018). Across time and space, insect vectors can naturally accumulate many viruses from different plants. In VEM, the insects are sampled from the plants, and metagenomic sequencing of viral nucleic acids is conducted (Ng et al., 2011). In the recent decade, this approach has been used in several studies to identify plant viruses within their insect vectors such as whiteflies (Ng et al., 2011; Rosario et al., 2015), leafhoppers (Fontenele et al., 2018; Kamali et al., 2017), aphids (Feng et al., 2017; Wamonje et al., 2017), psyllids (Nouri, Salem, Nigg, & Falk, 2016), and thrips (Thekke-Veetil et al., 2020), the predators of the vectors such as dragonflies (Rosario et al., 2013) and earwigs (Godinho, Paula, Varsani,

& Ribeiro, 2017), and herbivorous pests such as beetles (François et al., 2021). The majority of the sequences obtained in VEM are related to known viruses (Stobbe & Roossinck, 2016; Ng et al., 2011). Also, several numbers of unknown sequences have been identified in such studies (Ng et al., 2011).

Generally, our knowledge of insect (arthropod) viruses is biased towards species with economic and human health importance and, in a few cases, towards vectors of major crop diseases (François et al., 2021). On the other hand, while most studies are focused on all viruses inside the vector (virome), a term generally refers to those viruses replicating within a host (Virgin, 2014), vectorome studies in which the focus is on the viruses that use the vector for the transmission between hosts are mainly neglected. However, as described above, vectors can serve as outstanding reservoirs and potential nurseries for a wide variety of known and novel plant viruses (François et al., 2021), and vectorome studies are becoming an important part of diversity, biogeography, and evolution studies of plant viruses. The term 'vectorome' was first applied to describe replicating/non-replicating viruses carried by a viral vector that can be transmitted to non-homologous hosts but do not replicate in vector cells in many cases, e.g., non- or semi-persistent plant viruses (Valouzi, Hashemi, Wylie, Ahadiyat, & Golnaraghi, 2020). Three different strategies have been suggested to study the vectorome: (i) direct strategy in which the viruses within the vector are directly studied using different detection methods; (ii) semi-direct strategy involves transferring a possible vector to indicator (plant) hosts; and (iii) indirect strategy that can be conducted by putting indicator (plant) hosts in nature. In the two last strategies, the indicator (plant) hosts would subsequently be subjected to a variety of detection methods (Shahmohammadi & Golnaraghi, 2022). For all kinds of plant virus vectors, including fungi and arthropods, vectorome studies seem to be essential to extend our knowledge about possible species as vectors of plant viruses, the diversity of plant viruses, and potential risk of emerging or re-emerging viruses.

1.1.3 ENVIRONMENT

1.1.3.1 SOIL: A RICH RESERVOIR OF VIRAL DIVERSITY

Soil contains a vast diversity of viruses, some of which are probably unknown and capable of infecting plant hosts. New evidence shows that soils may contain various kinds of (RNA) viruses, most of which probably infect fungi.

They can also infect bacteria, plants, and animals (Starr, Nuccio, Pett-Ridge, Banfield, & Firestone, 2019). Few plant viruses in different genera can be found in and transmitted through the diverse world of soil. Soil-transmitted plant viruses are scattered throughout the world, and many of them cause significant economic losses to major crops. Soil-borne plant viruses always infect their hosts through roots or other underground structures, which may result in visible symptoms in the shoots due to pathological changes in root function (Roberts, 2014). The diversity of these viruses is severely underestimated and undersampled, although current measures of viral richness are higher for soils than for aquatic ecosystems (Starr, Nuccio, Pett-Ridge, Banfield, & Firestone, 2019; Williamson, Fuhrmann, Wommack, & Radosevich, 2017). Soil is critically important to terrestrial biodiversity and human civilization. However, information from aquatic systems suggests potential influences of viruses on nutrient cycles, food web interactions, and gene transfer. The impacts of viral activities on soil ecosystem services are poorly understood (Williamson, Fuhrmann, Wommack, & Radosevich, 2017). The abundance of viruses in soils can range from below detection limits in hot deserts to over 1 billion per gram in wetlands, which is strongly influenced by water availability and temperature (Williamson, Fuhrmann, Wommack, & Radosevich, 2017).

Soils encompass numerous microbes, including fungi and fungus-like organisms, which may serve as unknown vectors of known/unknown plant viruses. Soil-borne plant viruses can be classified into two categories: (i) viruses with abiotic transmission; and (ii) viruses with biotic transmission (Narayanasamy, 2011). While fungi, plasmodiophorids, or plant parasitic nematodes act as vectors for the spread of these viruses, in some cases, the virus may be released from infected roots and transmitted abiotically through the soil without the assistance of a vector. Typically, these viruses cannot rely on their vectors to transport them over long distances, usually ranging from millimeters to centimeters, and their distribution is often localized. The long-term stability of such viruses in the soil makes effective control almost impossible, and the soil itself can act as a reservoir of infection (Roberts, 2014). Soil movement during cultivation, caused by the wheels of agricultural machinery, water streams, or through planting material, is a significant means of spreading over large distances to new sites (Roberts, 2014; Pleše, Juretić, Mamula, Polák, & Krajačić, 1996). Rainfall leads to increased water percolation into the soil and raises groundwater levels, resulting in the washout of the soil surface layer and the transfer of viruses into environmental waters (Keswick & Gerba, 1980). The causal agent of rhizomania, beet necrotic yellow vein virus (BNYVV), can be transported during the washing of sugar beet roots harvested from infected sites and may contaminate water that is used to irrigate previously virus-free sites (Roberts, 2014), which may result in a higher virus disease incidence in sugar beet farms near sugar beet factories (Farzadfar, Pourrahim, Golnaraghi, & Ahoonmanesh, 2007). Rice yellow mottle virus (RYMV) has been readily detected in irrigation water flowing from infected rice plants using enzyme-linked immunosorbent assay (ELISA) and its infectivity confirmed by inoculation to susceptible rice cultivar. Inoculation of a susceptible rice cultivar has led to a high disease incidence of 86.6% (Traore et al., 2008), possibly showing the relative stability and high titers of the virus in the irrigation water. It has also been shown that transplantation of rice seedlings into RYMV-contaminated soil can lead to plant infection as well (Traore et al., 2008). The highly stable soil-borne viruses may freely remain in the soil and infect susceptible host plants without being transmitted by a vector (or their biological vectors for plant-to-plant are unknown) (Narayanasamy, 2011; Pleše, Juretić, Mamula, Polák, & Krajačić, 1996). Viruses can survive for weeks or sometimes for months in the soil and retain their infectivity depending on virus properties and environmental factors like soil type, level of moisture, pH reaction, temperature, the content of organic substances, and the biological activity of the medium (Nienhaus & Castello, 1989; Kegler, Kontzog, & Kleinhempel, 1991). Infectivity of tobamoviruses can persist for long periods in soil. Experiments showed that tomato mosaic virus (ToMV) infectivity persisted for two years in dry soil and root debris in soil for up to 22 months (Broadbent, 1976). However, RYMV survives for a short time in contaminated soil, and its infectivity decreases rapidly and is completely lost within 14 days (Traore et al., 2008).

Soil-born plant viruses identified from the soil include the families *Secoviridae* (*Nepovirus*, *Cheravirus*, *Sadwavirus*), *Potyviridae* (*Bymovirus*), *Ophioviridae* (*Ophiovirus*), *Tombusviridae* (*Carmovirus*, *Dianthovirus*, *Tombusvirus*, *Necrovirus*), *Virgaviridae* (*Furovirus*, *Pecluvirus*, *Pomovirus*, *Tobravirus*), *Benyviridae*, and the genus *Varicosavirus* (Narayanasamy, 2011; Lanter, McGuire, & Goode, 1982). Soil-borne viruses with abiotic transmission belong to the genera *Tobamovirus*, *Potexvirus*, *Tombusvirus*, and *Sobemovirus* which their stability and titer are extremely high (Narayanasamy, 2011). Members of the genus *Tombusvirus,* such as tomato bushy stunt virus (TBSV), cymbidium ringspot virus, and petunia asteroid mosaic virus (PetAMV), do not transmit through a vector and are exuded from infected roots into the soil, where they are transmitted abiotically. They are presumed to infect plants through microscopic root wounds or, in the case of cultivated crops, through cutting implements (Traore et al., 2008).

TMV and ToMV are the most stable viruses with a very wide host range (Komuro & Iwaki, 1969). Soil can play an important role in virus transmission (Lanter, McGuire, & Goode, 1982; Tan et al., 1997), and crops exposed to these two viruses are at great risk of infection when planted in areas where the viruses persist in soil sources (Agrios, 1988). Studies have shown that tomato roots become infected when seedlings are grown in soil that contains ToMV-infected plants left over from the previous crop (Pares, Gunn, & Keskula, 1996). Greenhouse experiments have confirmed the high level of abiotic soil transmission of ToMV, which can be detected in young plant shoots after successful transmission (Broadbent, 1965).

Another tobamovirus, tomato yellow fruit mosaic virus (TYFMV), has also been shown to be horizontally transmitted from soil to plants through wounds of delicate surface cells created by abrasion against gravel or other soil particles during normal root growth through soil (Jones, 2018).

Planting bait and test plants (*Chenopodium amaranticolor* and *C. quinoa*) in infested soils and detecting the virus serologically in soil and water in three forest districts in central Croatia showed that 40 out of 70 soil/root samples contained the TNV (Pleše, Juretić, Mamula, Polák, & Krajačić, 1996). Serological examination of the test plants also showed TMV and ToMV infection of the plants (Pleše, Juretić, Mamula, Polák, & Krajačić, 1996). In the baiting strategy for plant virus discovery, which is somewhat similar to the semi-direct strategy of vectorome studies (*see* Section 1.1.2.2), healthy seedlings will be transplanted into soils potentially contaminated with a particular virus. These plants may subsequently become infected with the virus through tiny root wounds made during planting (Jones, 2018). TMV, ToMV, and PVY (potyvirus) were identified in soils in forest ecosystems in Europe after cultivating bait plants (Büttner, & Nienhaus, 1989; Coutts & Jones, 2015). Several European forests are located where agricultural land once stood; therefore, viruses isolated from forest trees, soils, and waters from these areas might have originated from infected crop residues (Fillhart, Bachand, & Castello, 1998).

Previous studies have shown that certain viruses, such as the southern bean mosaic virus (SBMV; *Sobemovirus*) in common bean, cymbidium ringspot virus (*Tombusvirus*) in *N. clevelandii*, TBSV, and galinsoga mosaic virus (*Gallantivirus, Tombusviridae*) in *Celosia argentea* and *Galinsoga parviflora*, can be frequently transmitted abiotically in soil. This mode of transmission is likely more common than previously thought (Kleinhempel & Kegler,

1982; Teakle, 1986; Hollings, Komuro, & Tochihara, 1975; Shukla, Shanks, Teakle, & Behncken, 1979). A similar study was conducted for TMV, ToMV, and cucumber green mottle mosaic virus (CGMMV), where root infection from contaminated soil led to the infection of tobacco, tomato, and cucurbit crops, respectively (Broadbent, 1976; Hollings, Komuro, & Tochihara, 1975; Broadbent & Fletcher, 1963; Gooding & Todd, 1976). Transplanting cucurbit plants into soil infested with gayfeather mild mottle virus (GMMV) and cucumber fruit mottle mosaic virus (CFMMV) (*Tobamovirus*) resulted in root infection and subsequent infection of the plants (Antignus, Lachman, Pearlsman, & Koren, 2005; Li, Liu, & Gu, 2016).

1.1.3.2 WATER: PLANT VIRUSES IN THE CONNECTION PATH OF TERRESTRIAL AND AQUATIC ECOSYSTEMS

Viruses are the most abundant and diverse biological entities in environmental waters. They play a crucial role in the ecological networks of water and impact the global biochemical cycle and community composition dynamics. Previous studies have primarily focused on viruses that infect bacteria, with less attention given to plant, animal, and human-infecting viruses (Rastrojo & Alcamí, 2018). Through viromics, viral RNA and DNA of known or suspected terrestrial plant viral pathogens have been detected in various environments such as wastewaters (Aw, Howe, & Rose, 2014; Cantalupo et al., 2011), reclaimed waters (Rosario, Nilsson, Lim, Ruan, & Breitbart, 2009), freshwater lakes (Djikeng, Kuzmickas, Anderson, & Spiro, 2009), polar freshwater lakes (Aguirre De, López-Bueno, Pearce, & Alcamí, 2015; López-Bueno et al., 2009), rivers (Fernandez-Cassi et al., 2017), and seawater (Angly et al., 2006; Breitbart et al., 2002; Culley, Lang, & Suttle, 2006). Research has shown that plant viruses belonging to the Carmovirus (carnation mottle virus (CarMV)), Tombusvirideae (ToBSV, TNV, carnation ringspot virus (CRSV), PetAMV, grapevine Algerian latent virus (GALV)), Tobamovirus (TMV, ToMV, GGMMV), Potexvirus (PVXpotato virus X), and Furovirus families have been isolated from rivers, lakes, ditches, and sewage water, and have been found to be infectious (Sastry, 2013; Block, 1983; Tošić, & Tošić, 1984; Koenig, Rüdel, & Lesemann, 1989; Koenig, 1986; Piazzolla, Castellano, & De Stradis, 1986; Mehle & Ravnikar, 2012). Most of these viruses do not have known insect vectors and can apparently be transmitted to the roots of healthy plants without the aid of a vector (Sastry, 2013). These studies have reported the presence of 16 plant viruses representing seven genera (tobamoviruses, potexviruses, tombusviruses, carmoviruses, cucumoviruses, necroviruses, and dianthoviruses), as well as viroids, in freshwater ecosystems (Mehle & Ravnikar, 2012; Jeżewska, Zarzyńska-Nowak, & Trzmiel, 2018). A list of plant viruses detected in aqueous ecosystems is shown in Table 1.1.

TABLE 1.1 Infectious Plant Viruses Detected in Aquatic Environments

Virus Family	Virus Genus	Virus Species (Acronym)	Origin of Sample
Alphaflexiviridae	*Potexvirus*	Potato virus X (PVX)	Ponds, and creeks in forest area
Bromoviridae	*Cucumovirus*	Cucumber mosaic virus (CMV)	River
Tombusviridae	*Carmovirus*	Carnation mottle virus (CarMV)	River
–	*Dianthovirus*	Carnation ringspot virus (CRSV)	Canal near sewage plant; ditches, streams, drainage canals in agricultural areas; stream with no direct contact with fields.
–	*Necrovirus*	–	–
–	*Necrovirus*	Tobacco necrosis virus (TNV)	Ditches, streams, and drainage canals in agricultural areas; rivers, lakes.
–	*Tombusvirus*	Carnation Italian ringspot virus (CIRV)	Ponds, and creeks in forest area.
–	–	Grapevine Algerian latent virus (GALV)	Ditches, and streams in agricultural areas.
–	–	Petunia asteroid mosaic virus (PetAMV)	Ditches, drainage canals in agricultural area; seawater.
–	–	Tomato bushy stunt virus (TBSV)	Ponds, creeks in forest areas; rivers.
Virgaviridae	*Tobamovirus*	Tobacco mosaic virus (TMV)	Rivers, forest ditches, and brook.
–	–	Tomato mosaic virus (ToMV)	Ponds, creeks, brooks, streams, and lakes in forest area; rivers, gravel pits, clouds, and fog.

Source: Mehle, Gutierrez-Aguirre, Kutnjak, & Ravnikar (2018).

Rastrojo et al. (2018) showed that geminiviruses and nanoviruses, circular ssDNA viruses, and their satellites are abundant in polar lakes (135). Reverse-transcription polymerase (Pol) chain reaction (PCR) amplification detected various genotypes of TMV and ToMV in 1,40,000-year-old glacial ice subcores in Greenland (Castello et al., 1999). These viruses are also detected in clouds and fog; therefore, it is hypothesized that these viruses are present in ancient glacial ice because they are stable, airborne, and spread *via* the hydrologic cycle (Castello et al., 1995, 1999). In Europe, many studies on the identification of infectious plant viruses in waters have been performed. Infectious TNV (in Germany and England) and CRSV (in Germany) and CMV (in Italy) were confirmed to be present in ditches, streams, and canals (Koenig, Rüdel, & Lesemann, 1989; Piazzolla, Castellano, & De Stradis, 1986; Tomlinson, Faithfull, Webb, Fraser, & Seeley, 1983; Yi, Lesemann, Koenig, Rüdel, & Pfeilstetter, 1992). Melon necrotic spot virus (MNSV, Tombusviridae) has been detected in irrigation systems in some countries (Gosalvez et al., 2003; Boben et al., 2007). Vani & Varma (1993) have isolated CGMMV from the water of river Jamuna and showed high virus incidence in cucurbit crops irrigated with CGMMV-contaminated river water. Li et al. (2016); and Boben et al. (2007) also demonstrated transmission of CGMMV and ToMV to watermelon and tomato plants, respectively, by irrigation water contaminated with these viruses (Li, Liu, & Gu, 2016; Boben et al., 2007). Potato spindle tuber viroid (PSTVd) infection of potato tubers occurred when water containing PSTVd was used for tubers (Jones, 2018). Viruses obligatorily need their host for multiplication (Aw, Howe, & Rose, 2014). A majority of plant viruses occurring in waters are very stable and can survive outside of host cells for long periods. The occurrence of these viruses in aqueous environments has attracted great attention as waterborne viruses are able to infect their hosts even at very low concentrations (Sastry, 2013).

Regarding the survival of plant viruses in water, experiments have shown that TNV can survive in water for 46–80 hours (Yarwood, 1960). ToMV and pepper mild mottle virus (PMMoV) have been found to remain infectious in nutrient solution, whether stored in a glasshouse or at 4°C, for at least six months. This increases the risk of infection for plants grown in hydroponic systems (Pares, Gunn, & Cresswell, 1992). Pepino mosaic virus (PepMV) can remain infectious in water at 4°C for up to 3 weeks, and the PVY NTN strain for up to 1 week. When PVY NTN was stored at 4°C, it survived in the water for a much longer period (up to 10 weeks) (Mehle et al., 2014). Contaminated water can remain infectious for 7 weeks for PSTVd and 3 weeks for PepMV (Jones, 2018). Further research on the water-mediated transmission of viruses or viroids to plants and their survival is urgently required for managed systems like hydroponic cultures due to the continued use of water resources. Injured or decaying plant residues or infected roots of plants can be a source of releasing plant viruses into aqueous environments through washing away during irrigation or in hydroponic systems (Mehle et al., 2014; Schwarz et al., 2010; Stewart-Wade, 2011; Hong & Moorman, 2005). Running water can also wash out seeds, nematodes, and zoospores containing the virus in soil or drainage water (Koenig, 1986; Campbell, 1996; Matthews, 2012). Although waterborne transmission is not considered the most efficient method of virus transmission, it has been confirmed as the most likely route for certain plant viruses, such as PepMV, PVY, ToMV, and pelargonium flower break virus (PFBV; family *Tombusviridae*; genus *Alphacarmovirus*) (Mehle & Ravnikar, 2012). Additionally, horizontal transmission of waterborne viruses to healthy plants can occur through small wounds in the roots or even the upper parts of plants when infected water is used for irrigation (Mehle & Ravnikar, 2012; Mehle et al., 2014; Schwarz et al., 2010). The process of plant germination is another way in which waterborne viruses can infect plants (Teakle & Morris, 1981). Using quantitative PCR, the presence of PMMoV was detected in 76% of drinking water samples in Japan, with concentrations ranging from 2.03×10^3 to 2.90×10^6 copies/L (Haramoto et al., 2013). This virus was also found in seawater (Rosario, Symonds, Sinigalliano, Stewart, & Breitbart, 2009). PetAMV and CarMV were also isolated from the northeastern beach of the Isle of Helgoland and Baltic Seawater, respectively (Kontzog, Kleinhempel, & Kegler, 1988; Fuchs, Schlufter, & Kegler, 1996).

Aquatic plants were also found to be infected with plant viruses, suggesting that they are likely just as susceptible to plant pathogens as terrestrial plants (Zettler, Foxe, Hartman, Edwardson, & Christie, 1970). For example, it is proposed that a plant virus that infects *Gompherena globose* can also infect the related alligator weed (Zettler, Foxe, Hartman, Edwardson, & Christie, 1970). Virus infection can reduce or even completely eliminate seed production capacity in waterweeds (Sculthorpe, 1967). Proeseler et al. (1990) reported vein clearing in *Shinnersia rivularis* due to CMV infection (Proeseler, Stanarius, Kontzog, & Barth, 1990). Dasheen mosaic virus (DsMV) was identified in the aquatic ornamental plant *Cryptocoryne cordata* (Zettler, Foxe, Hartman, Edwardson, & Christie, 1970). On the contrary, artificial inoculation of *alfalfa mosaic virus* (AMV), TMV, TRSV, PVX, and PVY did not infect milfoil plants (Bayley, 1970). Virus vectors such *as Olpidium brassica* and species of dorylaimida nematodes (*Longidorus,*

Paralangidorus, and *Xiphinema*) are commonly found in aquatic environments (Grogan & Campbell, 1966; Harrison, 1964). Aphids, which are vectors of several viruses, have also been reported to colonize aquatic plants (Kennedy, Day, & Eastop, 1962).

Very little is known about plant viruses present in seawater and their diversity. Viromic studies of viral communities in the surface water of the East China Sea were measured to be 10^6 to 10^7 VLPs per ml (Wu et al., 2020). PMMoV was detected in seawater (Rosario et al., 2009). Mechanical inoculation of a resuspended pellet of seawater samples taken 200 m off the northeastern beach of the Isle of Helgoland caused lesions on *Chenopodium quinoa* (Rosario et al., 2009). Metagenomic analysis of coastal RNA virus communities detected *Nanoviridae, Tombusviridae, Comovirus, Sequivirus,* and *Umbravirus* (Culley, Lang, & Suttle, 2006).

1.1.3.3 WASTEWATER: THE UNDERESTIMATED WAY OF PLANT VIRUSES' DISTRIBUTION

Exploding human population and limited rainfall, especially within arid and semi-arid areas, increase water consumption, resulting in a perpetuation of a growing shortage of reliable water supplies (Rosario et al., 2009). Reclaimed or recycled water, which is the reusable end-product of wastewater treatment, is an alternative water supply that contributes to water conservation. It provides water for activities such as direct and indirect potable reuse, industrial use, agricultural irrigation, and recreational use that do not require drinking water quality standards (Young & York, 1996; Asano & Levine, 2004). While reclaimed water offers important benefits, its indiscriminate use has raised concerns regarding public health and the environment, including pathogen transfer (Rosario et al., 2009). Metagenomic studies have been widely conducted to identify plant viruses in reclaimed water. These studies have shown that plant viruses are abundant and highly diverse components of the wastewater virome and have even revealed emerging plant viruses that were not previously reported in the region (Bačnik et al., 2020). These viruses are including *Virigaviridae* (*Tobamovirus*), *Tymoviridae* (*Tymovirus* and *Maculavirus*), *Betaflexiviridae* (*Carlavirus*), *Alphaflexiviridae* (*Potexvirus*, *Allexivirus*), *Tombusviridae* (*Alphacarmovirus*, *gammacarmovirus*, *tombusvirus*, *alphanecrovirus*), *Sobemoviridae* (*Sobemovirus*), *Potyviridae* (*potyvirus*), *Secoviridae* (*Nepovirus*, *Fabavirus*, *Comovirus*), *Bromoviridae* (*Cucumovirus*, *Bromovirus*), *Luteoviridae* (*Luteovirus*), *Partitiviridae*

(*Deltapartitivirus*) (Cantalupo et al., 2011; Bačnik et al., 2020; Fernandez-Cassi et al., 2018). Among these viruses, *Secoviridae*, *Potyviridae*, *Tombusviridae,* and *Tobamovirus* have extremely stable virions that can resist high temperatures and are insensitive to organic solvents and non-ionic detergents (Fauquet, Mayo, Maniloff, Desselberger, & Ball, 2005). These studies also showed that ssDNA plant viruses, *Geminiviridae* and *Nanoviridae*, are dominant in wastewaters (Rosario, Nilsson, Lim, Ruan, & Breitbart, 2009). MNSV (*Tombusviridae*) and PMMoV, which have been detected in both raw sewage and reclaimed water metagenomes, were also the most abundant viruses in RNA viral metagenomes from healthy individuals (Zhang et al., 2006). This result indicates that these viruses are as stable and resistant to environmental degradation as they can reach irrigation systems, representing a potential problem for the agricultural sector (Rosario, Nilsson, Lim, Ruan, & Breitbart, 2009). Tobamoviruses (ToMV, TMGMV, PMMoV, and tomato black ring fruit virus (ToBRFV)) were found to be present in wastewater and remain infective even after conventional wastewater treatment (Bačnik et al., 2020; Dombrovsky & Smith, 2017). Outbreaks of these viruses through the application of treated wastewater as irrigation water could be causing major concerns for growers of Solanaceae and Cucurbitaceae worldwide (EPPO, 2019). The detection of nucleic acids of some viruses like ToBRFV in wastewater in Slovenia, which had never been reported in this country before, raises questions about their origin and suggests wastewater as a potential pathway for the dissemination and gradual globalization of plant viruses (Bačnik et al., 2020). In another study of deep sequencing of RNA and DNA plant viruses of raw sewage from different countries, including United States, Nigeria, Thailand, and Nepal, 11 plant viruses with RNA genome, e.g., *Alphaflexiviridae*, *Betaflexiviridae*, *Luteoviridae, Partitiviridae*, *Potyviridae*, *Sobemovirus*, *Secoviridae*, *Tombusviridae*, *Tymoviridae*, *Virgaviridae*, and *Umberavirus*) and two plant viruses with DNA genome, *Nanoviridae* and *Geminiviridae*, have been detected (Ng et al., 2012).

PMMoV was detected at concentrations ranging from 108 to 1,010 copies/L in raw sewage and treated wastewater samples collected throughout the USA (Rosario, Nilsson, Lim, Ruan, & Breitbart, 2009; Hamza, Jurzik, Überla, & Wilhelm, 2011). In another independent study, the relative abundance and occurrence of various viruses were investigated in two samples collected from wastewater treatment plants (WWTPs) in southern Arizona. The results indicated that among the 11 viruses tested (genogroup I, II, and IV noroviruses, sapovirus, enterovirus, group A rotavirus, Aichi virus, PMMoV, adenovirus, and JC and BK polyomaviruses), PMMoV was the most prevalent virus in both influent and effluent wastewater. The annual mean concentration of PMMoV in influent wastewater was 3.7–4.4 × 106 copies/L, while in effluent wastewater it was 4.6–6.3 × 105 copies/L (Kitajima, Iker, Pepper, & Gerba, 2014). PMMoV was found to be more abundant in wastewater samples compared to human adenoviruses (HuAdVs) and human polyomaviruses (HuPyVs), which were previously considered as potential viral indicators of human fecal contamination in environmental waters (Hamza, Jurzik, Überla, & Wilhelm, 2011; Bofill-Mas et al., 2006; McQuaig, Scott, Lukasik, Paul, & Harwood, 2009; Wong, Fong, Bibby, & Molina, 2012; Calgua et al., 2013). This virus can be introduced into wastewater through food items, as well as smoked and chewed tobacco products, and can subsequently be transmitted to humans (Pleše, Juretić, Mamula, Polák, & Krajačić, 1996). PMMoV was also detected in human feces at concentrations higher than 109 copies per gram of dry weight using qPCR (Zhang et al., 2006; Moore et al., 2015; Nakamura et al., 2009; Victoria et al., 2009). It is suggested that PMMoV has the potential to circulate between croplands and the environment. It can move from infected plants to wastewater through human and animal consumption and excreta, and then transmit back to plants through activities such as irrigation with reclaimed wastewater (Mehle, Gutierrez-Aguirre, Kutnjak, & Ravnikar, 2018). PMMoV has been found to be more persistent in wastewater compared to HuAdVs and HuPyVs (Hamza, Jurzik, Überla, & Wilhelm, 2011). As a result, this virus has been proposed as an indicator of fecal pollution in coastal, sub-tropical waters, untreated domestic wastewater (Wong, Fong, Bibby, & Molina, 2012), and tertiary-treated domestic wastewater (Rosario, Nilsson, Lim, Ruan, & Breitbart, 2009). In addition to PMMoV, a few reports have detected TMV in wastewaters (Piazzolla, Castellano, & De Stradis, 1986; Koenig & Lesemann, 1985). TMV is highly thermostable and resistant to tobacco manufacturing processes, and it can be present in chewing tobacco and cigars for extended periods of time (Kitajima, Iker, Pepper, & Gerba, 2014). Wastewater from plant processing factories, such as sugarcane factories that wash sugar beet roots, can contribute to the distribution of BNYVV (*see* Section 1.1.3.1).

Metagenomics studies of plant viruses in wastewater from different sources have provided a comprehension of the diversity of plant viruses, including emerging ones, circulating in a given environment. These studies need to be followed by systematic tests for specific viruses, especially in terms of their infectivity (Bačnik et al., 2020). The rapid transmission of pathogenic viruses to new areas through wastewater could be a consequence of the release of infective plant viruses into the environment, which cannot be reached by other vectors. It seems that such processes may speed up the global incidence and the possible impacts of viral diseases in agricultural ecosystems and a wide range of crops (Mehle, Gutierrez-Aguirre, Kutnjak, & Ravnikar, 2018). Although wastewater-based epidemiology has not been utilized yet, it has great potential hypothetically to prevent viral disease outbreaks by predicting critical locations and critical moments (Xagoraraki & O'Brien, 2020), similar to those performed recently to detect COVID-19 in wastewaters during the outbreak of the coronavirus SARS-CoV-2 and the Coronaceous period (Winn, 2020; Nazari, Sheykhrezaee, & Golnaraghi, 2021).

1.1.3.4 FECES AND GUANOS: PLANT VIRUSES ARE TRANSPORTING BY ANIMALS EVERYWHERE

Humans and animals, especially livestock, that consume fruits and leafy vegetables are exposed to plant viruses on a daily basis. Some of these viruses, such as PMMoV and TMV, have been found in food products and can still infect plants (Colson et al., 2010). As a result, the presence of plant viruses in human and animal feces is common (Aguado-García et al., 2020). Traditional medicine (WHO, n.d.) and smoking (Wahyuni, Hanapi, & Hartana, 2008; Wetter, 1975; Chyle, Chyle, Korb, & Papanek, 1971) are additional risk factors for human exposure to plant viruses. While plant viruses are generally thought to only infect plants and not cause diseases in humans and other vertebrates (Balique, Lecoq, Raoult, & Colson, 2015), there are instances where plant viruses can circulate and propagate in insect vectors (Balique, Lecoq, Raoult, & Colson, 2015). Humans have likely been exposed to plant viruses for thousands of years, suggesting that these viruses are generally safe for humans (Jiwaji et al., 2019). However, there are still many unanswered questions regarding the potential transmission of plant viruses from plants to humans. It is also unclear whether plant viruses benefit from intestinal cells or the microbiome to achieve stability (Balique, Lecoq, Raoult, & Colson, 2015). The Providence virus, an RNA insect virus, can infect both plants (cowpea) and mammalian cells (Jiwaji et al., 2019). There is evidence of immune responses to plant viruses or their genomes in invertebrates, non-human vertebrates, and humans after experimental exposure (Aguado-García et al., 2020).

Many studies have detected PMMoV in facial samples of human and non-human mammals. For example, early gut virus metagenome analysis by Zhang et al. (2006) discovered 35 plant viruses as the most prevalent RNA viruses in human feces, and PMMoV was the most abundant virus with a ratio of 10^9

virions per gram of dry weight of fecal matter (Zhang et al., 2006). Hamza et al. (2011) detected this virus by qRT-PCR in 7 out of 15 fecal samples from chickens, 1 out of 10 from geese, and 1 out of 6 from cows, at titers up to 3.1 $\log_{10}$ copies/mg (Hamza, Jurzik, Überla, & Wilhelm, 2011). Using real-time PCR in human feces, PMMoV was detected in 7.2% (22 out of 304) of adults (Colson et al., 2010). However, in another study, this virus was detected in 19 out of 20 healthy human fecal samples at titers up to 7.0 $\log_{10}$ viral RNA copies/mg (Hamza, Jurzik, Überla, & Wilhelm, 2011). Recently, a positive correlation has been shown between PMMoV virion content and some clinical signs, including fever and abnormal pain (Jiwaji et al., 2019; Kim, Yoon, Park, Kim, & Ryu, 2020). It has been suggested that this virus is present in two-thirds of healthy human populations in different geographic regions of the world (Zhang et al., 2006), and the high titer of the virus in human fecal samples might be the result of viral replication in the human gut (Kitajima, Sassi, & Torrey, 2018). Another tobamovirus, TMV, has also been detected in many studies. TMV RNA was retrieved in stools collected from South Asian children and from Japan (Nakamura et al., 2009; Victoria et al., 2009). The virus is resistant to manufacturing processes during the production of cigarettes, with a mean titer of 9.5 $\log_{10}$ RNA copies/cigarette and 3.8 $\log_{10}$ RNA copies/mL of smokers' saliva (Balique, Lecoq, Raoult, & Colson, 2015). TMV RNA was detected in all cigarettes from different brands, as well as in 45% of saliva samples from smokers. However, all saliva samples from non-smokers tested negative (Balique, Colson, & Raoult, 2012).

In addition to the gut of human adults, plant viruses have been frequently found in the gut and oropharynx of children during their first year of life, even when they are completely breastfed and before they initiate a solid food regime (Aguado-García et al., 2020). In this regard, 15 different species of tobamoviruses were identified, with tropical soda apple mosaic virus (TSAMV), PMMoV, and opuntia tobamovirus 2 (OpV2) being the most frequently detected species in 100%, 84%, and 61% of the fecal samples, respectively (Aguado-García et al., 2020). TMGMV and rattail cactus necrosis-associated virus (RCNaV) were other species commonly found in the samples (Aguado-García et al., 2020). In the oropharynx samples, 12 different families of plant viruses (*Virigaviridae*, *Geminiviridae*, *Secoviridae*, *Alfphaflexiviridae*, *Potyviridae*, *Tombusviridae*, *Bromoviridae*, *Betaflexiviridae*, *Luteoviridae*, *Endornaviridae* (*Alphaendornavirus*), *Procedovirinae*, and *Caulimoviridae*) were detected, and again, tobamoviruses were the most prevalent ones (Aguado-García et al., 2020). In another study, the fecal microbiome of eight healthy infants was characterized, and viruses from the *Virgaviridae, Alphaflexiviridae*, and *Tombusviridae* families were identified. The virus-positive samples were found in children older than 18 months. Geminiviral sequences were also detected in a single sample collected from a child a few days after birth (Lim et al., 2015).

Plant viruses have also been detected in non-human mammal's feces including mouse, vole, and rat (*Alphaflexiviridae, Luteoviridae, Secoviridae, Tombusviridae, Tymoviridae, Virigaviridae, Partitiviridae, Phycodnaviridae, Nanoviridae*, and *Geminiviridae*) (Rae et al., 2005; Phan et al., 2011), cow, donkey, and grass rat (*Sobemovirus, Tobamovirus*) (Van, 1988; Sarra & Peters, 2003), and sheep (*Sebemovirus*) (Sarra & Peters, 2003). For more details on genus and species, please refer to reference (Balique, Lecoq, Raoult, & Colson, 2015). These viruses were found to be stable and maintained their infectivity in the gastrointestinal tract. They could even be transmitted by mammals through their contaminated stools (Rae et al., 2005; Xu, Jones, & Rodgers, 1996). In a separate study, CGMMV was detected in cow manure three days after being fed with infected cucumbers (Van, 1988). Through a metagenomics approach, two viruses were recovered from caribou feces that had been frozen in a permanent ice patch for 700 years. One of the viruses was a DNA virus distantly related to plant-infecting geminiviruses and the fungi-infecting *Sclerotinia sclerotiorum* hypovirulence-associated DNA virus 1 (SsHADV-1), while the other was an RNA virus belonging to the insect-infecting Cripavirus genus (Ng et al., 2014). It has also been reported that modern caribou feces contain RNA viruses related to plant-infecting tombusviruses and sobemoviruses, as well as a DNA virus distantly related to fungi-infecting SsHADV-1 and plant-infecting geminiviruses.

Metagenomic studies have been conducted to identify the virome of wild rodents' stools. These studies have shown that 3.4% of the virome sequences belonged to plant viruses of *Nanoviridae, Geminiviridae, Alphaflexiviridae, Tymoviridae*, and *Secoviridae* (Phan et al., 2011; Li et al., 2010). Similarly, the identification of DNA and RNA viruses in pig feces using process-controlled deep sequencing revealed that 0.8% of the selected primary reads were plant virus sequences, including *Potyviridae, Luteoviridae, Bromoviridae*, and *Reoviridae* (Sachsenröder et al., 2012). Bats are known to host a variety of viruses (Li et al., 2010). In the USA, the characterization of the bat guano virome through metagenomics analysis showed that plant viruses constituted the second largest proportion of virus-related sequences. These included members of the viral families *Luteoviridae, Secoviridae, Tymoviridae*, and *Partitiviridae,* as well as the genus *Sobemovirus* and the ssDNA virus *Geminiviridae* (Li et al., 2010). Wu et al. (2012) also identified a range of viruses in bat species from China, including viruses from mammals, insects, fungi, plants, and phages (Wu et al., 2012).

Transmission of plant viruses through human and non-human animals has been reported. Ingestion of purified TBSV by human volunteers and inoculation of their feces on *Chenopodium quinoa* leaves induced signs of viral infection (Tomlinson, Faithfull, Flewett, & Beards, 1982; Tomlinson & Faithfull, 1984). Additionally, transmission of RYMV by cows, donkeys, and grass rats has been reported (Sarra & Peters, 2003). Subterranean clover mottle virus (SCMoV) could be transmitted by sheep (McKirdy, Jones, & Sivasithamparam, 1998). It has also been shown that cowpea mosaic virus (CPMV) can disseminate systemically in their bodies when orally administered to mice (Rae et al., 2005).

1.2 THE DIVERSITY OF PLANT VIRUSES IN UNKNOWN SOURCES: THE TIP OF THE ICEBERG

Our understanding of plant virus diversity has traditionally been limited to agricultural systems and commercial crops. However, recent studies have begun to uncover the vast diversity of viruses in other sources as well (Stobbe & Roossinck, 2016). Metagenomic studies are now being conducted on plant viruses in various new sources, beyond just crops, in order to gain a deeper understanding of their true diversity (Stobbe & Roossinck, 2016). It has been observed that virus families associated with wild plants can differ from those associated with crops (Hasiów-Jaroszewska, Boezen, & Zwart, 2021). While *Geminiviridae* and *Potyviridae* are the most abundant plant viruses, viruses from the *Totiviridae* and *Partitiviridae* families are commonly found in wild plants (Roossinck, 2012). Interestingly, these two families are typically associated with fungi and have not been previously identified as plant viruses (Roossinck, 2012). Non-plant-based metagenomics studies have also revealed a diverse array of plant viruses, which likely reflects the varying levels of virus stability (Roossinck, 2012).

Tobamoviruses are frequently found in environmental samples such as human feces, chimp feces, freshwater lakes, and reclaimed water, due to the high stability of their virions (Table 1.2). In fact, they are the most prevalent plant viruses in all environmental samples, except for wastewaters where ssDNA plant viruses, *Geminiviridae*, and *Nanoviridae* are dominant (Kim et al., 2008).

DNA viruses were the focus of the earliest studies of viruses in the sea, where phages were prevalently found (Roossinck, 2012). Although sequences distantly related to the *Nanoviridae* family have also been found in coastal seawaters (Culley, Lang, & Suttle, 2006). There have also been reports of these viruses in soil washing from rice paddies (Kim et al., 2008). ssDNA plant viruses are even part of the polar freshwater lake virome (Rastrojo & Alcamí, 2018; Aguirre De, López-Bueno, Pearce, & Alcamí, 2015).

Almost all the plant viruses that have been found in human feces metagenomics studies are known to infect wastewaters and remain infective in reclaimed waters. Table 1.2 summarizes the plant viral families/genera reported from different environments.

TABLE 1.2 A Summary of Plant Virus Genera/Families Reported from Different Environments

Environment		Virus Families/Genera	References
Soil samples	Soil	*Secoviridae, Potyviridae, Ophioviridae, Tombusviridae, Virgaviridae, Benyviridae, Varicosavirus*	Narayanasamy (2011); Lanter, McGuire, & Goode (1982)
	Rice paddies	*Nanoviridae, Geminiviridae*	Kim et al. (2008)
Aquatic environments	Coastal see waters	*Nanoviridae, Tombusviridae, Umbravirus*	Culley, Lang, & Suttle (2006)
	Freshwater	*Bromoviridae, Flexiviridae, Luteoviridae, Partitiviridae, Secoviridae, Tetraviridae, Tombusviridae, Tymoviridae, Sobemovirus, Umbravirus, Tobamovirus*	Djikeng, Kuzmickas, Anderson, & Spiro (2009)
	Wastewater/ reclaimed water	*Geminiviridae, Nanoviridae, Potyviridae, Reoviridae, Secoviridae, Virgaviridae, Tymoviridae, Betaflexiviridae, Alphaflexiviridae, Tombusviridae, Sobemoviridae, Bromoviridae, Luteoviridae, Partitiviridae*	Cantalupo et al. (2011); Rosario, Nilsson, Lim, Ruan, & Breitbart (2009); Bačnik et al. (2020); Fernandez-Cassi et al. (2018); Ng et al. (2012)
	Polar freshwater lake	*Geminiviridae, Nanoviridae*	Rastrojo & Alcamí (2018); Aguirre De, López-Bueno, Pearce, & Alcamí (2015); López-Bueno et al. (2009)
	River	*Tymoviridae, Virgaviridae*	Fernandez-Cassi et al. (2017); Vani & Varma (1993)

TABLE 1.2 *(Continued)*

Environment		Virus Families/Genera	References
Human and animal feces	Human feces	*Virgaviridae, Bromoviridae, Closteroviridae, Tombusviridae, Tymoviridae, Sobemovirus, Geminiviridae, Secoviridae, Alfphaflexiviridae, Potyviridae, Betaflexiviridae, Luteoviridae, Partitiviridae, Caulimoviridae*	Zhang et al. (2006); Hamza, Jurzik, Überla, & Wilhelm (2011); Nakamura et al. (2009); Victoria et al. (2009); Colson et al. (2010); Aguado-García et al. (2020); Lim et al. (2015)
	Wild-rodent feces	*Geminiviridae, Nanoviridae, Alphaflexiviridae, Luteoviridae, Virgaviridae, Phycodnaviridae, Partitiviridae, Secoviridae, Tombusviridae, Tymoviridae*	Rae et al. (2005); Phan et al. (2011)
	Cow/sheep/pig/donkey	*Sobemovirus, Virgaviridae, Potyviridae, Luteoviridae, Bromoviridae, Reoviridae*	Van (1988); Sarra & Peters (2003); Sachsenröder et al. (2012)
	Chimp feces	*Partitiviridae, Tobamovirus*	Victoria et al. (2009)
	Bat guano	*Luteoviridae, Partitiviridae, Secoviridae, Tymoviridae, Sobemovirus*	Li et al. (2010); Wu et al. (2012)

1.3 POSSIBLE IMPACTS OF UNKNOWN SOURCES OF PLANT VIRUSES ON AGRICULTURAL ECOSYSTEMS AND FOOD SECURITY

As discussed previously, our current understanding of the impacts of plant viruses is heavily biased towards economically important crops, with limited information available on emerging viruses that have not yet been detected on crops (Ng et al., 2011). However, undiscovered plant viruses from new sources play a crucial role in the complex microbial ecology of plants. They serve as reservoirs of biodiversity, allowing for recombination and reassortment with existing pathogenic plant viruses (Ng et al., 2011). This is particularly true for viruses that infect wild plants or are carried by vectors. The incidence rate of wild plant viruses is 60% (Roossinck, 2015), suggesting their potential as important ecological agents that may jump to crops in the future (Rodríguez-Nevado, Gavilán, & Pagán, 2020). Therefore, wild plants serve as permanent reservoir hosts for plant viruses, and their contribution to the emergence of plant viruses has been demonstrated in numerous cases (García-Arenal, & Zerbini, 2019; Gibbs, Trueman, & Gibbs, 2008; Nguyen et al., 2013). Additionally, plant viruses have been detected within the bodies of various arthropod species (Ng et al., 2011; Fontenele et al., 2018; Kamali et al., 2017; Feng et al., 2017; Wamonje et al., 2017; Nouri, Salem, Nigg, & Falk, 2016; Thekke-Veetil et al., 2020; Rosario et al., 2013, 2015; Godinho, Paula, Varsani, & Ribeiro, 2017; François et al., 2021), many of which are potential plant virus vectors or at least feed on plants. Vectors carrying viruses can increase the risk of emergence or reemergence of plant viruses by facilitating long-distance dispersal or acting as a pool of viral genetic information (Rojas & Gilbertson, 2008). Begomoviruses and tospoviruses are the most significant genera of plant viruses that have produced a remarkable number of new viral species in recent decades (Rojas & Gilbertson, 2008). In both cases, the increase in the distribution and population of their polyphagous insect vectors, the sweet potato whitefly (*Bemisia tabaci*) and the western flower thrips (WFT) [*Frankliniella occidentalis*], is considered as the main contributor to their emergence (Rojas & Gilbertson, 2008). With increasing global trade, the emergence of a virus may also be mediated by the introduction of infected plant materials (e.g., propagative materials or seeds) to a new geographical region (Rojas & Gilbertson, 2008). The introduction of PepMV to Europe and North America is an example of the emergence of a virus *via* seed dissemination (Rojas & Gilbertson, 2008; Gómez, Sempere, Elena, & Aranda, 2009). Additionally, TYLCV was introduced into the Dominican Republic in the early 1990s as the first emergence of an old world (OW) monopartite begomovirus in the New World (NW) with the imported tomato transplants (Salati et al., 2002).

Due to their great diversity and importance as genetic reservoirs in different environments, viruses can harbor ecological traits (Kimura, Jia, Nakayama, & Asakawa, 2008). The contamination of water sources with pathogenic microbes can negatively impact plant health and pose a threat to food security (Mehle & Ravnikar, 2012). Long-distance spread of viruses can occur through channels, lakes, and rivers. Almost immediately after the first plant viruses were detected in significant amounts in environmental waters (Koenig, 1986), the survival and spread of plant viruses by water became a concern (Mehle & Ravnikar, 2012). Presently, the extensive use of hydroponic systems and intensive irrigation in agriculture has highlighted the importance of water-mediated transmission of viruses (Mehle & Ravnikar, 2012). Many agricultural systems use surface water sources for irrigation, such as lakes, rivers, and reservoirs (Mehle & Ravnikar, 2012). These water sources usually harbor plant viruses (Rosario, Nilsson, Lim, Ruan, & Breitbart, 2009; Djikeng, Kuzmickas, Anderson, & Spiro, 2009; Culley, Lang, &

Suttle, 2006; Mehle & Ravnikar, 2012; Piazzolla, 1989), and therefore, can be an effective route for spreading the viruses. Tobamoviruses with exceptionally stable virions and a wide range of plant hosts are recovered from many water sources worldwide and can easily be released into the drainage waters. A number of studies have shown that these plant viruses can infect crops with contaminated irrigation water (Li, Liu, & Gu, 2016; Fernandez-Cassi et al., 2017; Boben et al., 2007; Vani & Varma, 1993). Most plant viruses collected from environmental waters have been shown to remain infectious. Viruses tend to associate with particulate matter present in water, such as clay minerals and organic matter. The long-term association of viruses with the material surfaces protects them against inactivation by offering physical and chemical protection from biotic and abiotic stresses (Hassard et al., 2016). The viral association is even stronger in sediments due to their finer texture, and the particles of CMV are shown to be protected by sediment (Piazzolla, 1989; Hassard et al., 2016). Roots are the main entry points for water-borne and soil-borne viruses to infect plants. Other ways are through leaves, other upper plant parts, or seeds at germination. The viral concentration in the water determines the severity of the symptoms (Mehle & Ravnikar, 2012). Plant viruses in contaminated reclaimed/wastewater are stable and resistant to environmental degradation. The viruses are infective even after wastewater treatment and can reach irrigation systems, representing a potential problem for agriculture, especially in Solanaceae and Cucurbitacea crops (Bačnik et al., 2020; Dombrovsky & Smith, 2017). In the same way as water sources, soil can play a role in virus transmission. By moving soil during cultivation, using agricultural machinery, or water streams, plant viruses can spread over large distances to new locations (Roberts, 2014). Planting crops in soil containing highly stable tobamoviruses always poses a significant risk of infection (Lanter, McGuire, & Goode, 1982). Transmission of plant viruses through human and non-human animals has been reported in several studies (Dombrovsky & Smith, 2017; Sarra & Peters, 2003; Tomlinson & Faithfull, 1984). Therefore, the contamination of irrigation water or planting soil with human and animal feces could also be a threat for agricultural systems. Infective plant viruses have also been recovered from other sources, e.g., PMMoV can survive in manufactured hot chili sauce, powdered chili, and human feces (Mehle & Ravnikar, 2012; Zhang et al., 2006). It has been suggested that people who are positive for PMMoV also experience pathogenic effects including fever, abdominal pain, and immune response, but the direct link between the virus and the symptom remains unclear (Colson et al., 2010; Jiwaji et al., 2019; Kim, Yoon, Park, Kim, & Ryu, 2020).

1.4 CONCLUSION

Plant viruses have been found everywhere, from domestic and wild plants and the interface between them, to vegetative and reproductive plant parts, vectors, and different environmental samples including soil, water, and feces (Roossinck, 2015). Innovations in metagenomics sequencing have been the key to most of these results, including enriching for VLPs, RNA or DNA specific to viruses, or extracting specific nucleic acids followed by next-generation deep sequencing and bioinformatics (Roossinck, 2015; Islam, Zhang, Adnan, Noman, Zaynab, & Wu, 2017). However, in most cases, these studies have revealed our surprising lack of understanding of plant viruses, their ecology, and evolution, indicating the need for more comprehensive research (Roossinck, 2015). Although metagenomic-related approaches have helped us gain a better understanding of plant viruses, it seems that integrative strategies, i.e., a combination of classic and modern approaches like those described earlier (Shahmohammadi, Mansourpour, & Golnaraghi, 2022), are necessary to study these viruses. For example, vectorome studies of air-borne, soil-borne, or water-borne vectors, planting indicator host plants in soils, and preparing viral concentrates from soil or water and inoculation on indicator hosts, followed by the discovery of viral sequences using metagenomics. Such strategies may hopefully help us learn more about the ecological/epidemiological roles of plant viruses in various systems. As discussed in the previous Section 1.3, plant viruses coming from different sources may cause serious problems in terms of food biosecurity and can threaten agricultural ecosystems. Therefore, we need to enhance our knowledge about plant viruses in other sources and their possible contribution to causing diseases that could cause great damage in the future (Islam, Zhang, Adnan, Noman, Zaynab, & Wu, 2017).

On the other hand, little is known about how viral abundance and diversity in resources other than domestic plants may affect the evolution of these entities. Viruses that disperse in different environments can serve as genetic material, allowing for modular evolution with a high degree of recombination and/or reassortment. Therefore, gaining a better understanding of these processes will provide us with a new perspective on virus evolution and disease emergence. Additionally, a majority of plant viruses are vectored by insects/fungi, and it is known that these organisms are abundant in various habitats such as soil and aquatic environments. It is possible that many of them are yet-unknown vectors of plant viruses, and their potential role in the distribution of plant viruses remains unclear.

KEYWORDS

- **diversity**
- **plant virome**
- **plant virus sources**
- **plant viruses**
- **seedvirome**
- **terrestrial and aquatic environment**
- **vectorome**
- **wild vs agricultural ecosystem**

REFERENCES

Agindotan, B. O., Shiel, P. J., & Berger, P. H., (2007). Simultaneous detection of potato viruses, PLRV, PVA, PVX and PVY from dormant potato tubers by TaqMan® real-time RT-PCR. *Journal of Virological Methods*, *142*(1, 2), 1–9.

Agrios, G., (1988). *Plant Pathology* (3rd edn.). Academic Press. San Diego.

Aguado-García, Y., Taboada, B., Morán, P., Rivera-Gutiérrez, X., Serrano-Vázquez, A., Iša, P., Rojas-Velázquez, L., et al., (2020). Tobamoviruses can be frequently present in the oropharynx and gut of infants during their first year of life. *Scientific Reports*, *10*(1), 13595.

Aguirre De, C. D., López-Bueno, A., Pearce, D. A., & Alcamí, A., (2015). Biodiversity and distribution of polar freshwater DNA viruses. *Science Advances*, *1*(5), e1400127.

Andika, I. B., Wei, S., Cao, C., Salaipeth, L., Kondo, H., & Sun, L., (2017). Phytopathogenic fungus hosts a plant virus: A naturally occurring cross-kingdom viral infection. *Proceedings of the National Academy of Sciences*, *114*(46), 12267–12272.

Angly, F. E., Felts, B., Breitbart, M., Salamon, P., Edwards, R. A., Carlson, C., Chan, A. M., et al., (2006). The marine viromes of four oceanic regions. *PLoS Biology*, *4*(11), e368.

Antignus, Y., Lachman, O., Pearlsman, M., & Koren, A., (2005). Note: Containment of *Cucumber fruit mottle mosaic virus* (CFMMV) infection through roots by planting into a virus-free intermediating medium. *Phytoparasitica*, *33*(1), 85–87.

Arif, M., Torrance, L., & Reavy, B., (1994). Improved efficiency of detection of *Potato mop-top furovirus* in potato tubers and in the roots and leaves of soil-bait plants. *Potato Research*, *37*(4), 373–381.

Asano, T., & Levine, A., (2004). Recovering sustainable waste from wastewater. *Environ. Sci. Technol.*, *38*, 201A–208A.

Aw, T. G., Howe, A., & Rose, J. B., (2014). Metagenomic approaches for direct and cell culture evaluation of the virological quality of wastewater. *Journal of Virological Methods*, *210*, 15–21.

Bačnik, K., Kutnjak, D., Pecman, A., Mehle, N., Žnidarič, M. T., Aguirre, I. G., & Ravnikar, M., (2020). Viromics and infectivity analysis reveal the release of infective plant viruses from wastewater into the environment. *Water Research*, *177*, 115628.

Balique, F., Colson, P., & Raoult, D., (2012). *Tobacco mosaic virus* in cigarettes and saliva of smokers. *Journal of Clinical Virology*, *55*(4), 374–376.

Balique, F., Lecoq, H., Raoult, D., & Colson, P., (2015). Can plant viruses cross the kingdom border and be pathogenic to humans? *Viruses*, *7*(4), 2074–2098.

Barragan, C. E., & Guzmán-Barney, M., (2014). Molecular detection of *Potato yellow vein virus* in the natural whitefly vector *Trialeurodes vaporariorum*, Westwood. *Revista de Protección Vegetal*, *29*(3), 168–176.

Bayley, S. E. M., (1970). *The Ecology and Disease of Eurasian Water Milfoil (Myriophyllum Spicatum L.) in the Chesapeake Bay*. The Johns Hopkins University.

Beijerinck, M., (1898). Concerning a *Contagium viwm fluidum* as cause of the spot disease of tobacco leaves. *Phytopathol Classics*, *7*, 33–52.

Beuch, U., Persson, P., Edin, E., & Kvarnheden, A., (2014). Necrotic diseases caused by viruses in Swedish potato tubers. *Plant Pathology*, *63*(3), 667–674.

Bhat, A. I., & Rao, G. P., (2020). Virus transmission through pollen. In: *Characterization of Plant Viruses* (pp. 61–64). Springer.

Blanc, S., Uzest, M., & Drucker, M., (2011). New research horizons in vector-transmission of plant viruses. *Current Opinion in Microbiology*, *14*(4), 483–491.

Block, J. C., (1983). Viruses in environmental waters. In: Berg, G., (ed.), *Viral Pollution of the Environment* (pp. 117–145). CRC Press.

Boben, J., Kramberger, P., Petrovič, N., Cankar, K., Peterka, M., Štrancar, A., & Ravnikar, M., (2007). Detection and quantification of *Tomato mosaic virus* in irrigation waters. *European Journal of Plant Pathology*, *118*(1), 59–71.

Bofill-Mas, S., Albinana-Gimenez, N., Clemente-Casares, P., Hundesa, A., Rodriguez-Manzano, J., Allard, A., et al., (2006). Quantification and stability of human adenoviruses and polyomavirus JCPyV in wastewater matrices. *Applied and Environmental Microbiology*, *72*(12), 7894–7896.

Breitbart, M., Salamon, P., Andresen, B., Mahaffy, J. M., Segall, A. M., Mead, D., Azam, F., & Rohwer, F., (2002). Genomic analysis of uncultured marine viral communities. *Proceedings of the National Academy of Sciences*, *99*(22), 14250–14255.

Broadbent, L., & Fletcher, J., (1963). The epidemiology of tomato mosaic: IV. Persistence of virus on clothing and glasshouse structures. *Annals of Applied Biology*, *52*(2), 233–241.

Broadbent, L., (1965). The epidemiology of tomato mosaic VIII. Virus infection through tomato roots. *Annals of Applied Biology*, *55*(1), 57–66.

Broadbent, L., (1976). Epidemiology and control of *Tomato mosaic virus*. *Annual Review of Phytopathology*, *14*(1), 75–96.

Büttner, C., & Nienhaus, E., (1989). Virus contamination of soils in forest ecosystems of the federal republic of Germany. *European Journal of Forest Pathology*, *19*(1), 47–53.

Calgua, B., Fumian, T., Rusinol, M., Rodriguez-Manzano, J., Mbayed, V. A., Bofill-Mas, S., Miagostovich, M., & Girones, R., (2013). Detection and quantification of classic and emerging viruses by skimmed-milk flocculation and PCR in river water from two geographical areas. *Water Research*, *47*(8), 2797–2810.

Campbell, R., (1996). Fungal transmission of plant viruses. *Annual Review of Phytopathology*, *34*(1), 87–108.

Cantalupo, P. G., Calgua, B., Zhao, G., Hundesa, A., Wier, A. D., Katz, J. P., Grabe, M., et al., (2011). Raw sewage harbors diverse viral populations. *MBio*, *2*(5), e00180–11.

Card, S., Pearson, M., & Clover, G., (2007). Plant pathogens transmitted by pollen. *Australasian Plant Pathology*, *36*(5), 455–461.

Castello, J. D., Rogers, S. O., Starmer, W. T., Catranis, C. M., Ma, L., Bachand, G. D., Zhao, Y., & Smith, J. E., (1999). Detection of tomato mosaic *Tobamovirus* RNA in ancient glacial ice. *Polar Biology*, *22*(3), 207–212.

Castello, J., Lakshman, D., Tavantzis, S., Rogers, S., Bachand, G., Jagels, R., Carlisle, J., & Liu, Y., (1995). Detection of infectious tomato mosaic *Tobamovirus* in fog and clouds. *Phytopathology*, *85*(11), 1409–1412.

Chauhan, R. P., Hamon, H. F., Rajakaruna, P., Webb, M. A., Payton, M., & Verchot, J., (2015). Reliable detection for *Bean yellow mosaic virus*, *Canna yellow streak virus*, and *Canna yellow mottle virus* in canna varieties with red foliage. *Plant Disease*, *99*(2), 188–194.

Chikoti, P. C., Mulenga, R. M., Tembo, M., & Sseruwagi, P., (2019). Cassava mosaic disease: A review of a threat to cassava production in Zambia. *Journal of Plant Pathology*, *101*(3), 467–477.

Chyle, P., Chyle, M., Korb, J., & Papanek, M., (1971). Virus tabakove mozaiky, polyfenoly a kancerogenita tabakoveho navyku. *Cesk Epidemiol. Mikrobiol. Imunol.*, (20), 32–42.

Claverie, S., Bernardo, P., Kraberger, S., Hartnady, P., Lefeuvre, P., Lett, J. M., Galzi, S., et al., (2018). From spatial metagenomics to molecular characterization of plant viruses: A geminivirus case study. *Advances in Virus Research*, *101*, 55–83.

Colson, P., Richet, H., Desnues, C., Balique, F., Moal, V., Grob, J. J., Berbis, P., et al., (2010). *Pepper mild mottle virus,* a plant virus associated with specific immune responses, fever, abdominal pains, and pruritus in humans. *PloS One*, *5*(4), e10041.

Cooper, I., & Jones, R. A., (2006). Wild plants and viruses: Under-investigated ecosystems. *Advances in Virus Research*, *67*, 1–47.

Coutts, B., & Jones, R., (2015). *Potato virus Y*: Contact transmission, stability, inactivation, and infection sources. *Plant Disease*, *99*(3), 387–394.

Culley, A. I., Lang, A. S., & Suttle, C. A., (2006). Metagenomic analysis of coastal RNA virus communities. *Science*, *312*(5781), 1795–1798.

Dáder, B., Then, C., Berthelot, E., Ducousso, M., Ng, J. C., & Drucker, M., (2017). Insect transmission of plant viruses: Multilayered interactions optimize viral propagation. *Insect Science*, *24*(6), 929–946.

Davey, T., Carnegie, S., Saddler, G., & Mitchell, W. J., (2014). The importance of the infected seed tuber and soil inoculum in transmitting *Potato mop-top virus* to potato plants. *Plant Pathology*, *63*(1), 88–97.

Diaz-Lara, A., Stevens, K., Klaassen, V., Golino, D., & Al Rwahnih, M., (2020). Comprehensive real-time RT-PCR assays for the detection of fifteen viruses infecting *Prunus* spp. *Plants*, *9*(2), 273.

Djikeng, A., Kuzmickas, R., Anderson, N. G., & Spiro, D. J., (2009). Metagenomic analysis of RNA viruses in a fresh water lake. *PLOS One*, *4*(9), e7264.

Dolja, V. V., & Koonin, E. V., (2018). Metagenomics reshapes the concepts of RNA virus evolution by revealing extensive horizontal virus transfer. *Virus Research*, *244*, 36–52.

Dombrovsky, A., & Smith, E., (2017). Seed transmission of *Tobamoviruses*: Aspects of global disease distribution. In: Jimenez-Lopez, J. C., (ed.), *Advances in Seed Biology* (pp. 233–260). InTech.

EPPO, (2019). *EPPO Reporting Service* (p. 1).

Fabre, F., Kervarrec, C., Mieuzet, L., Riault, G., Vialatte, A., & Jacquot, E., (2003). Improvement of *Barley yellow dwarf virus*-PAV detection in single aphids using a fluorescent real time RT-PCR. *Journal of Virological Methods*, *110*(1), 51–60.

Farzadfar, S., Golnaraghi, A., & Pourrahim, R., (2002). *Plant Viruses of Iran*. Saman Co.: Tehran (Iran).

Farzadfar, S., Pourrahim, R., Golnaraghi, A., & Ahoonmanesh, A., (2007). Surveys of *Beet necrotic yellow vein virus*, *Beet soil borne virus*, *Beet virus Q* and *Polymyxa betae* in sugar beet fields in Iran. *Journal of Plant Pathology*, 277–281.

Farzadfar, S., Pourrahim, R., Golnaraghi, A., & Ahoonmanesh, A., (2005). Occurrence of *Cauliflower mosaic virus* in different cruciferous plants in Iran. *Plant Pathology*, *54*(6).

Farzadfar, S., Pourrahim, R., Golnaraghi, A., & Ahoonmanesh, A., (2006). Distribution and incidence of some aphid and leafhopper transmitted viruses infecting sugar beets in Iran. *Plant Disease*, *90*(3), 252–258.

Fauquet, C. M., Mayo, M. A., Maniloff, J., Desselberger, U., & Ball, L. A., (2005). *Virus Taxonomy: VIIIth Report of the International Committee on Taxonomy of Viruses*. Academic Press.

Feng, Y., Krueger, E. N., Liu, S., Dorman, K., Bonning, B. C., & Miller, W. A., (2017). Discovery of known and novel viral genomes in soybean aphid by deep sequencing. *Phytobiomes*, *1*(1), 36–45.

Fernandez-Cassi, X., Timoneda, N., Gonzales-Gustavson, E., Abril, J., Bofill-Mas, S., & Girones, R., (2017). A metagenomic assessment of viral contamination on fresh parsley plants irrigated with fecally tainted river water. *International Journal of Food Microbiology*, *257*, 80–90.

Fernandez-Cassi, X., Timoneda, N., Martínez-Puchol, S., Rusiñol, M., Rodriguez-Manzano, J., Figuerola, N., Bofill-Mas, S., et al., (2018). Metagenomics for the study of viruses in urban sewage as a tool for public health surveillance. *Science of the Total Environment*, *618*, 870–880.

Fetters, A. M., Cantalupo, P. G., Wei, N., Robles, M. T. S., Stanley, A., Stephens, J. D., Pipas, J. M., & Ashman, T. L., (2022). The pollen virome of wild plants and its association with variation in floral traits and land use. *Nature Communications*, *13*(1), 1–11.

Fillhart, R. C., Bachand, G. D., & Castello, J. D., (1998). Detection of infectious tobamoviruses in forest soils. *Applied and Environmental Microbiology*, *64*(4), 1430–1435.

Fontenele, R. S., Alves-Freitas, D. M., Silva, P. I., Foresti, J., Silva, P., Godinho, M. T., Varsani, A., & Ribeiro, S. G., (2018). Discovery of the first maize-infecting mastrevirus in the Americas using a vector-enabled metagenomics approach. *Archives of Virology*, *163*(1), 263–267.

François, S., Antoine-Lorquin, A., Kulikowski, M., Frayssinet, M., Filloux, D., Fernandez, E., Roumagnac, P., et al., (2021). Characterization of the viral community associated with the alfalfa weevil (*Hypera postica*) and its host plant, Alfalfa (*Medicago sativa*). *Viruses*, *13*(5), 791.

Fuchs, E., Schlufter, C., & Kegler, H., (1996). Occurrence of a Plant Virus in the Northern Sea. *Archives of Phytopathology and Plant Protection, 30,* 365–366.

Garcés-Ruiz, M., Senés-Guerrero, C., Declerck, S., & Cranenbrouck, S., (2019). Community composition of arbuscular mycorrhizal fungi associated with native plants growing in a petroleum-polluted soil of the Amazon region of Ecuador. *Microbiology Open*, *8*(4), e00703.

García-Arenal, F., & Zerbini, F. M., (2019). Life on the edge: Geminiviruses at the interface between crops and wild plant hosts. *Annual Review of Virology*, *6*, 411–433.

Gibbs, A. J., Trueman, J., & Gibbs, M. J., (2008). The bean common mosaic virus lineage of potyviruses: Where did it arise and when? *Archives of Virology*, *153*(12), 2177–2187.

Gillaspie, Jr. A., Hopkins, M., & Pinnow, D. (1995). Seedborne viruses in preintroduction cowpea seed lots and establishment. *Plant Disease 79*(4).

Godinho, M. T., Paula, D. P., Varsani, A., & Ribeiro, S. G., (2017). Genome sequence of *Cauliflower mosaic virus* identified in earwigs (*Doru luteipes*) through a metagenomic approach. *Genome Announcements*, *5*(11), e00043–17.

Golnaraghi, A., Shahraeen, N., Pourrahim, R., Farzadfar, S., & Ghasemi, A., (2004). Occurrence and relative incidence of viruses infecting soybeans in Iran. *Plant Disease*, *88*(10), 1069–1074.

Golnaraghi, A., Shahraeen, N., Pourrahim, R., Farzadfar, S., & Ghasemi, A., (2002). First report of the natural occurrence of eight viruses affecting soybeans in Iran. *Plant Pathology*, *51*(6), 794–794.

Gómez, P., Sempere, R., Elena, S. F., & Aranda, M. A., (2009). Mixed infections of *Pepino mosaic virus* strains modulate the evolutionary dynamics of this emergent virus. *Journal of Virology*, *83*(23), 12378–12387.

Gooding, Jr. G., & Todd, F., (1976). Soil-borne *Tobacco mosaic virus* as an inoculum source for flue-cured tobacco. *Tob. Sci.*, *20*, 133–135.

Gosalvez, B., Navarro, J., Lorca, A., Botella, F., Sanchez-Pina, M., & Pallas, V., (2003). Detection of *Melon necrotic spot virus* in water samples and melon plants by molecular methods. *Journal of Virological Methods*, *113*(2), 87–93.

Grogan, R., & Campbell, R., (1966). Fungi as vectors and hosts of viruses. *Annual Review of Phytopathology*, *4*(1), 29–52.

Groves, C., German, T., Dasgupta, R., Mueller, D., & Smith, D. L., (2016). Seed transmission of soybean vein necrosis virus: The first *tospovirus* implicated in seed transmission. *PLoS One*, *11*(1), e0147342.

Hamza, I. A., Jurzik, L., Überla, K., & Wilhelm, M., (2011). Evaluation of pepper mild mottle virus, human picobirnavirus and *Torque teno virus* as indicators of fecal contamination in river water. *Water Research*, *45*(3), 1358–1368.

Haramoto, E., Kitajima, M., Kishida, N., Konno, Y., Katayama, H., Asami, M., & Akiba, M., (2013). Occurrence of pepper mild mottle virus in drinking water sources in Japan. *Applied and Environmental Microbiology*, *79*(23), 7413–7418.

Harrison, B., (1964). The transmission of plant viruses in soil. *Plant Virology*, 118–147.

Hasanvand, V., Kamali, M., Heydarnejad, J., Massumi, H., Kvarnheden, A., & Varsani, A., (2018). Identification of a new turncurtovirus in the leafhopper *Circulifer haematoceps* and the host plant species *Sesamum indicum*. *Virus Genes*, *54*(6), 840–845.

Hasiów-Jaroszewska, B., Boezen, D., & Zwart, M. P., (2021). Metagenomic studies of viruses in weeds and wild plants: A powerful approach to characterize variable virus communities. *Viruses*, *13*(10), 1939.

Hassard, F., Gwyther, C. L., Farkas, K., Andrews, A., Jones, V., Cox, B., Brett, H., et al., (2016). Abundance and distribution of enteric bacteria and viruses in coastal and estuarine sediments – A review. *Frontiers in Microbiology*, *7*, 1692.

Hollings, M., Komuro, Y., & Tochihara, H., (1975). *Descriptions of Plant Viruses No. 154* (p. 4). Wellesbourne, UK, p AAB.

Hong, C., & Moorman, G., (2005). Plant pathogens in irrigation water: Challenges and opportunities. *Critical Reviews in Plant Sciences*, *24*(3), 189–208.

Hull, R., (2014). *Plant Virology*. Academic Press.

Islam, W., Zhang, J., Adnan, M., Noman, A., Zaynab, M., & Wu, Z., (2017). Plant virus ecology: A glimpse of recent accomplishments. *Appl. Ecol. Environ. Res.*, *15*(1), 691–705.

Jeżewska, M., Zarzyńska-Nowak, A., & Trzmiel, K., (2018). Detection of infectious *Tobamoviruses* in irrigation and drainage canals in Greater Poland. *Journal of Plant Protection Research*.

Jiwaji, M., Matcher, G. F., De Bruyn, M. M., Awando, J. A., Moodley, H., Waterworth, D., Jarvie, R. A., & Dorrington, R. A., (2019). Providence virus: An animal virus that replicates in plants or a plant virus that infects and replicates in animal cells? *PLoS One*, *14*(6), e0217494.

Johansen, E., Edwards, M. C., & Hampton, R. O., (1994). Seed transmission of viruses: Current perspectives. *Annual Review of Phytopathology*, *32*(1), 363–386.

Jones, R. A., & Naidu, R. A., (2019). Global dimensions of plant virus diseases: Current status and future perspectives. *Annual Review of Virology*, *6*, 387–409.

Jones, R. A., (2018). Plant and insect viruses in managed and natural environments: Novel and neglected transmission pathways. *Advances in Virus Research*, *101*, 149–187.

Kaiser, W. J., (1980). Use of thermotherapy to free potato tubers of alfalfa mosaic, potato leaf roll and *Tomato black ring viruses*. *Phytopathology*, *70*(11), 1119–1122.

Kamali, M., Heydarnejad, J., Pouramini, N., Masumi, H., Farkas, K., Kraberger, S., & Varsani, A., (2017). Genome sequences of *Beet curly top Iran virus*, *Oat dwarf virus*, *Turnip curly top virus*, and *Wheat dwarf virus* identified in leafhoppers. *Genome Announcements*, *5*(8), e01674–16.

Kateřina, S., Helena, S., & Karel, D., (2017). Distribution of viruses in the shallot germplasm collection of the Czech Republic–short communication. *Horticultural Science*, *44*(1), 49–52.

Katoch, M., Abdin, M., Ram, R., & Zaidi, A., (2003). An overview of diagnostics for viruses infecting gladiolus. *Crop Protection*, *22*(1), 153–156.

Kegler, H., Kontzog, H. G., & Kleinhempel, H., (1991). Das verhalten pflanzenpathogener viren im boden (übersichtsarbeit). *Archives of Phytopathology & Plant Protection*, *27*(5), 339–346.

Kennedy, J. S., Day, M. F., & Eastop, V. F., (1962). A Conspectus of Aphids as Vectors of Plant Viruses. Commonwealth Institute of Entomology, 114 pp. Published June, 1962. The Eastern Press Ltd., London, UK.

Keswick, B. H., & Gerba, C. P., (1980). Viruses in groundwater. *Environmental Science & Technology*, *14*(11), 1290–1297.

Khankhum, S., (2016). Persistent RNA Viruses of Common Bean (Phaseolus vulgaris): Distribution and Interaction with the Host and Acute Plant Viruses. LSU Doctoral Dissertations. 838. https://repository.lsu.edu/gradschool_dissertations/838.

Khelifa, M., (2019). Detection and quantification of *Potato virus Y* genomes in single aphid stylets. *Plant Disease*, *103*(9), 2315–2321.

Kil, E. J., Park, J., Choi, H. S., Kim, C. S., & Lee, S., (2017). Seed transmission of *Tomato yellow leaf curl virus* in white soybean (Glycine max). *The Plant Pathology Journal*, *33*(4), 424.

Kim, J. S., Yoon, S. J., Park, Y. J., Kim, S. Y., & Ryu, C. M., (2020). Crossing the kingdom border: Human diseases caused by plant pathogens. *Environmental Microbiology*, *22*(7), 2485–2495.

Kim, J., Cha, D. J., Kwon, M., & Maharjan, R., (2016). *Potato virus Y* (PVY) detection in a single aphid by one-step RT-PCR with boiling technique. *Entomological Research*, *46*(4), 278–285.

Kim, K. H., Chang, H. W., Nam, Y. D., Roh, S. W., Kim, M. S., Sung, Y., Jeon, C. O., et al., (2008). Amplification of uncultured single-stranded DNA viruses from rice paddy soil. *Applied and Environmental Microbiology*, *74*(19), 5975–5985.

Kimura, M., Jia, Z. J., Nakayama, N., & Asakawa, S., (2008). Ecology of viruses in soils: Past, present, and future perspectives. *Soil Science and Plant Nutrition*, *54*(1), 1–32.

Kitajima, E. W., (2020). An annotated list of plant viruses and viroids described in Brazil (1926–2018). *Biota Neotropica*, *20*(2).

Kitajima, M., Iker, B. C., Pepper, I. L., & Gerba, C. P., (2014). Relative abundance and treatment reduction of viruses during wastewater treatment processes—Identification of potential viral indicators. *Science of the Total Environment*, *488*, 290–296.

Kitajima, M., Sassi, H. P., & Torrey, J. R., (2018). Pepper mild mottle virus as a water quality indicator. *NPJ Clean Water*, *1*(1), 1–9.

Kleinhempel, H., & Kegler, G., (1982). Transmission of *Tomato bushy stunt virus* without vectors. *Acta Phytopathologica Academiae Scientiarum Hungaricae*, *17*(1, 2), 17–21.

Klerks, M., Leone, G., Verbeek, M., Van, D. H. J., & Schoen, C., (2001). Development of a multiplex AmpliDet RNA for the simultaneous detection of *Potato leafroll virus* and *Potato virus Y* in potato tubers. *Journal of Virological Methods*, *93*(1, 2), 115–125.

Koenig, R., & Lesemann, D. E., (1985). Plant viruses in German Rivers and Lakes: I. tombusviruses, a potexvirus and carnation mottle virus. *Journal of Phytopathology*, *112*(2), 105–116.

Koenig, R., (1986). Plant viruses in rivers and lakes. *Advances in Virus Research*, *31*, 321–333.

Koenig, R., Rüdel, M., & Lesemann, D. E., (1989). Detection of petunia asteroid mosaic, carnation ringspot and tobacco necrosis viruses in ditches and drainage canals in a grapevine-growing area in West Germany. *Journal of Phytopathology*, *127*(2), 169–172.

Komuro, Y., & Iwaki, M., (1969). Presence of *Tobacco mosaic virus* in roots of field-grown tomato plants healthy in appearance. *Japanese Journal of Phytopathology*, *35*(4), 294–298.

Kontzog, H., Kleinhempel, H., & Kegler, H., (1988). Detection of plant pathogenic viruses in waters. *Archiv für Phytopathologie und Pflanzenschutz*, *24*(2), 171, 172.

Kumar, P. L., Cuervo, M., Kreuze, J. F., Muller, G., Kulkarni, G., Kumari, S. G., Massart, S., et al., (2021). Phytosanitary interventions for safe global germplasm exchange and the prevention of transboundary pest spread: The role of CGIAR germplasm health units. *Plants*, *10*(2), 328.

Kyrychenko, A., Shcherbatenko, I., & Kovalenko, A., (2021). Viruses of wild plants and current metagenomic methods for their investigation. *Cytology and Genetics*, *55*(3), 248–255.

Lacroix, C., Renner, K., Cole, E., Seabloom, E. W., Borer, E. T., & Malmstrom, C. M., (2016). Methodological guidelines for accurate detection of viruses in wild plant species. *Applied and Environmental Microbiology*, *82*(6), 1966–1975.

Lal, A., Pant, M., & Rani, A., (2015). The who's who of plant viruses: A cognitive approach. *Asian Journal of Pharmaceutical and Clinical Research*, 60–68.

Lanter, J. M., McGuire, J., & Goode, M., (1982). Persistence of *Tomato mosaic virus* in tomato debris and soil under field conditions. *Plant Disease*, *66*(7), 552–555.

Latvala-Kilby, S., Aura, J. M., Pupola, N., Hannukkala, A., & Valkonen, J. P., (2009). Detection of *Potato mop-top virus* in potato tubers and sprouts: Combinations of RNA2 and RNA3 variants and incidence of symptomless infections. *Phytopathology*, *99*(5), 519–531.

Lee, C. H., Lee, G. B. V., Wang, G. J., & Jan, F. J., (2021). Development of novel techniques to detect orchid viruses. In: *Orchid Biotechnology IV* (pp. 207–218). World Scientific.

Li, J. X., Liu, S. S., & Gu, Q. S., (2016). Transmission efficiency of *Cucumber green mottle mosaic virus* via seeds, soil, pruning and irrigation water. *Journal of Phytopathology*, *164*(5), 300–309.

Li, L., Victoria, J. G., Wang, C., Jones, M., Fellers, G. M., Kunz, T. H., & Delwart, E., (2010). Bat guano virome: Predominance of dietary viruses from insects and plants plus novel mammalian viruses. *Journal of Virology*, *84*(14), 6955–6965.

Lim, E. S., Zhou, Y., Zhao, G., Bauer, I. K., Droit, L., Ndao, I. M., Warner, B. B., et al., (2015). Early life dynamics of the human gut virome and bacterial microbiome in infants. *Nature Medicine*, *21*(10), 1228–1234.

López-Bueno, A., Tamames, J., Velázquez, D., Moya, A., Quesada, A., & Alcamí, A., (2009). High diversity of the viral community from an Antarctic Lake. *Science*, *326*(5954), 858–861.

Maclot, F., Candresse, T., Filloux, D., Malmstrom, C. M., Roumagnac, P., Van, D. V. R., & Massart, S., (2020). Illuminating an ecological blackbox: Using high throughput sequencing to characterize the plant virome across scales. *Frontiers in Microbiology*, *11*, 578064.

Mascia, T., Labarile, R., Doohan, F., & Gallitelli, D., (2019). *Tobacco mosaic virus* infection triggers an RNAi-based response in phytophthora infestans. *Scientific Reports*, *9*(1), 1–13.

Mascia, T., Nigro, F., Abdallah, A., Ferrara, M., De Stradis, A., Faedda, R., Palukaitis, P., & Gallitelli, D., (2014). Gene silencing and gene expression in phytopathogenic fungi using a plant virus vector. *Proceedings of the National Academy of Sciences*, *111*(11), 4291–4296.

Mascia, T., Vučurović, A., Minutillo, S., Nigro, F., Labarile, R., Savoia, M., Palukaitis, P., & Gallitelli, D., (2019). Infection of *Colletotrichum acutatum* and *Phytophthora infestans* by taxonomically different plant viruses. *European Journal of Plant Pathology*, *153*(4), 1001–1017.

Matthews, R. E. F., (2012). *Fundamentals of Plant Virology*. Academic Press.

McKirdy, S., Jones, R., & Sivasithamparam, K., (1998). Determining the effectiveness of grazing and trampling by livestock in transmitting white clover mosaic and subterranean clover mottle viruses. *Annals of Applied Biology*, *132*(1), 91–105.

McQuaig, S. M., Scott, T. M., Lukasik, J. O., Paul, J. H., & Harwood, V. J., (2009). Quantification of human polyomaviruses JC virus and BK virus by TaqMan quantitative PCR and comparison to other water quality indicators in water and fecal samples. *Applied and Environmental Microbiology*, *75*(11), 3379–3388.

Mehle, N., & Ravnikar, M., (2012). Plant viruses in aqueous environment–survival, water mediated transmission and detection. *Water Research*, *46*(16), 4902–4917.

Mehle, N., Gutierrez-Aguirre, I., Kutnjak, D., & Ravnikar, M., (2018). Water-mediated transmission of plant, animal, and human viruses. *Advances in Virus Research*, *101*, 85–128.

Mehle, N., Gutiérrez-Aguirre, I., Prezelj, N., Delić, D., Vidic, U., & Ravnikar, M., (2014). Survival and transmission of *Potato virus Y*, *Pepino mosaic virus*, and potato spindle tuber viroid in water. *Applied and Environmental Microbiology*, *80*(4), 1455–1462.

Min, B. E., Feldman, T. S., Ali, A., Wiley, G., Muthukumar, V., Roe, B. A., Roossinck, M., et al., (2012). Molecular characterization, ecology, and epidemiology of a novel tymovirus in *Asclepias viridis* from Oklahoma. *Phytopathology*, *102*(2), 166–176.

Moore, N. E., Wang, J., Hewitt, J., Croucher, D., Williamson, D. A., Paine, S., Yen, S., et al., (2015). Metagenomic analysis of viruses in feces from unsolved outbreaks of gastroenteritis in humans. *Journal of Clinical Microbiology*, *53*(1), 15–21.

Nagata, T., Inoue-Nagata, A. K., Ávila, A. C. D., & Giordano, L. D. B., (2004). Print-capture PCR for detection of tomato *Begomoviruses* from plants and whiteflies. *Fitopatologia Brasileira*, *29*, 91–93.

Nakamura, S., Yang, C. S., Sakon, N., Ueda, M., Tougan, T., Yamashita, A., Goto, N., et al., (2009). Direct metagenomic detection of viral pathogens in nasal and fecal specimens using an unbiased high-throughput sequencing approach. *PloS One*, *4*(1), e4219.

Narayanasamy, P., (2011). Detection of viral and viroid plant pathogens in the environment. In: *Microbial Plant Pathogens-Detection and Disease Diagnosis* (pp. 221–249). Springer.

Nault, L., (1997). Arthropod transmission of plant viruses: A new synthesis. *Annals of the Entomological Society of America*, *90*(5), 521–541.

Navarro, J. A., Sanchez-Navarro, J. A., & Pallas, V., (2019). Key checkpoints in the movement of plant viruses through the host. *Advances in Virus Research*, *104*, 1–64.

Nazari, K., Sheykhrezaee, Z., & Golnaraghi, A., (2021). *Man in the Coronaceous Period (1): How Long Will Corona Vaccination Last in the World?* Iranian Students's News Agency, Report No. 1400020503193.

Nerva, L., Varese, G., Falk, B., & Turina, M., (2017). Mycoviruses of an endophytic fungus can replicate in plant cells: Evolutionary implications. *Scientific Reports*, *7*(1), 1–11.

Ng, T. F. F., Chen, L. F., Zhou, Y., Shapiro, B., Stiller, M., Heintzman, P. D., Varsani, A., et al., (2014). Preservation of viral genomes in 700-y-old caribou feces from a subarctic ice patch. *PNAS*, *111*(47), 16842–16847.

Ng, T. F. F., Duffy, S., Polston, J. E., Bixby, E., Vallad, G. E., & Breitbart, M., (2011). Exploring the diversity of plant DNA viruses and their satellites using vector-enabled metagenomics on whiteflies. *PloS One*, *6*(4), e19050.

Ng, T. F. F., Marine, R., Wang, C., Simmonds, P., Kapusinszky, B., Bodhidatta, L., Oderinde, B. S., et al., (2012). High variety of known and new RNA and DNA viruses of diverse origins in untreated sewage. *Journal of Virology*, *86*(22), 12161–12175.

Nguyen, H. D., Tomitaka, Y., Ho, S. Y., Duchêne, S., Vetten, H. J., Lesemann, D., Walsh, J. A., et al., (2013). Turnip mosaic potyvirus probably first spread to *Eurasian brassica* crops from wild orchids about 1000 years ago. *PloS One*, *8*(2), e55336.

Nibert, M. L., Vong, M., Fugate, K. K., & Debat, H. J., (2018). Evidence for contemporary plant mitoviruses. *Virology*, *518*, 14–24.

Nienhaus, F., & Castello, J., (1989). Viruses in forest trees. *Annual Review of Phytopathology*, *27*(1), 165–186.

Nouri, S., Salem, N., Nigg, J. C., & Falk, B. W., (2016). Diverse array of new viral sequences identified in worldwide populations of the Asian citrus psyllid (*Diaphorina citri*) using viral metagenomics. *Journal of Virology*, *90*(5), 2434–2445.

Ohshima, K., Nomiyama, R., Mitoma, S., Honda, Y., Yasaka, R., & Tomimura, K., (2016). Evolutionary rates and genetic diversities of mixed potyviruses in narcissus. *Infection, Genetics and Evolution*, *45*, 213–223.

Olmos, A., Bertolini, E., Gil, M., & Cambra, M., (2005). Real-time assay for quantitative detection of non-persistently transmitted *Plum pox virus* RNA targets in single aphids. *Journal of Virological Methods*, *128*(1, 2), 151–155.

Pares, R., Gunn, L., & Cresswell, G., (1992). *Tomato mosaic virus* infection in a recirculating nutrient solution. *Journal of Phytopathology*, *135*(3), 192–198.

Pares, R., Gunn, L., & Keskula, E., (1996). The role of infective plant debris, and its concentration in soil, in the ecology of *Tomato mosaic tobamovirus*—A non-vectored plant virus. *Journal of Phytopathology*, *144*(3), 147–150.

Parizad, S., Dizadji, A., Habibi, M. K., Winter, S., Kalantari, S., Movi, S., Tendero, C. L., et al., (2019). The effects of geographical origin and virus infection on the saffron (*Crocus sativus* L.) quality. *Food Chemistry*, *295*, 387–394.

Parizad, S., Dizadji, A., Koohi, H. M., Winter, S., Kalantari, S., Movi, S., García-Arenal, F., & Ayllón, M. A., (2018). Description and genetic variation of a distinct species of potyvirus infecting saffron (*Crocus sativus* L.) plants in major production regions in Iran. *Annals of Applied Biology*, *173*(3), 233–242.

Phan, T. G., Kapusinszky, B., Wang, C., Rose, R. K., Lipton, H. L., & Delwart, E. L., (2011). The fecal viral flora of wild rodents. *PLoS Pathogens*, *7*(9), e1002218.

Piazzolla, P., (1989). Partial stabilization of *Cucumber mosaic virus* (CMV) induced by a clay mineral. *Journal of Phytopathology*, *124*(2), 97–100.

Piazzolla, P., Castellano, M., & De Stradis, A., (1986). Presence of plant viruses in some rivers of southern Italy. *Journal of Phytopathology*, *116*(3), 244–246.

Pleše, N., Juretić, N., Mamula, D., Polák, Z., & Krajačić, M., (1996). Plant viruses in soil and water of forest ecosystems in Croatia. *Phyton (Horn)*, *36*(1), 135–143.

Pourrahim, R., Farzadfar, S., Golnaraghi, A., & Ahoonmanesh, A., (2007). Incidence and distribution of important viral pathogens in some Iranian potato fields. *Plant Disease*, *91*(5), 609–615.

Prendeville, H. R., (2010). *Ecological Effects of Virus-Resistant Transgenic Squash*. University of Nebraska Lincoln, Lincoln, Nebraska.

Přibylová, J., Lenz, O., Fránová, J., Koloniuk, I., & Špak, J., (2020). Comparison of the incidence of viruses within germplasm, orchards, and wild cherry trees in the Czech Republic. *Annals of Applied Biology*, *176*(2), 138–146.

Proeseler, G., Stanarius, A., Kontzog, H., & Barth, H., (1990). Virus infections of aquatic plants. *Archiv für Phytopathologie und Pflanzenschutz*, *26*(1), 19–24.

Rae, C. S., Khor, W., Wang, Q., Destito, G., Gonzalez, M. J., Singh, P., Thomas, D. M., et al., (2005). Systemic trafficking of plant virus nanoparticles in mice via the oral route. *Virology*, *343*(2), 224–235.

Rastrojo, A., & Alcamí, A., (2018). Viruses in polar lake and soil ecosystems. *Advances in Virus Research*, *101*, 39–54.

Roberts, A. G., (2014). Plant viruses: Soil-borne. *eLS*.

Rodríguez-Nevado, C., Gavilán, R. G., & Pagán, I., (2020). Host abundance and identity determine the epidemiology and evolution of a generalist plant virus in a wild ecosystem. *Phytopathology*, *110*(1), 94–105.

Rojas, M. R., & Gilbertson, R. L., (2008). Emerging plant viruses: A diversity of mechanisms and opportunities. In: *Plant Virus Evolution* (pp. 27–51). Springer.

Roossinck, M. J., & García-Arenal, F., (2015). Ecosystem simplification, biodiversity loss and plant virus emergence. *Current Opinion in Virology*, *10*, 56–62.

Roossinck, M. J., (2010). Lifestyles of plant viruses. *Phil. Trans. R. Soc. B*, *365*(1548), 1899–1905.

Roossinck, M. J., (2011). The big unknown: Plant virus biodiversity. *Current Opinion in Virology*, *1*(1), 63–67.

Roossinck, M. J., (2012). Plant virus metagenomics: Biodiversity and ecology. *Annual Review of Genetics*, *46*, 359–369.

Roossinck, M. J., (2015). Metagenomics of plant and fungal viruses reveals an abundance of persistent lifestyles. *Frontiers in Microbiology*, *5*, 767.

Roossinck, M. J., (2015). Plants, viruses, and the environment: Ecology and mutualism. *Virology*, *479*, 271–277.

Roossinck, M. J., (2018). Evolutionary and ecological links between plant and fungal viruses. *New Phytologist*, *221*(1), 86–92.

Rosario, K., Nilsson, C., Lim, Y. W., Ruan, Y., & Breitbart, M., (2009). Metagenomic analysis of viruses in reclaimed water. *Environmental Microbiology*, *11*(11), 2806–2820.

Rosario, K., Padilla-Rodriguez, M., Kraberger, S., Stainton, D., Martin, D. P., Breitbart, M., & Varsani, A., (2013). Discovery of a novel mastrevirus and alphasatellite-like circular DNA in dragonflies (*Epiprocta*) from Puerto Rico. *Virus Research*, *171*(1), 231–237.

Rosario, K., Seah, Y. M., Marr, C., Varsani, A., Kraberger, S., Stainton, D., Moriones, E., et al., (2015). Vector-enabled metagenomic (VEM) surveys using whiteflies (*Aleyrodidae*) reveal novel *Begomovirus* species in the new and old worlds. *Viruses*, *7*(10), 5553–5570.

Rosario, K., Symonds, E. M., Sinigalliano, C., Stewart, J., & Breitbart, M., (2009). *Pepper mild mottle virus* as an indicator of fecal pollution. *Applied and Environmental Microbiology*, *75*(22), 7261–7267.

Sachsenröder, J., Twardziok, S., Hammerl, J. A., Janczyk, P., Wrede, P., Hertwig, S., & Johne, R., (2012). Simultaneous identification of DNA and RNA viruses present in pig faeces using process-controlled deep sequencing. *PloS One*, *7*(4), e34631.

Salati, R., Nahkla, M. K., Rojas, M. R., Guzman, P., Jaquez, J., Maxwell, D. P., & Gilbertson, R. L., (2002). *Tomato yellow leaf curl virus* in the Dominican Republic: Characterization of an infectious clone, virus monitoring in whiteflies, and identification of reservoir hosts. *Phytopathology*, *92*(5), 487–496.

Salem, N., Ehlers, J., Roberts, P., & Ng, J., (2010). Biological and molecular diagnosis of seedborne viruses in cowpea germplasm of geographically diverse Sub-Saharan origins. *Plant Pathology*, *59*(4), 773–784.

Sarra, S., & Peters, D., (2003). *Rice yellow mottle virus* is transmitted by cows, donkeys, and grass rats in irrigated rice crops. *Plant Disease*, *87*(7), 804–808.

Sarwar, M., Aslam, M., Sarwar, S., & Iftikhar, R., (2020). Different nematodes and plasmodiophorids as vectors of plant viruses. In: *Applied Plant Virology* (pp. 275–290). Elsevier.

Sastry, K. S., & Zitter, T. A., (2014). Management of virus and viroid diseases of crops in the tropics. In: *Plant Virus and Viroid Diseases in the Tropics* (pp. 149–480). Springer.

Sastry, K. S., (2013). Plant virus transmission through vegetative propagules (asexual reproduction). In: *Seed-Borne Plant Virus Diseases* (pp. 285–305). Springer.

Schwarz, D., Beuch, U., Bandte, M., Fakhro, A., Büttner, C., & Obermeier, C., (2010). Spread and interaction of *Pepino mosaic virus* (PepMV) and *Pythium aphanidermatum* in a closed nutrient solution recirculation system: Effects on tomato growth and yield. *Plant Pathology*, *59*(3), 443–452.

Sculthorpe, C. D., (1967). Biology of Aquatic Vascular Plants. Edward Arnold Ltd., London, UK.

Shahmohammadi, N., Mansourpour, M., & Golnaraghi, A., (2022). Current challenges and future perspectives on detection of geminiviruses. In: Gaur, R., Sharma, P., & Czosnek, H., (eds.), *Geminivirus: Detection, Diagnosis and Management* (pp. 3–24). Elsevier.

Shukla, D., Shanks, G., Teakle, S., & Behncken, G., (1979). Mechanical transmission of *Galinsoga mosaic virus* in soil. *Australian Journal of Biological Sciences*, *32*(2), 267–276.

Singh, D., Verma, N., & Varma, A., (2008). The fungal transmitted viruses. In: *Mycorrhiza* (pp. 485–503). Springer.

Starr, E. P., Nuccio, E. E., Pett-Ridge, J., Banfield, J. F., & Firestone, M. K., (2019). Metatranscriptomic reconstruction reveals RNA viruses with the potential to shape carbon cycling in soil. *Proceedings of the National Academy of Sciences*, *116*(51), 25900–25908.

Stein, A., Levy, S., & Loebenstein, G., (1988). In: *Detection of Viruses in Gladioli Corms, VII International Symposium on Virus Diseases of Ornamental Plants* (Vol. 234, pp. 275–280).

Stewart-Wade, S. M., (2011). Plant pathogens in recycled irrigation water in commercial plant nurseries and greenhouses: Their detection and management. *Irrigation Science*, *29*(4), 267–297.

Stobbe, A., & Roossinck, M. J., (2016). Plant virus diversity and evolution. In: *Current Research Topics in Plant Virology* (pp. 197–215). Springer.

Susi, H., Filloux, D., Frilander, M. J., Roumagnac, P., & Laine, A. L., (2019). Diverse and variable virus communities in wild plant populations revealed by metagenomic tools. *PeerJ*, *7*, e6140.

Sutela, S., Poimala, A., & Vainio, E. J., (2019). Viruses of fungi and oomycetes in the soil environment. *FEMS Microbiology Ecology*, *95*(9), fiz119.

Tan, S. H., Nishiguchi, M., Sakamoto, W., Ogura, Y., Murata, M., Ugaki, M., Tomiyama, M., & Motoyoshi, F., (1997). Molecular analysis of the genome of an attenuated strain of *Cucumber green mottle mosaic virus*. *Japanese Journal of Phytopathology*, *63*(6), 470–474.

Teakle, D., & Morris, T., (1981). Transmission of *Southern bean mosaic virus* from soil to bean seeds. *Plant Disease*, *65*(7), 599, 600.

Teakle, D., (1986). Abiotic transmission of *Southern bean mosaic virus* in soil. *Australian Journal of Biological Sciences*, *39*(4), 353–360.

Teimoori-Boghsani, Y., Ganjeali, A., Cernava, T., Müller, H., Asili, J., & Berg, G., (2020). Endophytic fungi of native salvia abrotanoides plants reveal high taxonomic diversity and unique profiles of secondary metabolites. *Frontiers in Microbiology*, *10*, 3013.

Thekke-Veetil, T., Lagos-Kutz, D., McCoppin, N. K., Hartman, G. L., Ju, H. K., Lim, H. S., & Domier, L., (2020). Soybean thrips (*Thysanoptera: Thripidae*) harbor highly diverse populations of arthropod, fungal and plant viruses. *Viruses*, *12*(12), 1376.

Thomas, J., (2015). *Technical Guidelines for the Safe Movement of Musa Germplasm*. Bioversity International.

Tomlinson, J., & Faithfull, E. M., (1984). Studies on the occurrence of *Tomato bushy stunt virus* in English rivers. *Annals of Applied Biology*, *104*(3), 485–495.

Tomlinson, J., Faithfull, E. M., Webb, M., Fraser, R., & Seeley, N., (1983). Chenopodium necrosis: A distinctive strain of *Tobacco necrosis virus* isolated from river water. *Annals of Applied Biology*, *102*(1), 135–147.

Tomlinson, J., Faithfull, E., Flewett, T., & Beards, G., (1982). Isolation of infective *Tomato bushy stunt virus* after passage through the human alimentary tract. *Nature*, *300*(5893), 637, 638.

Tošić, M., & Tošić, D., (1984). Occurrence of tobacco mosaic virus in water of the Danube and Sava rivers. *Journal of Phytopathology*, *110*(3), 200–202.

Traore, M., Traoré, V., Galzi-Pinel, A., Fargette, D., Konate, G., Traore, A., & Traore, O., (2008). Abiotic transmission of *Rice yellow mottle virus* through soil and contact between plants. *Pakistan Journal of Biological Sciences: PJBS*, *11*(6), 900–904.

Valouzi, H., Hashemi, S. S., Wylie, S. J., Ahadiyat, A., & Golnaraghi, A., (2020). *Wisteria vein mosaic virus* detected for the first time in Iran from an unknown host by analysis of aphid vectors. *The Plant Pathology Journal*, *36*(1), 87.

Valouzi, H., Shahmohammadi, N., Golnaraghi, A., Moosavi, M. R., & Ohshima, K., (2021). Genetic diversity and evolutionary analyses of potyviruses infecting narcissus in Iran. *Journal of Plant Pathology*, *104*(1), 237–250.

Van, D. H. J. F., Hogenhout, S. A., & Van, D. W. F., (1999). Recognition and receptors in virus transmission by arthropods. *Trends in Microbiology*, *7*(2), 71–76.

Van, D. H., (1988). Surface water as source in the spread of cucumber *Green mottle mosaic virus*. *Netherlands Journal of Agricultural Science*, *36*(3), 291–299.

Vani, S., & Varma, A., (1993). Properties of *Cucumber green mottle mosaic virus* isolated from water of river Jamuna. *Indian Phytopathology*, *46*(2), 118–122.

Victoria, J. G., Kapoor, A., Li, L., Blinkova, O., Slikas, B., Wang, C., Naeem, A., et al., (2009). Metagenomic analyses of viruses in stool samples from children with acute flaccid paralysis. *Journal of Virology*, *83*(9), 4642–4651.

Virgin, H. W., (2014). The virome in mammalian physiology and disease. *Cell*, *157*(1), 142–150.

Wahyuni, W. S., Hanapi, M., & Hartana, I., (2008). The presence of *Tobacco mosaic virus* in the compost extract of cigar tobacco debris. *HAYATI Journal of Biosciences*, *15*(3), 118–122.

Wamonje, F. O., Michuki, G. N., Braidwood, L. A., Njuguna, J. N., Musembi, M. J., Djikeng, A., Harvey, J. J., & Carr, J. P., (2017). Viral metagenomics of aphids present in bean and maize plots on mixed-use farms in Kenya reveals the presence of three dicistroviruses including a novel *Big Sioux River virus*-like dicistrovirus. *Virology Journal*, *14*(1), 1–13.

Wetter, C., (1975). *Tobacco mosaic virus* and *Para-tobacco mosaic virus* in cigarettes. *Die Naturwissenschaften*, *62*(11), 533–533.

WHO, *World Health Organization Media Centre Fact Sheet*. URL: http://www.who.int/mediacentre/factsheets/fs394/en/ (accessed on 02 March 2024).

Williamson, K. E., Fuhrmann, J. J., Wommack, K. E., & Radosevich, M., (2017). Viruses in soil ecosystems: An unknown quantity within an unexplored territory. *Annu. Rev. Virol.*, *4*(1), 201–219.

Winn, Z., (2020). *MIT Begins Testing Wastewater to Help Detect COVID-19 on Campus*. https://news.mit.edu/2020/testing-wastewater-covid-19-1002 (accessed on 02 March 2024).

Wong, K., Fong, T. T., Bibby, K., & Molina, M., (2012). Application of enteric viruses for fecal pollution source tracking in environmental waters. *Environment International*, *45*, 151–164.

Wu, S., Zhou, L., Zhou, Y., Wang, H., Xiao, J., Yan, S., & Wang, Y., (2020). Diverse and unique viruses discovered in the surface water of the East China Sea. *BMC Genomics*, *21*(1), 1–15.

Wu, Z., Ren, X., Yang, L., Hu, Y., Yang, J., He, G., Zhang, J., et al., (2012). Virome analysis for identification of novel mammalian viruses in bat species from Chinese provinces. *Journal of Virology*, *86*(20), 10999–11012.

Xagoraraki, I., & O'Brien, E., (2020). Wastewater-based epidemiology for early detection of viral outbreaks. In: *Women in Water Quality* (pp. 75–97). Springer.

Xu, F., Jones, T., & Rodgers, P., (1996). Potential of chimaeric plant virus particles as novel, stable vaccines. *Developments in Biological Standardization*, *87*, 201–205.

Yarwood, C., (1960). Release and preservation of virus by roots. *Phytopathology*, *50*(2).

Yazdani-Khameneh, S., & Golnaraghi, A., (2017). The status of *Begomoviruses* in Iran. In: *Begomoviruses: Occurrence and Management in Asia and Africa* (pp. 229–253). Springer.

Yazdani-Khameneh, S., Golnaraghi, A., & Wylie, S., (2019). Diverse endophytic fungi colonize indigenous grasses in the Hyrcanian forest of Iran. In: *The 18th Congress of European Mycologists*. Warsaw, Poland.

Yi, L., Lesemann, D. E., Koenig, R., Rüdel, M., & Pfeilstetter, E., (1992). Isometric plant viruses in ditches and streams in agricultural areas: Recovery of previously found viruses and identification of hitherto unrecorded carmoand *Tombusviruses* including grapevine *Algerian latent virus*. *Journal of Phytopathology*, *134*(2), 121–132.

Young, H. W., & York, D. W., (1996). Reclaimed water reuse in Florida and the South Gulf Coast. *Florida Water Resour. J.*, 32–36.

Zettler, F., Foxe, M., Hartman, R., Edwardson, J., & Christie, R., (1970). Filamentous viruses infecting dasheen and other araceous plants. *Phytopathology*, *60*(6), 983–987.

Zhang, T., Breitbart, M., Lee, W. H., Run, J. Q., Wei, C. L., Soh, S. W. L., Hibberd, M. L., et al., (2006). RNA viral community in human feces: Prevalence of plant pathogenic viruses. *PLoS Biology*, *4*(1), e3.

Zhang, Y., Gao, Y. L., He, W. Q., Wang, Y. Q., Qian, Y. J., Zhou, X. P., & Wu, J., (2020). Monoclonal antibody-based serological detection of *Potato virus* M in potato plants and tubers. *Journal of Integrative Agriculture*, *19*(5), 1283–1291.

CHAPTER 2

Wild Plants: A Source of Emerging Viruses and Their Impact on Agriculture and Food Security

RODOLFO TORRES-DE LOS SANTOS,[1]
MA. GUADALUPE BUSTOS-VÁZQUEZ,[1] HERMILO LUCIO-CASTILLO,[1]
VÍCTOR ALBORES-FLORES,[2] R. IVONNE TORRES-ACOSTA,[1]
DANIEL TRUJILLO-RAMÍREZ,[1] and HUMBERTO MARTÍNEZ-MONTOYA[3]

[1]*Mante-Centro Multidisciplinary Academic Unit, Autonomous University of Tamaulipas, Cd. Mante, Tamaulipas, Mexico*

[2]*Institute of Biosciences, Autonomous University of Chiapas, Tapachula, Chiapas, Mexico*

[3]*Reynosa-Aztlán Multidisciplinary Academic Unit, Autonomous University of Tamaulipas, Reynosa, Tamaulipas, Mexico*

ABSTRACT

Wild plants are part of an agricultural ecosystem and can be a reservoir for emerging disease viruses that damage agricultural crops around the world. In this sense, the diversity of viruses that exist in nature, their genetic variability, and their agro-ecological distribution are unknown, especially at the interfaces between growing areas and neighboring areas; in addition to their impact on world food security.

New and Emerging Plant Viruses: The Threat to Food Security.
Alireza Golnaraghi and Rajarshi Kumar Gaur (Eds.)

2.1 INTRODUCTION

2.1.1 IMPACT OF VIRAL DISEASES ON AGRICULTURE AND FOOD SECURITY

It is common for agriculture to be constantly threatened by plant viruses, which results in economic losses for producers. Annually, losses due to viral diseases reach figures around 60 million dollars. Therefore, it is crucial to control these viruses to maintain an abundant and high-quality food production (Abdelkhalek & Hafez, 2020). According to Salaudeen et al. (2016), climate change can affect the suitability of plants and crops. However, integrated management systems, such as developing virus-resistant varieties, applying traditional or transgenic techniques, and implementing biological control of virus vectors, can help reduce these impacts to some extent.

Phytopathogenic viruses, like other types of viruses, have mechanisms that facilitate their acquisition, retention, and transmission by vectors. Recent studies have shown that plant viruses can manipulate plant phenotypes and vector behavior to ensure their own transmission (Ray & Casteel, 2022). The emergence of new viral genotypes that cause previously unknown diseases in plants is often associated with ecological changes or agricultural practices that facilitate new interactions and interspecies jumps between viral agents, vectors, and hosts (Elena et al., 2011).

To become an emergent viral agent, the emergence of a virus in a new species population can be divided into three stages. First, a single initial infection occurs in a new host with no onward transmission. Secondly, successful spillovers cause local chains of transmission in the new host population before the third stage, which involves epidemic transmission among individuals in the new population (Parrish et al., 2008). The initial step involves the molecular mechanisms needed to trigger inter-species jumping. Initial exposure of the virus to the new host sparks new host colonization. However, in most cases, viral replication and transmission in the new host are not efficient and will not be established, and only viral agents with advantageous genetic mutations could persist in the new population (Elena et al., 2011; Parrish et al., 2008).

2.2 PLANT VIRUSES

Viruses are considered the smallest parasites that interact with all living organisms. Their size ranges between 0.02 and 0.3 micrometers, depending on the type, genus, and species. Some viruses can measure 1 micrometer.

Viruses rely on a cellular host (microorganisms, plants, or humans) to reproduce and are inactive outside of cells. Generally, viruses have a protein outer shell called capsule or capsid, although lipids can also be present. Inside, they contain genetic material such as DNA or RNA, depending on the virus type. Viruses can have various shapes and may distribute their genetic material across multiple particles (Rivera & Wright, 2020). Unlike living systems, viruses only replicate within the cells of their hosts, utilizing the host's protein synthesis machinery. During this process, numerous copies of viral nucleic acids and coat proteins (CPs) are produced. The coat proteins form the capsid, which surrounds and stabilizes the nucleic acid, facilitates adherence to the host, and enables penetration into susceptible cells upon contact (Jawetz et al., 2022). In plant research, the most important viruses are often related to phytopathology, particularly those belonging to the Riboviria empire, Orthornavirae kingdom. Examples include the genera Potyvirus, Cucumovirus, Ilarvirus, Closterovirus, Tobamovirus, Portexvirus, Trichovirus, Polerovirus, Luteovirus, Orthotospovirus, Ophiovirus, and Fijivirus.

On the other hand, some viruses have been discovered that infect plants, causing positive reactions and favoring resistance to stress and infections by microbial pathogens. Some viruses are associated with tripartite virus-fungus-plant interactions, which allow plant development in extreme environments, such as thermotolerance (Márquez et al., 2007; Roossinck, 2015). The behavior induced in plants by virus infection is the result of an evolutionary process over time, influenced by the specific interaction formed, the interaction with vector insects, the condition of the host plant and its species, along with the environmental conditions (Aguilar & Lozano-Duran, 2022; García et al., 2020; Susi et al., 2019; Roossinck, 2015).

2.3 VIRUS VECTORS

Viruses can migrate from one host to another or from one location to another through a vector or transportation system, which can be physical or biological. Based on interactions within ecosystems involving living and non-living organisms (specifically viruses), arthropods are considered the most efficient biological vectors of viruses (Catto et al., 2022; Ghosh et al., 2021; Wu et al., 2022; Bhattacharjee et al., 2022).

Among arthropods, mites and insects are the most common vectors of phytopathogenic viruses. These arthropods visit plants of different crops depending on the host range and the type of phytophagy they exhibit. Thrips, aphids, and whiteflies are the main vectors among insects, and they seek access to plant sap through their jaws or mouthparts (such as chopper-sucker, sponge sucker, siphon sucker, scraper-sucker). There are also reports of viruses infecting plants being transported by nematodes, bacteria, or fungi that are associated with plants (Ziegler-Graff et al., 2020; Di Mattia et al., 2020; Kloth et al., 2020; van Munster, 2020; Wu et al., 2020; Belval et al., 2019).

Virus vector insects may be attracted to infected host plants to extract resources such as sap, bark, or pollen. Some species of *Trigona* spp. visiting *Jatropha curcas* have tested positive for begomovirus infection. Therefore, it is suggested that they have a dual ecological function, transporting viruses and pollinating cultivated and non-cultivated plants. Additionally, some vector insects use cultivated or uncultivated plants as resting areas during the transfer or migration to other crops (Torres de Los Santos et al., 2016; Gallegos-Gómez et al., 2018; Carboney-Mejía et al., 2020).

2.4 WEEDS AND THEIR RELATIONSHIP WITH AGRICULTURAL CROPS

From an ecological perspective, a weed can be defined as any plant that grows in an area cultivated by humans for various purposes, including crop establishment. A wild plant is considered a weed when it lacks an agricultural or economic function in an agroecosystem and is deemed to have a negative impact on it. Plant species that grow as weeds and compete with crops usually possess ecological attributes that enable them to effectively compete for resources such as water, light, and space (Kaur et al., 2018). They also exhibit accelerated growth rates, high reproduction rates, and the ability to tolerate changing environments and diverse conditions. Consequently, they can exploit the niches created during land cultivation and compete, and even outpace, the growth rates of the crop (Bagavathiannan & Davis, 2018; Bourgeois et al., 2019). These weed species can originate from the original seed bank of the site, which refers to all the propagules accumulated in the cultivated area that possess the necessary attributes to compete with the crop.

Also, they can originate from the arrival of seeds from sites adjacent to the cultivation area and, in addition, have the necessary attributes to be considered as weeds. This introduction from nearby sites is due to environmental transport, agricultural practices, animals, or contaminated materials (Bagavathiannan & Davis, 2018). Although many plant species can be found in both local and neighboring seed reservoirs, only those that can negatively interact with the crop through competition, allelopathic effects, or acting as reservoirs of pests or diseases are considered weeds. Generally, weeds are native or introduced wild plants that arrive as part of the surrounding vegetation and take advantage of the land conditions to establish themselves, either by competing or coexisting with the crop, causing a deterioration in yield (Korav et al., 2018).

Weeds also have an important impact on the agroecosystem, which is related to their potential as reservoirs for pests and phytopathogens (Gerling et al., 2022). Phytopathogens, such as fungi, bacteria, viruses, and viroids, can interfere with plant functionality and cause diseases, resulting in economic losses. Therefore, weed control and removal are necessary to reduce the presence of pathogens and minimize the risk to crops. Proper identification of weeds and their phylogenetic relationship with the target crop is crucial to understand the potential impact of phytopathogens that could be transmitted and cause disease. This helps to assess the risks associated with shared vectors, genetic similarities, or related physiological processes. Some botanical families, such as Amaranthaceae, Cucurbitaceae, Fabaceae, Solanaceae, Brassicaceae, and Poaceae, are known reservoirs of viruses (Byron et al., 2019).

A deeper understanding of the biology, taxonomy, and physiology of weeds and crops provides valuable insights to assess the risk of phytopathogen propagation between them. Additionally, the presence of insect vectors further complicates the dynamics of disease transmission between weeds and crops. Important vector orders include Hemiptera, Thysanoptera, Coleoptera, Diptera, and Hymenoptera (Purcell & Almida, 2005; Sarwar, 2020; Budziszewska et al., 2021).

It is well known that wild plants are reservoirs of viruses, which can persist in weeds that are genetic relatives of food crops, where they do not cause apparent damage. However, they do replicate and can pass from these weeds to cultivated plants. To colonize new hosts and persist in populations, plant viruses require vectors that facilitate horizontal transmission. Plant viruses are transmitted horizontally by vector insects and associate with them in a non-persistent, semi-persistent, or persistent mode (Fiallo-Olive et al., 2020; Torres de Los Santos et al., 2022).

This is a common phenomenon, although the real impact is still unknown. Often, there are no symptoms in the weeds, or they are mistaken for mineral deficiencies. Additionally, the weed may not be correctly identified, or effective detection methods may be lacking or inaccessible (Suproniene et al., 2019; Susi et al., 2019; Hasiów-Jaroszewska et al., 2021; Pandey et al., 2022). Therefore, management strategies for weeds that have the potential to be disease reservoirs must consider multiple control points to be successful.

2.5 EMERGING DNA AND RNA VIRUSES

The so-called cryptic plant viruses (Boccardo et al., 1987) have a persistent lifestyle, remaining with their hosts for many generations and transmitting strictly vertically at a rate close to 100%. In wild plants, this type of virus is the most abundant, with the Partitiviridae family being the most representative (Roossinck, 2012). However, they are also commonly found in some agricultural plants, such as bean (*Phaseolus vulgaris*) crops. These viruses have been reported to be genetically related to fungi (Roossinck, 2003), and partitiviruses and endornaviruses have occasionally been reported to be horizontally transmitted between plants and fungi (Roossinck et al., 2015b).

Susi et al. (2019) evaluated the diversity and prevalence of viruses infecting the wild plant *Plantago lanceolata* across the Åland Islands, South-West Finland. They collected 64 plants from 12 populations, regardless of symptoms, and 90 plants collected from five populations showing conspicuous viral symptoms. In addition to the latent *Plantago lanceolata* virus (PlLV), typical of the wild plant analyzed, they found four potentially new virus species belonging to the Caulimovirus, Betapartitivirus, Enamovirus, and Closterovirus genera. Virus prevalence and diversity varied among sampled host populations. Only a single virus was detected in a single population, while five of the populations supported between two and five of the virus species studied. Additionally, 20% of the plants were co-infected. Interestingly, virus infections were detected in asymptomatic plants (44%). In general, a wide diversity of viruses was found using recently developed tools and protocols that offer interesting opportunities for future studies on the eco-evolutionary dynamics of viruses that infect wild plants in nature.

The Mexican physic nut (*Jatropha curcas* L.) is a non-edible wild plant with great potential for the biofuel and biojet industry worldwide due to the quality of the oil obtained from its seeds. It is a non-cultivated plant and is used by farmers as green or biological barriers between the agricultural fields of one producer and another. Gallegos-Gómez et al. (2018) found the presence of begomoviruses, DNA viruses, associated with this plant, with a prevalence of 88% in all the populations evaluated, although the presence of *Bemisia tabaci*, an insect vector of begomoviruses, has not been reported associated with *J. curcas*. In this sense, Carboney-Mejía et al. (2020) report the presence of begomoviruses in native bees, *Trigona* spp., collected from populations of *J. curcas*. They suggest the possibility of bees, *Trigona spp*., carrying the virus, although their role as a begomovirus vector has not been proven. It is interesting to continue with new investigations of this tripartite interaction to understand the ecological function of both the pollinating function of the bee and the function of transporting the virus from one plant to another.

Wild plant communities may harbor a wide diversity of unknown viruses, as virus research focuses on agricultural crops of economic interest. Currently, the development of high-level technology, especially massive sequencing and metagenomic techniques, allows for the search of small nucleic acids, such as cytoplasmic and organelle RNAs. This will provide knowledge about the viruses present in the wild plants surrounding agricultural crops, also known as wild plants (Susi et al., 2019). Small RNA sequencing in hosts is a powerful method for discovering virus diversity in wild plants (Roossinck et al., 2015b). The method is based on the plant resistance mechanism of RNA silencing, where virus-specific double-stranded RNA (dsRNA) generated in most virus infections is cleaved into 21–24 nucleotide fragments (Susi et al., 2019). By sequencing these small RNAs, novel viruses from both insects and plants have been discovered (Wu et al., 2010). This approach will enable us to understand the distribution of new viruses in all ecosystems and wild plants (Susi et al., 2019).

RNA viruses are the most common viruses affecting plants and are classified into broad taxonomic groups that also contain animal or insect viruses (Roossinck, 2003). Currently, viral emergent infections in plants are caused by members of the RNA virus families *Potyviridae*, *Bunyaviridae*, *Closteroviridae*, *Betaflexiviridae*, *Secoviridae*, and the DNA virus family *Geminiviridae* (Ertunc, 2020).

Potyviridae is the most prominent family of plant-infecting single-stranded positive-sense RNA viruses. It is composed of 12 genera: *Arepavirus*, *Bevemovirus*, *Brambyvirus*, *Bymovirus*, *Celavirus*, *Ipomovirus*, *Macluravirus*, *Poacevirus*, *Potyvirus*, *Roymovirus*, *Rymovirus*, and *Tritimovirus*. *Potyvirus* comprises about 150 viruses, and the genera *Macluravirus* are transmitted by aphids, the genus Ipomovirus is transmitted by whiteflies, and *Poacevirus, Rymovirus*, and *Tritimovirus* are transmitted by mites (Valli et al., 2015). *Potyviruses* have a worldwide distribution and can be transmitted by vectors (aphids), seeds, and in some cases, pollen. Most *Potyviridae* are restricted to a small host species, but a few could infect up to 30 plant families (Inoue-Nagata et al., 2022). Together, members of this family account for more than half of the viral crop damage worldwide (Qin et al., 2020).

The probability of new infection emergence could occur in previously known free areas, although most members are restricted to several host ranges. Several *Potyviridae* have weeds as alternative hosts, and this capacity could increase the incidence of infection by these viruses and ultimately affect commercial crops. In recent years, novel RNA viral outbreaks caused by *Potyviruses* have been identified in weeds associated with commercial crops, such as the papaya ringspot virus (PRSV) and the pennnisetum alopecuroides mosaic virus (PalMV) in whitegrass species (*P. alopercuroides*) (Bermudez-Guzman et al., 2018; Liu et al., 2021). Other major outbreaks caused by *Potyviridae* include potato necrotic ringspot caused by potato virus Y (PVY), sweet potato feathery mottle virus (SPFMV), and plum pox virus (PPV) which can spread to commercial or industrial crops through wild weeds (Jones, 2021).

Within the insect-borne *Bunyaviridae* family, the genus *Tospovirus* is the only one associated with plant infections, transmitted by thrips. Although this virus genus is not as abundant as the plant-affecting members of Potyviridae, it is one of the most devastating for agriculture due to its broad host range, including weeds and commercial crop species from over 1,000 plant species from around 80 families, including both dicots and monocots (Kormelink et al., 2011). The tomato spotted wilt virus (TSWV) is the most well-known member of the *Tospovirus* genus, causing significant losses to commercial crops due to the nature of the vector thrips and the wide range of host plants (Rotenberg et al., 2015). Currently, the genus *Tospovirus* includes 29 species vectored by thrips (Rotenberg & Whitfield, 2018). Due to its broad host range, including weeds, Tospoviruses can "jump" between hosts and affect large areas of crops. Recently, several outbreaks and reports of new hosts affected by TSWV include solanaceous and cucurbitaceous crops in India (Kumari et al., 2022) and the perennial herb butterbur (*Petasites japonicas*) in Korea (Kwak et al., 2020).

The family of plant single-stranded positive-sense RNA virus, *Closteroviridae*, is restricted to the plant phloem. The family comprises four genera: *Closterovirus, Ampelovirus, Crinivirus*, and *Velarivirus* (Candresse & Fuchs, 2020). The members of this genus have a range of vectors. Closteroviruses are vectored by aphids, while ameloviruses use pseudococcid malybugs, and criniviruses are transmitted by whiteflies from the genera *Trialeurodes* and *Bemisia*. Within this family, Crinivirus is a genus that has emerged as a threat to agriculture due to the establishment and naturalization of its whitefly vectors in several temperate climate regions around the world (Tzanetakis et al., 2013).

Recent outbreaks caused by *Closteroviridae* members include the sweet potato chlorotic stunt virus (SPCSV), a crinivirus that affects several species of bindweed (*Convolvulaceae: Ipomoea* spp.), ultimately functioning as a pathway to affect sweet potato crops through its whitefly vector (Jones, 2021). The beet pseudo-yellows virus (BPYV) is another virus with high invasive potential. It is transmitted by the vector *Trialeurodes vaporariorum*, which in recent years has been found in more regions due to the movement of plant material (Tzanetakis et al., 2013).

The important citrus tristeza virus (CTV) is a *Closterovirus* that mainly affects citrus, causing significant economic losses. However, to date, there have been no reports associating this virus with wild weeds or hosts outside of citrus and related species.

Betaflexiviridae is a family of plant-infecting, single-stranded, positive-sense, polyadenylated RNA viruses. This family consists of three genera: *Carlavirus, Foveavirus*, and *Robigovirus*, as well as three unassigned species (Ryu & Song, 2020). In 2012, a new monotypic genus was proposed within this family, represented by the potato virus T (PVT). PVT has the ability to infect not only potato species but also herbaceous hosts through mechanical inoculation, although the vector is unknown. Transmission primarily occurs through propagating material, seeds, and pollen (Rubino et al., 2012). *Carlavirus* is a genus that includes approximately 52 species, which primarily infect woody hosts. However, knowledge of vector transmission is limited, and the whitefly (*B. tabaci*) is the only known vector for these viruses.

Secoviridae is a family of non-enveloped, single-stranded, positive-sense RNA viruses that infect dicots and are related to picornaviruses. It includes several genera such as *Comovirus, Fabavirus, Nepovirus, Cheravirus, Sadwavirus, Torradovirus, Sequivirus,* and *Waikavirus* (Sanfacon et al., 2009; Thompson et al., 2014, 2017). Among these genera, Torradovirus is well-known for causing torrado disease in tomatoes, transmitted by whiteflies and aphids. Recent research has shown that Torradovirus can also naturally infect several weed species, increasing the risk of emergent infections in economically important crops (Alfaro-Fernández et al., 2008). *Fabavirus* and *Nepovirus* have a wide range of hosts, including monocots and dicots, transmitted by aphids and longidorus nematodes, respectively (Thompson et al., 2017). The transmission mechanism of *Sadwavirus* is still unknown, despite having many known hosts. Lastly, *Sequivirus* has multiple plant hosts, but its transmission depends on the presence of a helper virus from the *Waikavirus* genus.

2.6 CONCLUSION

The main studies in plant virology focus on analyzing the diversity of viruses in plants of agronomic or ornamental interest. Recent studies are also exploring the discovery of new viruses in wild or non-cultivated plants using metagenomic studies (Roossinck, 2012; Roossinck et al., 2015). The relationship between wild plants and cultivated plants is closely related to the symbiosis between the transmitting insect or vector, the virus, and the host plant. This mutualistic symbiosis allows the virus to associate with different plants depending on the environmental conditions of the host (Roossinck, 2015).

Until 2012, the International Committee for the Taxonomy of Viruses reported that most of the known plant viruses, less than 1,000 different species, originated from cultivated plants. However, new studies have revealed a high prevalence of new viruses in wild plants from the tropics to Antarctica (Roossinck, 2012; Hopkins et al., 2014). Additionally, plant virus-like nucleic sequences have been found in various aquatic environments (Mehle & Ravnikar, 2012; Tamaki et al., 2012). This undiscovered diversity implies an evolution of viruses and their association with hosts. It also highlights the importance of conducting new studies on the diversity of plant viruses using innovative methodological approaches, especially in environments where exotic species are frequently introduced and can serve as new hosts (Bao & Roossinck, 2013).

KEYWORDS

- **agricultural crops**
- **DNA and RNA viruses**
- **food security**
- **plant viruses**
- **viral diseases**
- **virus vectors**
- **weeds**
- **wild plants**

REFERENCES

Abdelkhalek, A., & Hafez, E., (2020). Plant viral diseases in Egypt and their control. In: *Cottage Industry of Biocontrol Agents and Their Applications* (pp. 403–421). Springer: Cham.

Aguilar, E., & Lozano, D. R., (2022). Plant viruses as probes to engineer tolerance to abiotic stress in crops. *Stress Biology*. doi.org/10.1007/s44154-022-00043-4.

Alfaro-Fernández, A., Córdoba-Sellés, C., Cebrián, M. C., Herrera-Vásquez, J. A., Sánchez-Navarro, J. A., Juárez, M., Espino, A., et al., (2008). First report of *Tomato torrado virus* on weed hosts in Spain. *Plant Disease, 92*, 831–831.

Bagavathiannan, M. V., & Davis, A. S., (2018). An ecological perspective on managing weeds during the great selection for herbicide resistance. *Pest Management Science*, *74*, 2277–2286.

Bao, X., & Roossinck, M. J., (2013). A life history view of mutualistic viral symbioses: Quantity or quality for cooperation? *Curr. Op. Microbiol.*, *16*, 514–518.

Belval, L., Marmonier, A., Schmitt-Keichinger, C., Gersch, S., Andret-Link, P., Komar, V., Vigne, E., et al., (2019). From a movement-deficient *Grapevine fanleaf virus* to the identification of a new viral determinant of nematode transmission. *Viruses*, *11*, 1146.

Bermudez-Guzman, M. J., Guzman-Gonzalez, S., Lara-Reyna, J., Palmeros-Suarez, P. A., Lopez-Muraira, I. G., & Gomez-Leyva, J. F., (2018). Presence of *Papaya ringspot virus* (PRS) in weed associated with *Carica papaya* in Colima. Mexico. *Revista Mexicana de Fitopatología*, *36*, 1–15.

Bhattacharjee, B., & Hallan, V., (2022). Geminivirus-derived vectors as tools for functional genomics. *Front. Microbiol.*, *13*, 799345. doi: 10.3389/fmicb.2022.799345.

Boccardo, G., Lisa, V., Luisoni, E., & Milne, R. G., (1987). **Cryptic plant viruses.** *Advances in Virus Research*, *32*, 171–214.

Bourgeois, B., Munoz, F., Fried, G., Mahaut, L., Armengot, L., Denelle, P., & Violle, C., (2019). What makes a weed a weed? A large-scale evaluation of arable weeds through a functional lens. *American Journal of Botany*, *106*, 90–100.

Budziszewska, M., Frąckowiak, P., & Obrępalska-Stęplowska, A., (2021). Analysis of the role of *Bradysia impatiens* (*Diptera: Sciaridae*) as a vector transmitting peanut stunt virus on the model plant *Nicotiana benthamiana*. *Cells*, *10*, 1546.

Byron, M., Treadwell, D., & Dittmar, P., (2019). *Weeds as Reservoirs of Plant Pathogens Affecting Economically Important Crops.* University of Florida, Gainesville, Florida, USA.

Candresse, T., & Fuchs, M., (2020). Closteroviridae. In: *eLS*. John Wiley & Sons, Ltd: Chichester.

Carboney-Mejía, I. G., Ruíz-Toledo, J., Hernández-Reyes, J., Lucio-Castillo, H., Hernandez-Robledo, V., Torres-Castillo, J. A., & Torres, De. L. S. R., (2020). Prevalence of *Begomoviruses* in *Trigona fuscipennis* and *Trigona fulviventris* visitors of *Jatropha curcas*. *Journal of Apicultural Research*. doi: 10.1080/00218839.2020.1756181.

Catto, A. M., Mugerwa, H., Myers, B. K., Pandey, S., Dutta, B., & Srinivasan, R., (2022). A review on transcriptional responses of interactions between insect vectors and plant viruses. *Cells*, *11*, 693.

Di Mattia, J., Ryckebusch, F., Vernerey, M. S., Pirolles, E., Sauvion, N., Peterschmitt, M., Zeddam, J. L., & Blanc, S., (2020). Co-acquired nanovirus and geminivirus exhibit a contrasted localization within their common aphid vector. *Viruses*, *12*, 299.

Elena, S. F., Bedhomme, S., Carrasco, P., Cuevas, J. M., De La Iglesia, F., Lafforgue, G., Lalić, J., et al., (2011). The evolutionary genetics of emerging plant RNA viruses. *Mol. Plant Microbe. Interact*, *24*, 287–293.

Ertunc, F., (2020). Emerging plant viruses. In: *Emerging and Reemerging Viral Pathogens* (1st edn.), USA: Academic Press.

Fiallo-Olive, E., Pan, L. L., Liu, S. S., & Navas-Castillo, J., (2020). Transmission of *Begomoviruses* and other whitefly-borne viruses: Dependence on the vector species. *Phytopathology*, *110*, 10–17.

Gallegos-Gómez, E. H., Salvador-Figueroa, M., & Torres-De, L. S. R., (2018). Detección de *begomovirus* en plantas de *Jatropha curcas* L., en Chiapas, México. *Agroproductividad*, *11*, 3–7.

García, Y., Sierra-Mejia, A., Gutiérrez, P. A., & Marín, M. M., (2020). Prevalencia de cinco virus de ARN en tubérculos – semilla de papa cultivados en antioquia (Colombia). *Biotecnología En el Sector Agropecuario E Industrial*, *19*, 66–78.

Gerling, M., Pätzig, M., Hempel, L., Büttner, C., & Müller, M. E., (2022). Arable weeds at the edges of kettle holes as overwintering habitat for phytopathogenic fungi. *Agronomy*, *12*, 823.

Ghosh, A., Mandal, B., & Dietzgen, R. G., (2021). Progression of *Watermelon bud necrosis virus* infection in its vector, *Thrips Palmi. Cell*, *10*, 392.

Hasiów-Jaroszewska, B., Boezen, D., & Zwart, M. P., (2021). Metagenomic studies of viruses in weeds and wild plants: A powerful approach to characterize variable virus communities. *Viruses*, *13*, 1939.

Hopkins, D. W., Swanson, M. M., & Taliansky, M. E., (2014). What do we know about viruses in terrestrial antarctica? In: Cowan, D., (ed.), *Antarctic Terrestrial Microbiology* (p. 12). Springer: Berlin.

Inoue-Nagata, A. K., Jordan, R., Kreuze, J., Li, F., Lopez-Moya, J. J., Makinen, K., Ohshima, K., et al., (2022). ICTV virus taxonomy profile: *Potyviridae* 2022. *J. Gen. Virol.*, *103*. 10.1099/jgv.0.001738.

Jawetz, E., Melnick, J. L., & Aldeberg, E. A., (2022). *Propiedades Generales de Los Virus: Microbiología Médica* (28th edn.). USA: Mc.Graw Hill.

Jones, R. A. C., (2021). Global plant virus disease pandemics and epidemics. *Plants (Basel)*, 10.

Kaur, R., Mudgal, R., Narwal, M., & Tomar, S., (2018). Development of an ELISA assay for screening inhibitors against divalent metal ion dependent alphavirus capping enzyme. *Virus Res*, *256*, 209–218.

Kloth, K. J., & Kormelink, R., (2020). Defenses against virus and vector: A phloem-biological perspective on RTM and SLI1- mediated resistance to potyviruses and aphids. *Viruses*, *12*, 129. doi.org/10.3390/v12020129.

Korav, S., Dhaka, A. K., Singh, R., Premaradhya, N., & Chandramohan, R. G., (2018). A study on crop weed competition in field crops. *Journal of Pharmacognosy and Phytochemistry*, *7*, 3235–3240.

Kormelink, R., Garcia, M. L., Goodin, M., Sasaya, T., & Haenni, A. L., (2011). Negative-strand RNA viruses: The plant-infecting counterparts. *Virus Res.*, *162*, 184–202.

Kumari, S., Nagendran, K., Manimurugan, C., Pandey, K. K., & Singh, J., (2022). Diversity of *tospoviruses* infecting vegetable crops of eastern Indo-Gangetic plains of Uttar Pradesh and its impact on cowpea seed health. *European Journal of Plant Pathology*, *162*, 403–414.

Kwak, H. R., Park, G., Choi, H. Y., Go, W. R., Baek, E., Kim, J. E., Kim, M., & Choi, H. S., (2020). First report of *Tomato spotted wilt virus* in *Petasites japonicus* in Korea. *Plant Dis.*, 1235.

Liu, X., Chen, X., Liu, S., Du, K., Wang, P., Jiang, T., Cao, M., et al., (2021). A new potyvirus isolated from *Pennisetum alopecuroides* with the potential to infect cereal crops. *Phytopathology Research*, *3*, 6.

Márquez, L. M., Redman, R. S., Rodriguez, R. J., & Roossinck, M. J., (2007). A virus in a fungus in a plant: Three-way symbiosis required for thermal tolerance. *Science, 315*, 513–515.

Mehle, N., & Ravnikar, M., (2012). Plant viruses in aqueous environment-survival, water mediated transmission and detection. *Water Res.*, *46*, 4902–4917.

Pandey, S., Girish, T. R., Basavaraj, S., Padmaja, A. S., & Nagaraju, N., (2022). RCA-based detection of *Begomoviruses* in weed genera associated with legumes in Southern Karnataka. *Legume Research-An International Journal*, *1*, 8.

Parrish, C. R., Holmes, E. C., Morens, D. M., Park, E. C., Burke, D. S., Calisher, C. H., Laughlin, C. A., et al., (2008). Cross-species virus transmission and the emergence of new epidemic diseases. *Microbiol. Mol. Biol. Rev.*, *72*, 457–470.

Purcell, A. H., & Almeida, R. P., (2005). Insects as vectors of disease agents. *Encyclopedia of Plant and Crop Science*, *10*, 1–5.

Qin, L., Shen, W., Tang, Z., Hu, W., Shangguan, L., Wang, Y., Tuo, D., et al., (2020). A newly identified virus in the family potyviridae encodes two leader cysteine proteases in tandem that evolved contrasting RNA silencing suppression functions. *J. Virol.*, 95.

Ray, S., & Casteel, C. L., (2022). Effector-mediated plant-virus-vector interactions. *Plant Cell*, *34*, 1514–1531.

Rivera, M. C., & Wright, E. R., (2020). *Apuntes de Patología Vegetal: Fundamentos y Prácticas Para la Salud de las Plantas* (1st edn.). BA: Editorial Facultad de Agronomía.

Roossinck, M. J., (2003). Plant RNA virus evolution. *Curr. Opin. Microbiol.*, *6*, 406–409.

Roossinck, M. J., (2012). Plant virus metagenomics: Biodiversity and ecology. *Ann. Rev. Genet.*, *46*, 357–367.

Roossinck, M. J., (2015). Plants, viruses, and the environment: Ecology and mutualism. *Virology*, *480*, 271–277.

Roossinck, M. J., Martin, D. P., & Roumagnac, P., (2015b). Plant virus metagenomics: Advances in virus discovery. *Phytopathology*, *105*, 716–727.

Rotenberg, D., & Whitfield, A. E., (2018). Molecular interactions between *Tospoviruses* and thrips vectors. *Curr. Opin. Virol.*, *33*, 191–197.

Rotenberg, D., Jacobson, A. L., Schneweis, D. J., & Whitfield, A. E., (2015). Thrips transmission of *tospoviruses*. *Curr. Opin. Virol., 15*, 80–89.

Rubino, L., Russo, M., De Stradis, A., & Martelli, G. P., (2012). Tepovirus, a novel genus in the family *Betaflexiviridae*. *Arch. Virol.*, *157*, 1629–1633.

Ryu, K. H., & Song, E. G., (2020). Quinviruses (*Betaflexiviridae*). In: *Encyclopedia of Virology* (1st edn., Vol. 3). London: Elsevier Academic Press.

Salaudeen, M. T., Adama, C. J., Abdullahi, A. A., Ayeleke, D. A., & Ibrahim, A. D., (2016). Climate change and viral diseases in relation to crop productivity and food security: A review. *International Journal of Applied Biological Research, 7*, 56–65.

Sanfacon, H., Wellink, J., Le Gall, O., Karasev, A., Van, D. V. R., & Wetzel, T., (2009). *Secoviridae*: A proposed family of plant viruses within the order *Picornavirales* that combines the families *Sequiviridae* and *Comoviridae*, the unassigned genera *Cheravirus* and Sadwavirus, and the proposed genus torradovirus. *Arch. Virol.*, *154*, 899–907.

Sarwar, M., (2020). Insects as transport devices of plant viruses. In: *Applied Plant Virology* (1st edn.). USA: Academic Press.

Suproniene, S., Kadziene, G., Irzykowski, W., Sneideris, D., Ivanauskas, A., Sakalauskas, S., & Jedryczka, M., (2019). Asymptomatic weeds are frequently colonized by pathogenic species of *fusarium* in cereal-based crop rotations. *Weed Research*, *59*, 312–323.

Susi, H., Filloux, D., Frilander, M. J., Roumagnac, P., & Laine, A. L., (2019). Diverse and variable virus communities in wild plant populations revealed by metagenomic tools. *PeerJ*, *7*, e6140.

Tamaki, H., Zhang, R., Angly, F. E., Nakamura, S., Hong, P. Y., Yasunaga, T., Kamagata, Y., & Liu, W. T., (2012). **Metagenomic analysis of DNA viruses in a wastewater treatment plant in tropical climate.** *Environ. Microbiol.*, *14*, 441–452.

Thompson, J. R., Dasgupta, I., Fuchs, M., Iwanami, T., Karasev, A. V., Petrzik, K., Sanfaçon, H., et al., (2017). ICTV report consortium. ICTV virus taxonomy profile: Secoviridae. *The Journal of General Virology*, *98*, 529–531.

Thompson, J. R., Kamath, N., & Perry, K. L., (2014). An evolutionary analysis of the *Secoviridae* family of viruses. *PLoS One*, *9*, e106305.

Torres-De, L. S. R., Arévalo-Monterrubio, L. D., Torres-Acosta, R. I., & Martínez-Montoya, H., (2022). The other face of pollinating insects and their relationship to geminivirus transmission. In: *Geminivirus: Detection, Diagnosis and Management* (1st edn.). United Kingdom: Elsevier.

Torres-De, L. S. R., Salvador-Figueroa, M., Gallegos-Gómez, E. H., Arévalo-Monterrubio, L. D., Rincón-Rabanales, M., & Grajales-Conesa, J., (2016). Detection of *Begomovirus* in the stingless bee *Trigona fuscipennis* visiting *Jatropha curcas* in the South of Mexico. *Journal of Apicultural Research*, *55*, 185–186.

Tzanetakis, I. E., Martin, R. R., & Wintermantel, W. M., (2013). Epidemiology of criniviruses: An emerging problem in world agriculture. *Frontiers in Microbiology*, *4*, 119.

Valli, A., García, J. A., & Lopez-Moya, J. J., (2015). Potyviridae. In: *eLS*. John Wiley & Sons, Ltd: Chichester.

Van, M. M., (2020). Impact of abiotic stresses on plant virus transmisión by aphids. *Viruses*, *12*, 216. doi.org/10.3390/v12020216.

Wu, W., Shan, H. W., Li, J. M., Zhang, C. X., Chen, J. P., & Mao, Q., (2022). Roles of bacterial symbionts in transmission of plant virus by hemipteran vectors. *Front. Microbiol.*, *13*, 805352. doi: 10.3389/fmicb.2022.805352.

Wu, X., & Ye, J., (2020). Manipulation of jasmonate signaling by plant viruses and their insect vectors. *Viruses*, *12*, 148. doi.org/10.3390/v12020148.

Ziegler-Graff, V., (2020). Molecular insights into host and vector manipulation by plant viruses. *Viruses*, *12*, 263. doi.org/10.3390/v12030263.

CHAPTER 3

Tobacco and Weeds as Hidden Sources of Plant Viruses Threatening Vegetable Production of Solanaceae Crops

NIKOLAY MANCHEV PETROV,[1] MARIYA IVANOVA STOYANOVA,[2] RAJARSHI KUMAR GAUR,[3] MILENA GEORGIEVA BOZHILOVA-SAKOVA,[4] and IVONA VASSILEVA DIMITROVA[5]

[1]*New Bulgarian University, Sofia, Bulgaria*

[2]*Institute of Soil Science, Agrotechnologies and Plant Protection (ISSAPP) "Nikola Pushkarov," Agricultural Academy, Sofia, Bulgaria*

[3]*Deen Dayal Upadhyaya Gorakhpur University, Gorakhpur, Uttar Pradesh, India*

[4]*Institute of Animal Science, Department of "Genetics, Breeding, Selection, Reproduction, and Biotechnologies of Farm Animals," Kostinbrod, Bulgaria*

[5]*University of Forestry, Agronomy Faculty, Department of Plant Protection, Sofia, Bulgaria*

ABSTRACT

Solanaceae crops, including vegetables such as potatoes, tomatoes, peppers, and eggplants, are among the most widely produced food plants in the world. Plant viruses affecting these crops pose a significant threat to both yield and quality. Tobacco, as a mass-produced industrial crop, and weeds from the *Solanaceae* family serve as hidden sources and reservoirs for the spread of viruses to these plants. Therefore, it is crucial to identify specific weeds

New and Emerging Plant Viruses: The Threat to Food Security.
Alireza Golnaraghi and Rajarshi Kumar Gaur (Eds.)

and tobacco cultivars that act as reservoirs for particular viruses in order to effectively control viral infections.

3.1 INTRODUCTION

The *Solanaceae* family includes a lot of species, the most cultivated of which are horticultural crops such as potatoes, tomatoes, and peppers, as well as the industrial crop tobacco. The family also includes some widely spread weeds, such as petunia, *Physalis* sp., *Datura stramonium, Solanum nigrum*, etc., which are the most common viral reservoirs. Several viruses that infect tomatoes and peppers can also infect *D. stramonium* and use the plant as a weed reservoir. For example, potato virus X (PVX) and tomato mosaic virus (ToMV) (Alemu et al., 2002). Potato (*Solanum tuberosum*), being the most important cultivated species of this family, ranks as the third most important food crop in the world (Devaux et al., 2014). A major limiting factor in potato production is viral diseases, which can cause significant losses in production and a decline in tuber quality (Loebenstein, 2009). Controlling potato virus diseases is crucial to minimize losses and increase the yield and quality of produced and planting material (Harahagazwe et al., 2018).

Global environmental and climate changes are among the main reasons for the emergence and spread of virus species in new geographical areas. On the other hand, the globalization of world markets and new methods of rapid transportation of goods provide opportunities for easier and faster spread of viruses over long distances (Rojas & Gilbertson, 2008; Jones, 2014, 2016). The density of virus vector populations, which is affected by annual temperature changes, is also one of the main factors for the more efficient spread of viruses in *Solanaceae* species. With the expansion of agricultural areas and the implementation of new agricultural practices in agroproduction, new opportunities for contact between cultivated species and native natural flora are created, leading to the spread of invasive weed species as virus reservoirs (Jones, 2014). Consequently, viruses introduced by new cultivated species can utilize native wild species as new hosts (Cooper & Jones, 2006; Jones, 2015).

3.2 PLANT VIRUSES INFECTING VEGETABLES AND WEEDS FROM *SOLANACEAE* FAMILY

Cultivated *Solanaceous* species, grown for specific purposes by mankind, exist within a complex agroecosystem where the boundaries between cultivated plants and natural wild populations are not clearly distinguishable. Viruses have the ability to infect both cultivated species and wild weeds through vectors (Burdon & Thrall, 2008; Hančinský et al., 2020). The methods of viral transmission among hosts can vary depending on the virus. In *Solanaceae* plants, viruses are primarily transmitted horizontally by insect vectors such as aphids (Ng & Perry, 2004; Mondal et al., 2017), thrips (for Tospoviruses) (Rotenberg et al., 2015), or nematodes for tobacco rattle virus (TRV) (Riga et al., 2009). Vertical transmission of viruses to progeny through seeds or pollen has been observed in various *Solanaceae* species for tobamoviruses (Dombrovsky & Smith, 2017), pepino mosaic virus (PepMV) (Hanssen et al., 2010), and tomato torrado virus (ToTV) (Pospieszny et al., 2019).

Most research on plant viruses focuses on crops and their interactions with cultivated hosts. There is limited research on the preservation and spread of viruses in native wild species and weeds. Studies on virus transmission from cultivated to weedy wild species are scarce (Shates et al., 2019). Understanding the natural life cycles of viruses in wild host populations is crucial for predicting their evolution and future behavior (Pagán & Holmes, 2010; Gibbs et al., 2015). The introduction of new agro-industrial practices by humans globally alters natural plant-virus interactions in cultivated fields (Anderson et al., 2004). It is suggested that Potyviruses rapidly evolved and became the largest genus of plant viruses around 4,610 years BC, during the dawn of agriculture (Gibbs et al., 2008). This information is vital for preventing disease epidemics and ensuring food supply. Therefore, studying viruses in natural wild plant populations is necessary to understand their natural life cycles (Roossinck, 2015; Hančinský et al., 2020). Human activities aimed at increasing agricultural production, such as altering virus cycles and promoting monocultures with low genetic diversity, can lead to the emergence of new virus species and strains (Roossinck & García-Arenal, 2015; Hančinský et al., 2020). On the other hand, in natural ecosystems without human intervention, phytoviruses are subjected to diversified pressure due to the stress from biotic/abiotic factors and the more limited accessibility to their relevant vectors and hosts (Fraile et al., 1997). Infection of plants with certain plant viruses, such as cucumber mosaic virus (CMV), can result in increased accumulation of osmoprotectants, leading to enhanced drought tolerance in the host plants and creating better conditions for virus reproduction (Xu et al., 2008; Gorovits et al., 2022). Infection with tomato yellow leaf curl virus (TYLCV) (a Begomovirus) causes mild symptoms and promotes host survival under high temperatures and drought through several mechanisms, including down-regulation of heat-shock transcription factors

(HSFA) transcription, induction of protein complexes capturing HSFA2, accumulation of osmoprotectants, mitigation of stress-induced increase in metabolites, and their reallocation from shoots to roots (Anfoka et al., 2016; Shteinberg et al., 2021; Gorovits et al., 2022).

Solanaceous plants are naturally infected by viruses from more than 20 genera (Table 3.1). Deeper information on viruses infecting cultivated and wild plant species, as well as the interactions among them, will provide valuable data for future outbreak prevention (McLeisch et al., 2019; Hančinský et al., 2020). It has been demonstrated that the weed *Solanum nigrum*, belonging to the *Solanaceae* family, is the most efficient source for ToCV transmission (Orfanidou et al., 2016; Hančinský et al., 2020).

A summary of viruses infecting both cultivated and native wild weed species, along with their modes of transmission, is presented in Table 3.1 (ICTV, 2019; VIDE Database, 2020; NCBI Virus Portal, 2020; Sastry et al., 2019; Hančinský et al., 2020).

Within native wild plants from the *Solanaceae* family, the phenomenon of synergism among viruses can emerge. Virus synergism can lead to the breakage of plant resistance (Hassani-Mehraban et al., 2009). In this study, Bag et al. (2012) demonstrated synergism between ambisense viruses like TSWV (tomato spotted wilt virus) and IYSV (iris yellow spot virus), infecting the solanaceous weed *D. stramonium*. *D. stramonium* is permissive to TSWV, meaning the virus infects the plant in a systemic way, while the spread of IYSV within the plant is restricted only to the infected leaves in the case of single infections. However, in the case of a dual infection with TSWV and IYSV, datura plants displayed increased symptom severity and a significantly shortened period for symptom apparition. Additionally, the silencing suppressor gene NSs (nonstructural protein gene) of IYSV was also found in the new leaves of the plants (Bag et al., 2012).

ToLCNDV (Tomato leaf curl New Delhi begomovirus) exhibits a high level of genetic diversity, and its viral population dynamics indicate a higher mutation rate in the *Solanaceae* weed *D. stramonium* compared to cultivated hosts. As a result, it can serve as a potential source of inoculum in *D. stramonium* for other cultivated plants from different families (Juárez et al., 2019).

The exchange of viruses between cultivated crops and their wild relatives plays a significant role in the development of virus populations, and this transfer can occur in both directions (Ma et al., 2020). Other Ilaviruses closely related to ToLCNDV, such as PMoV (Parietaria mottle virus) and ToNSV (tomato necrotic spot virus), have caused outbreaks of infection in Europe and the USA, respectively. It is reasonable to assume that the weed

TABLE 3.1 List of Viruses Naturally Infecting Cultivated and Weed Species of the *Solanaceae* Family

Virus Genera	Virus Species	Transmission	Natural Cultivated Host	Natural Wild Host
Alfamovirus	Alfalfa mosaic virus	Vector (aphids) and seeds	*Capsicum annuum*, *Solanum lycopersicum*, *Solanum tuberosum*, *Solanum melongena*, *Solanum muricatum*, *Solanum betaceum*	*Petunia* sp., *Solanum betaceum*
Alphanucleorhabdovirus	Eggplant mottled dwarf virus	Vector (leafhopper)	*Solanum melongena*, *Solanum lycopersicum*, *Solanum tuberosum*, *Capsicum annuum*	*Nicotiana tabacum*
	Potato yellow dwarf virus	Vector (leafhopper)	*Solanum tuberosum*, *Capsicum annuum*, *Solanum lycopersicum*	*Nicotiana benthamiana* and *Nicotiana rustica*
Betanecrovirus	Tobacco necrosis virus D	Vector (*Olpidium brassicae*)	*Nicotiana tabacum*	*Nicotiana tabacum*
Betanucleorhabdovirus	Datura yellow vein nucleorhabovirus	–	*Solanum lycopersicum*	*Datura stramonium*
Begomovirus	Tomato leaf curl New Delhi virus, tomato yellow leaf curl virus	Vector (*Bemisia tabaci*)	*Physalis philadelphica*, *Capsicum annuum*, *Solanum lycopersicum*, *Solanum melongena*	*Petunia* sp., *Datura stramonium*, *Solanum nigrum*, *Nicotiana tabacum*
Carlavirus	Cowpea mild mottle virus	Vector (*Bemisia tabaci*)	*Solanum lycopersicum*, *Solanum melongena*	*Browallia speciosa*
	Potato virus M	Vector (aphids)	*Solanum lycopersicum*, *Solanum tuberosum*	*Solanum dulcamara*, *Solanum muricatum*
	Potato virus S	Vector (aphids)	*Solanum tuberosum*, *Solanum lycopersicum*	*Solanum nigrum*
Closterovirus	Tobacco virus 1	Vector (aphids)	–	*Nicotiana tabacum*

TABLE 3.1 *(Continued)*

Virus Genera	Virus Species	Transmission	Natural Cultivated Host	Natural Wild Host
Crinivirus	Potato yellow vein virus	Vector (whiteflies)	*Solanum tuberosum*, *Solanum lycopersicum*	*Solanum nigrum*
	Tomato chlorosis virus	Vector (whiteflies)	*Capsicum annuum*, *Solanum tuberosum*, *Solanum lycopersicum*, *Solanum melongena*	*Nicotiana tabacum*, *Datura stramonium*, *Physalis species*, *Physalis peruviana*, *Physalis philadelphica*, *Solanum aethiopicum*
Cucumovirus	Cucumber mosaic virus	Vector (aphids)	*Capsicum annuum*, *Solanum tuberosum*, *Solanum lycopersicum*	*Physalis* sp., *Solanum nigrum*, *Nicotiana tabacum*
	Peanut stunt virus	Vector (aphids)	*Capsicum annuum*, *Solanum lycopersicum*	*Nicotiana tabacum*
	Tomato aspermy virus	Vector (aphids)	*Capsicum annuum*, *Solanum lycopersicum*	*Petunia* sp.
Fabavirus	Broad bean wilt virus	Vector (aphids)	*Capsicum annuum*, *Solanum melongena*	*Petunia* sp., *Nicotiana tabacum*
Ilarvirus	Tobacco streak virus	Vector (thrips); pollen and seeds	*Solanum tuberosum*, *Solanum lycopersicum*	*Nicotiana tabacum*, *Petunia* sp., *Datura stramonium*, *Physalis* species
Nepovirus	Arabis mosaic virus	Vector (nematodes)	*Solanum tuberosum*, *Solanum lycopersicum*	*Petunia* sp., *Solanum nigrum*, *Solanum betaceum*, *Physalis peruviana*
	Artichoke yellow ringspot virus	Vector (nematodes)	–	*Nicotiana tabacum*
	Cherry leaf roll virus	Vector (nematodes)	*Solanum tuberosum*	*Solanum berthaultii*, *Petunia* sp., *Solanum jamesii*
	Tomato black ring virus	Vector (nematodes)	*Solanum tuberosum*, *Solanum lycopersicum*	*Petunia* sp.

TABLE 3.1 *(Continued)*

Virus Genera	Virus Species	Transmission	Natural Cultivated Host	Natural Wild Host
	Tomato ringspot virus	Vector (nematodes)	*Solanum tuberosum*, *Solanum lycopersicum*, *Solanum melongena*	*Nicotiana tabacum*, *Petunia* sp., *Solanum betaceum*
	Tobacco ringspot virus	Vector (nematodes)	*Solanum lycopersicum*, *Solanum melongena*	*Petunia* sp., *Nicotiana tabacum*
Orthotospovirus	Tomato spotted wilt virus	Vector (thrips)	*Solanum tuberosum*, *Solanum lycopersicum*, *Solanum melongena*, *Capsicum annuum*	*Nicotiana tabacum*, *Solanum nigrum*, *Petunia*, *Physalis peruviana*, *Physalis philadelphica*, *Solanum dulcamara*, *Solanum betaceum*, *Datura stramonium*
–	Impatiens necrotic spot virus	Vector (thrips)	*Solanum tuberosum*, *Solanum lycopersicum*, *Solanum melongena*, *Capsicum annuum*	*Nicotiana tabacum*, *Solanum nigrum*, *Petunia*, *Physalis peruviana*, *Physalis philadelphica*, *Solanum dulcamara*, *Solanum betaceum*, *Datura stramonium*
–	Iris yellow spot virus	Vector (thrips)	*Solanum tuberosum*, *Solanum lycopersicum*, *Solanum melongena*, *Capsicum annuum*	*Nicotiana tabacum*, *Solanum nigrum*, *petunia*, *Physalis peruviana*, *Physalis philadelphica*, *Solanum dulcamara*, *Solanum betaceum*, *Datura stramonium*
Polerovirus	Potato leafroll virus	Vector (aphids)	*Solanum tuberosum*, *Solanum lycopersicum*	*Datura stramonium*, *Solanum nigrum*, *Solanum betaceum*
	Beet western yellows virus	Vector (aphids)	*Capsicum annuum*, *Solanum tuberosum*	*Solanum nigrum*
	Pepper vein yellows virus 1	Vector (aphids)	*Capsicum annuum*	*Nicotiana tabacum*
	Tobacco vein distorting virus	Vector (aphids)	–	*Nicotiana tabacum*

TABLE 3.1 *(Continued)*

Virus Genera	Virus Species	Transmission	Natural Cultivated Host	Natural Wild Host
Potexvirus	Potato virus X	Mechanical contact	*Solanum tuberosum*, *Solanum lycopersicum*, *Solanum melongena*	*Physalis peruviana*
	Pepino mosaic virus	Mechanical contact and seeds.	*Solanum tuberosum*, *Solanum lycopersicum*, *Solanum melongena*	*Nicotiana tabacum*, *Solanum nigrum*, *Solanum muricatum*
	Potato aucuba mosaic virus	Vector (aphids)	*Solanum tuberosum*	*Solanum betaceum*
Potyvirus	Potato virus Y	Vector (aphids)	*Solanum tuberosum*, *Solanum lycopersicum*, *Solanum melongena*, *Capsicum annuum*	*Nicotiana tabacum*, *Petunia* sp., *Physalis peruviana*, *Datura stramonium*, *Solanum nigrum*, *Solanum betaceum*
	African eggplant mosaic virus	Vector (aphids)	*Solanum melongena*	*Solanum aethiopicum*
	Colombian datura virus	Vector (aphids)	–	*Datura stramonium*, *Petunia* sp., *Physalis peruviana*, *Solanum muricatum*
	Chili veinal mottle virus	Vector (aphids)	*Capsicum annuum*, *Solanum lycopersicum*	*Nicotiana tabacum*, *Datura stramonium*, *Solanum aethiopicum*
	Datura shoestring virus	Vector (aphids)	–	*Datura stramonium*
	Henbane mosaic virus	Vector (aphids)	*Solanum lycopersicum*	*Nicotiana tabacum*, *Datura stramonium*, *Hyoscyamus niger*
	Pepper mottle virus	Vector (aphids)	*Capsicum annuum*, *Solanum lycopersicum*	*Physalis pruinosa*
	Peru tomato mosaic virus	Vector (aphids)	*Solanum lycopersicum*	*Physalis peruviana*, *Solanum nigrum*, *Solanum betaceum*

TABLE 3.1 *(Continued)*

Virus Genera	Virus Species	Transmission	Natural Cultivated Host	Natural Wild Host
	Potato virus A	Vector (aphids)	*Solanum tuberosum*	*Nicotiana tabacum*
	Pepper veinal mottle virus	Vector (aphids)	*Solanum lycopersicum*, *Solanum melongena*	*Nicotiana tabacum*, *Solanum nigrum*, *Datura stramonium*, *Physalis pruinosa*
	Potato virus V	Vector (aphids)	*Solanum tuberosum*, *Solanum lycopersicum*	*Solanum betaceum*
	Tobacco etch virus	Vector (aphids)	*Capsicum annuum*, *Solanum lycopersicum*	*Datura stramonium*, *Physalis* sp., *Nicotiana tabacum*, *Petunia* sp.
	Tamarillo leaf malformation virus.	Vector (aphids)	–	*Solanum betaceum*
	Tobacco vein banding mosaic virus.	Vector (aphids)	*Solanum tuberosum*	*Datura stramonium*, *Nicotiana tabacum*
	Tobacco vein mottling virus	Vector (aphids)	–	*Nicotiana tabacum*
Sobemovirus	*Solanum nodiflorum* mottle virus	Vector (beetles)	–	*Solanum nodiflorum*
Solendovirus	Tobacco vein clearing virus	Seeds	–	*Nicotiana tabacum*
Tombusvirus	Moroccan pepper virus	–	*Capsicum annuum*, *Solanum lycopersicum*	*Datura stramonium*
	Petunia asteroid mosaic virus	–	–	*Petunia* sp.
Tobamovirus	Tobacco mosaic virus	Mechanical contact and seeds	*Solanum tuberosum*, *Solanum lycopersicum*, *Solanum melongena*, *Capsicum annuum*	*Physalis peruviana*, *Solanum nigrum*, *Petunia* sp., *Nicotiana tabacum*

TABLE 3.1 *(Continued)*

Virus Genera	Virus Species	Transmission	Natural Cultivated Host	Natural Wild Host
	Tomato mosaic virus	Mechanical contact and seeds.	*Solanum tuberosum*, *Solanum lycopersicum*, *Solanum melongena*, *Capsicum annuum*	*Petunia* sp., *Solanum betaceum*
	Brugmansia mild mottle virus	Mechanical contact	–	*Brugmansia* spp.
	Obuda pepper virus	Mechanical contact	*Capsicum annuum*	*Nicotiana tabacum*
	Pepper mild mottle virus	Mechanical contact and seeds	*Capsicum annuum*	*Nicotiana tabacum*, *Datura stramonium*, *Petunia* sp., *Physalis peruviana*
	Ribgrass mosaic virus	Mechanical contact	–	*Nicotiana tabacum*
	Tobacco latent virus	Mechanical contact and seeds	–	*Nicotiana tabacum*
	Tobacco mild green mosaic virus.	–	*Capsicum annuum*, *Solanum lycopersicum*	*Nicotiana tabacum*, *Petunia* sp.
Tobravirus	Tobacco rattle virus	Vector (nematodes)	*Solanum tuberosum*, *Capsicum annuum*	*Nicotiana tabacum*
Torradovirus	Tomato torrado virus	Vector (whiteflies)	*Solanum lycopersicum*	*Solanum nigrum*
Tymovirus	Belladonna mottle virus	Vector (beetles)	–	*Physalis pruinosa*, *Atropa belladonna*
	Dulcamara mottle virus	Vector (beetles)	–	*Solanum dulcamara*

TABLE 3.1 *(Continued)*

Virus Genera	Virus Species	Transmission	Natural Cultivated Host	Natural Wild Host
	Eggplant mosaic virus	Vector (beetles)	*Solanum lycopersicum*, *Solanum melongena*	*Nicotiana tabacum*
	Petunia vein banding virus	Vector (beetles)	–	*Petunia* sp.
	Physalis mottle virus	Vector (beetles)	–	*Physalis philadelphica*
Umbravirus	Tobacco bushy top virus	Vector (aphids)	*Solanum lycopersicum*, *Capsicum annuum*	Nicotiana tabacum

Solanum nigrum acts as a reservoir, supplying viruses to neighboring tomato plants (Carpino et al., 2019). SnIV 1 (*Solanum nigrum* ilarvirus 1) can infect both the weed *Solanum nigrum* and tomato plants, but it is predominantly prevalent in *Solanum nigrum* populations (Ma et al., 2020). PVY (potato virus Y) was found to prevail in *Solanum nigrum* populations only in areas with tomato plantations, suggesting a spillover effect of PVY from tomato to weeds (Carpino et al., 2019). Conversely, BBWV (broad bean wilt virus) was only detected in *Solanum nigrum* plants, but not in tomatoes, indicating the presence of an unidentified ecological barrier that hinders the efficient spread of BBWV from *Solanum nigrum* to tomatoes under natural conditions (Carpino et al., 2019). *Solanaceae* weeds (*D. stramonium, Physalis minima, S. nigrum, Withania somnifera*) growing near potato fields have been identified as potential sources of inoculum for PLRV (potato leaf roll virus) (Singh, 2016; Hobbs et al., 2000; Barreto et al., 2013).

More than 40 viruses can infect potato and tobacco plants. Yield losses can be caused mainly by PVY and PLRV (Kerlan, 2009; Loebenstein, 2009; Kreuze et al., 2020). PVY has become the most important pathogen in Solanaceous crops, resulting in yield losses of up to 80% (Dabijev et al., 2005; Kerlan, 2009; Kostiw, 2011). It was first reported in England (Smith, 1931) and has a wide range of plant host species belonging to 72 genera. The virus is transmitted by nearly 70 aphid species (Kerlan, 2009; Kerlan & Moury, 2009; Loebenstein, 2009). The spread of the virus is restricted at altitudes above 700 m due to unfavorable conditions for the aphids (Petrov, 2015). Symptoms depend on the virus strain, host, and vary significantly (Jones et al., 2009). The virus was divided into five major strain groups: PVY^C, PVY^O, PVY^Z, PVY^N, and PVY^E (Singh et al., 2008; Kerlan & Moury, 2009).

3.3 TOBACCO CULTIVARS AS SOURCES OF VIRAL INFECTION IN VEGETABLES FROM *SOLANACEAE* FAMILY

Tobacco (*Nicotiana tabacum*) is one of the most significant technical field crops in the Balkan Peninsula. It is primarily cultivated in deficient, mountainous, dry, and poor soil areas. Tobacco is susceptible to a wide range of viruses (Shew & Lucas, 1991). The recurrent viruses, along with the invasion of emerging viruses, have a serious impact on tobacco production, leading to symptoms such as yellowing, mosaic patterns, mottling, deformation, venial necrosis, ring spots, and stunting. The pathogens responsible belong to the Cucumovirus, Tobamovirus, Polerovirus, and Potyvirus families, with the most predominant virus species being PVY, CMV, TMV (tobacco mosaic virus), TVBMV (tobacco vein banding mosaic virus), PeMV (pepper mottle virus), Brassica yellow virus, CVMV (chili venial mottle virus), and BBWV (Ding et al., 2011; Tian et al., 2011; Dong et al., 2010; Wang et al., 2016). Tobacco crops act as virus reservoirs for other *Solanaceae* crops (Shew & Lucas, 1991). The most important virus is PVY, followed by CMV, TMV, TSWV, AMV (*alfalfa mosaic virus*), TRV (tobacco rattle virus), AYRV (Artichoke yellow ringspot virus), and EMDV (eggplant mottled dwarf virus) (Katis et al., 1993, 2000).

3.4 CONTROL MEASURES AND MANAGEMENT

One of the important goals of research in plant protection against viruses is to control and reduce the spread of viruses in cultivated plant populations, with the aim of reducing losses and increasing quality (Roselló et al., 1996). Infections in plants caused by viruses cannot be treated, so it is essential to use diagnostic methods to correctly identify the virus and prevent its further spread. Prevention of phytoviruses mainly includes suppression of the vector population and agrotechnical measures (Lapidot et al., 2014). In potato varieties, different methods such as meristem cultures, thermotherapy, and electrotherapy of tubers can be applied individually or in combination to produce virus-free planting material (Gong et al., 2019).

For some viruses, such as PLRV, that are persistently transmitted (which takes longer for the virus to remain in the vector before being transmitted to other plants), the use of insecticides as a preventative measure has been successful. However, for viruses such as PVY, which are transmitted non-persistently by aphids, insecticides are ineffective (Jones, 2014).

Choosing the right diagnostic method for identifying viral infections is extremely important to ensure that different viral strains and mutants are not missed. The use of at least two different methods, such as ELISA and PCR, improves the accuracy of virus diagnosis (Meena et al., 2017; Jeevalatha et al., 2016). Loop mediated isothermal amplification (LAMP) is an innovative method with high sensitivity and specificity for detecting various viruses in *Solanaceae* species, such as PVY (Treder et al., 2018), PLRV (Ahmadi et al., 2013), and PVX (Jeong et al., 2015).

3.4.1 CROSS PROTECTION

Viruses that cause very mild symptoms or do not affect the quantity and quality of agricultural production can be used to prevent infection with more dangerous virus species or strains. Attenuated or asymptomatic variants of such viruses can be applied in pre-inoculation of plants to prevent further virus disease (Nicaise, 2014; Agüero et al., 2018). Various viruses have been used for the species of the *Solanaceae* family. The ToMV strain Fukushima was applied in tomatoes to protect the plants against the highly virulent ToMV strains (Pechinger et al., 2019; Matsumoto et al., 2002). Similarly, a PepMV KD mutant (Chewachong et al., 2015) and weak strain CH_2 (Hanssen et al., 2010; Hasiów-Jaroszewska et al., 2014; Agüero et al., 2018) have been used to protect the plants from secondary infection with dangerous PepMV strains. Cross-protection against PVM was achieved with *Datura stramonium* as a host plant (Grupa & Syller, 2016).

3.4.2 INDUCERS OF SYSTEMIC RESISTANCE (SAR)

Plant mechanisms for protection from pathogens are based on defense, tolerance, or resistance to biotic and abiotic stress factors. The activation of protection is based on different pathways according to the specific type of pathogen. System-acquired resistance (SAR) results in resistance to subsequent pathogen attacks (Mejía-Teniente et al., 2010) and is triggered by inducer molecules. The outcomes of such research show the potential to replace toxic fungicides (Holopainen et al., 2009). Inducer substances based on salicylic acid activate SAR against PVY (Petrov & Andonova, 2012). The application of BION (benzothiadiazole) prior to inoculation with PVY in tomato plants prevented the development of viral infection in 85% of the plants, depending on the variety (Petrov & Andonova, 2012; Petrov, 2014a). Treatments with EXIN showed 58% resistant plants against PVY. The combined application of EXIN and BION resulted in full protection against PVY in the tested plants (Petrov & Andonova, 2012). BION also reduced CMV infection in tomato plants by 85% (Petrov, 2014b).

3.4.3 PLANT EXTRACTS WITH ANTIVIRAL ACTIVITY

Different ethanol and methanol extracts from medicinal plants and bacteria have shown antiviral activity against PVY, ToMV (Petrov et al., 2015, 2016b), and CMV (Petrov et al., 2016b).

3.4.4 INDUCTION OF POSTTRANSCRIPTIONAL GENE SILENCING (PTGS) THROUGH SPECIFIC SMALL INTERFERING RNAS (SIRNAS)

Posttranscriptional gene silencing (PTGS) is a protective mechanism in plants known as RNA gene silencing (Baulcombe, 1999; Vance & Vaucheret, 2001). PTGS was induced in potato cultivars by silencing essential PVY fragments from the viral genome of different strains using specific dsRNAs and siRNAs (Petrov et al., 2015a, b, 2019, 2022).

3.5 CONCLUSIONS

The *Solanaceae* family comprises a wide variety of cultivated and wild plants with varying levels of intraspecific genetic variation. Plant species of the *Solanacea*e family, including important cultivated and weedy species, serve as an important model for virological studies due to their ecological and climatic flexibility and wide distribution (Gebhardt, 2016; Goodin et al., 2008). These cultivated species, such as *Nicotiana tabacum* and *N. benthamiana*, and weeds exhibit specific susceptibility to many plant viruses and can provide valuable information for adjusting plant protection practices in a constantly changing environment.

KEYWORDS

- **control measures**
- **cross protection**
- **posttranscriptional gene silencing**
- **systemic resistance**
- **tobacco cultivars**
- **viruses of Solanaceae crops**
- **weed reservoirs**

REFERENCES

Agüero, J., Gómez-Aix, C., Sempere, R. N., García-Villalba, J., García-Núñez, J., Hernando, Y., & Aranda, M. A., (2018). Stable and broad spectrum cross-protection against *Pepino mosaic virus* attained by mixed infection. *Front. Plant Sci., 9*, 1810.

Ahmadi, S., Almasi, M. A., Fatehi, F., Struik, P. C., & Moradi, A., (2013). Visual detection of *Potato leafroll virus* by one-step reverse transcription loop-mediated isothermal amplification of DNA with hydroxynaphthol blue dye. *Journal of Phytopathology, 161*(2), 120–124.

Alemu, T., Hamacher, J., & Dehne, H. W., (2002). The role of some weeds as hosts of *Capsicum viruses* in the rift valley parts of Ethiopia. *Meded Rijksuniv Gent Fak Landbouwkd Toegep Biologische Wetenschappen, 67*, 283–289.

Anderson, P. K., Cunningham, A. A., Patel, N. G., Morales, F. J., Epstein, P. R., & Daszak, P., (2004). Emerging infectious diseases of plants: Pathogen pollution, climate change and agrotechnology drivers. *Trends Ecol. Evol., 19*, 535–544.

Anfoka, G., Moshe, A., Fridman, L., Amrani, L., Rotem, O., Kolot, M., Zeidan, M., Czosnek, H., & Gorovits, R., (2016). *Tomato yellow leaf curl virus* infection mitigates the heat stress response of plants grown at high temperatures. *Sci. Rep., 6*, 19715. https://doi.org/10.1038/srep19715.

Bag, S., Mitter, N., Eid, S., & Pappu, H. R., (2012). Complementation between two *Tospoviruses* facilitates the systemic movement of a plant virus silencing suppressor in an otherwise restrictive host. *PLoS One, 7*, e44803.

Barreto, S. S., Hallwass, M., Aquino, O. M., & Inoue-Nagata, A. K., (2013). A study of weeds as potential inoculums sources for a tomato-infecting *Begomovirus* in Central Brazil. *Phytopathology, 103*, 436–444.

Baulcombe, D. C., (1999). Gene silencing: RNA makes no protein. *Curr. Biol., 9,* R599–R601.

Bratsch, S. A., Grinstead, S., Creswell, T. C., Ruhl, G. E., & Mollov, D., (2019). Characterization of *Tomato necrotic spot virus*, a subgroup 1 ilarvirus causing necrotic foliar, stem, and fruit symptoms in tomatoes in the United States. *Plant Dis., 103*, 1391–1396.

Burdon, J. J., & Thrall, P. H., (2008). Pathogen evolution across the agro-ecological interface: Implications for disease management. *Evol. Appl., 1*, 57–65.

Carpino, C., Elvira-González, L., Rubio, L., Peri, E., Davino, S., & Galipienso, L., (2019). A comparative study of viral infectivity, accumulation and symptoms induced by broad *Bean wilt virus* 1 isolates. *J. Plant Pathol., 101*, 275–281.

Chewachong, G., Miller, S., Blakeslee, J., Francis, D., Morris, T., & Qu, F., (2015). Generation of an attenuated, cross-protective *Pepino mosaic virus* variant through alignment-guided mutagenesis of the viral capsid protein. *Phytopathology, 105*, 126–134.

Cooper, I., & Jones, R. A., (2006). Wild plants and viruses: Under-investigated ecosystems. *Adv. Virus Res., 67*, 1–47.

Dabijev, J., Mihova, T., & Muletarova, S., (2005). Investigation of the virus status of the potatoes in some regions producing high quality seed potatoes in Bulgaria. *Biotechnology & Biotechnological Equipment, 19*(Suppl. 1), 52–55.

Devaux, A., Kromann, P., & Ortiz, O., (2014). Potatoes for sustainable global food security. *Potato Research, 57*, 85–199.

Ding, M., Yang, C., Zhang, L., Jiang, Z. L., Fang, Q., Qin, X. Y., & Zhang, Z. K., (2011). Occurrence of *Chilli veinal mottle virus* in *Nicotiana tabacum* in Yunnan. *China Plant Disease, 95*, 357.

Dombrovsky, A., & Smith, E., (2017). Seed transmission of tobamoviruses: Aspects of global disease distribution. In: Jimenez-Lopez, J. C., (ed.), *Advances in Seed Biology* (pp. 234–260). Intech Open: London, UK.

Dong, J. H., Yin, Y. Y., Xu, X. Y., Duan, Y. M., & Zhang, Z. K., (2010). First report of *Tomato spotted wilt virus* in tomato and tobacco in China. *J. Plant Pathol., 92*, S121.

Forsius, M., Anttila, S., Arvola, L., Bergström, I., Hakola, H., Heikkinen, H. I., Helenius, J., Hyvärinen, M., Jylhä, K., Karjalainen, J., et al., (2013). Impacts and adaptation options of climate change on ecosystem services in Finland: A model-based study. *Curr. Opin. Environ. Sustain., 5*, 26–40.

Fraile, A., Alonso-Prados, J. L., Aranda, M. A., Bernal, J. J., Malpica, J. M., & García-Arenal, F., (1997). Genetic exchange by recombination or reassortment is infrequent in natural populations of a tripartite RNA plant virus. *J. Virol., 71*, 934–940.

Gebhardt, C., (2016). The historical role of species from the *Solanaceae* plant family in genetic research. *Theor. Appl. Genet., 129*, 2281–2294.

Gibbs, A. J., Ohshima, K., Phillips, M. J., & Gibbs, M. J., (2008). The prehistory of potyviruses: Their initial radiation was during the dawn of agriculture. *PLoS One, 3*, 2523.

Gibbs, A. J., Wood, J., García-Arenal, F., Ohshima, K., & Armstrong, J. S., (2015). Tobamoviruses have probably co-diverged with their eudicotyledonous hosts for at least 110 million years. *Virus Evol., 1*, vev 019.

Gong, H., Igiraneza, C., & Dusengemungu, L., (2019). Major *in vitro* techniques for *Potato virus* elimination and post eradication detection methods: A review. *Am. J. Potato Res., 96*, 379–389.

Goodin, M. M., Zaitlin, D., Naidu, R. A., & Lommel, S. A., (2008). *Nicotiana benthamiana*: Its history and future as a model for plant-pathogen interactions. *Mol. Plant Microbe Interact., 21*, 1015–1026.

Gorovits, R., Shteinberg, M., Anfoka, G., & Czosnek, H., (2022). Exploiting virus infection to protect plants from abiotic stresses: Tomato protection by a *Begomovirus*. *Plants, 11*, 2944. https://doi.org/10.3390/plants11212944.

Grupa, A., & Syller, J., (2016). Cross-protection between a naturally occurring mild isolate of *Potato virus M* (PVM) and a more virulent isolate in *Datura metel* plants. *J. Phytopathol., 164*, 69–73.

Hančinský, R., Mihálik, D., Mrkvová, M., Candresse, T., & Glasa, M., (2020). Plant viruses infecting *Solanaceae* family members in the cultivated and wild environments: A review. *Plants (Basel)., 9*(5), 667. doi: 10.3390/plants9050667. PMID: 32466094.

Hanssen, I. M., Gutierrez-Aguirre, I., Paeleman, A., Goen, K., Wittemans, L., Lievens, B., Vanachter, A. C. R. C., et al., (2010). Cross-protection or enhanced symptom display in greenhouse tomato co-infected with different *Pepino mosaic virus* isolates. *Plant Pathol., 59*, 13–21.

Hanssen, I. M., Mumford, R., Blystad, D., Cortez, I., Hasiów-Jaroszewska, B., Hristova, D., Pagán, I., Pereira, A. M., Peters, J., Pospieszny, H., et al., (2010). Seed transmission of *Pepino mosaic virus* in tomato. *Eur. J. Plant Pathol., 126*, 145–152.

Harahagazwe, D., Condor, B., Barreda, C., Bararyenya, A., Byarugaba, A. A., Kude, D. A., Lung'aho, C., et al., (2018). How big is the potato (*Solanum tuberosum* L.) yield gap in Sub-Saharan Africa and why? A participatory approach. *Open Agriculture, 3*(1), 180–189.

Hasiów-Jaroszewska, B., Minicka, J., & Pospieszny, H., (2014). Cross-protection between different pathotypes of *Pepino mosaic virus* representing Chilean 2 genotype. *Acta Sci. Pol. Hortorum Cultus., 13*, 177–185.

Hassani-Mehraban, A., Brenkman, A., Broek, N., Goldbach, R., & Kormelink, R., (2009). RNAi-mediated transgenic tospovirus resistance broken by intraspecies silencing suppressor protein complementation. *Mol. Plant Microbe Interact., 22*, 1250–1257.

Hobbs, H. A., Eastburn, D. M., D'Arcy, C. J., Kindhart, J. D., Masiunas, J. B., Voegtlin, D. J., Weinzierl, R. A., & McCoppin, N. K., (2000). Solanaceous weeds as possible sources of *Cucumber mosaic virus* in southern Illinois for aphid transmission to pepper. *Plant Dis., 84*, 1221–1224.

Holopainen, J., Heijari, J., Nerg, A., Vuorinen, M., & Kainulainen, P., (2009). Potential for the use of exogenous chemical elicitors in disease and insect pest management of conifer seedling production. *The Open Forest Science Journal, 2*, 17–24.

ICTV Virus Taxonomy: 2019 Release. Available online: https://talk.ictvonline.org/taxonomy (accessed on 02 March 2024).

Jeevalatha, A., Kaundal, P., Kumar, A., Guleria, A., Sundaresha, S., Pant, R. P., Sridhar, J., et al., (2016). SYBR green based duplex RT-qPCR detection of a *Begomovirus*, *Tomato leaf curl New Delhi virus*-[potato] along with *Potato virus X* and *Potato leaf roll virus* in potato. *Potato Journal, 43*, 125–137.

Jeong, J., Cho, S. Y., Lee, W. H., Lee, K. J., & Ju, H. J., (2015). Development of a rapid detection method for *Potato virus X* by reverse transcription loop-mediated isothermal amplification. *Plant Pathology Journal, 31*(3), 219.

Jones, R. A. C., & Coutts, B. A., (2015). Spread of introduced viruses to new plants in natural ecosystems and the threat this poses to plant biodiversity. *Mol. Plant Pathol., 16*, 541–545.

Jones, R. A. C., (2014). Virus disease problems facing potato industries worldwide: Viruses found, climate change implications, rationalizing virus strain nomenclature and addressing the *Potato virus Y* issue. In: Navarre, R., & Pavek, M. J., (eds.), *The Potato: Botany, Production and Uses* (1st edn., pp. 202–224). CABI: Wallingford, UK.

Jones, R. A. C., (2016). Future scenarios for plant virus pathogens as climate change progresses. *Adv. Virus Res., 95*, 87–147.

Jones, R. A., Charkowski, A., Fribourg, C. E., Stevenson, W. R., & Slack, S. A., (2009). *Potato virus* and virus-like diseases. In: *Virus Diseases of Plants: Grape, Potato, and Wheat Image Collection and Teaching Resource CD-Rom* (pp. 1–102). APS Press, St. Paul, MN 5512, USA.

Juárez, M., Rabadán, M. P., Martínez, L. D., Tayahi, M., Grande-Pérez, A., & Gómez, P., (2019). Natural hosts and genetic diversity of the emerging *Tomato leaf curl New Delhi virus* in Spain. *Front. Microbiol., 10,* 140.

Katis, N. I., Chatzivassiliou, E. K., Avgelis, A. D., Manoussopoulos, I., & Lecoq, H., (2000). Occurrence of *Eggplant mottled dwarf nucleorhabdovirus* (EMDV) in tobacco and cucumber in Greece. *Phytopathologia Mediterranea, 39*, 318.

Katis, N., Chryssohoou, A., & Woods, R., (1993). Tobacco viruses in Greece. *Coresta 1992 Information Bulletin* (p. 169). Jerez de la Frontera, Spain.

Kerlan, C., & Moury, B., (2009). *Potato virus Y*. In: Mahy, B. W., & Van, R. M. H., (eds.), *Desk Encyclopedia of Plant and Fungal Virology* (p. 613). London: Elsevier Academic Press, UK.

Kerlan, C., (2009). Potato viruses. In: Mahy, B. W., & Van, R. M. H., (eds.), *Desk Encyclopedia of Plant and Fungal Virology* (pp. 458–471, 613). London: Elsevier Academic Press, UK.

Kostiw, M., (2011). The occurrence of major potato viruses in Poland. *Journal of Plant Protection Research, 51,* 204–209.

Kreuze, J. F., Souza-Dias, J. A. C., Jeevalatha, A., Figueira, A. R., Valkonen, J. P. T., & Jones, R. A., (2020). Chapter "viral diseases of potato." In: Campos, H., & Ortiz, O., (eds.), *The Potato Crop*. ISBN 978-3-030-28683-5.

Lapidot, M., Legg, J. P., Wintermantel, W. M., & Polston, J. E., (2014). Management of whitefly-transmitted viruses in open-field production systems. *Adv. Virus Res., 90*, 147–206.

Loebenstein, G., (2009). Plant virus diseases: Economic aspects. In: Mahy, B. W., & Van, R. M. H., (eds.), *Desk Encyclopedia of Plant and Fungal Virology* (pp. 426–430, 613). London: Elsevier Academic press, UK.

Lovisolo, O., & Bartels, R., (1970). On a new strain of *Henbane mosaic virus* from *Physalis alkekengi*. *J. Phytopathol., 69*, 189–201.

Ma, Y., Marais, A., Lefebvre, M., Faure, C. H., & Candresse, T., (2020). Metagenomic analysis of virome cross-talk between cultivated *Solanum lycopersicum* and wild *Solanum nigrum*. *Virology, 540,* 38–44.

Matsumoto, T., Nara, Y., Furuya, H., Takahashi, H., Tairako, K., & Yamamoto, H., (2002). Characteristics for practical use of attenuated isolate UA-Fukushima of *Tomato mosaic virus*. *J. Gen. Plant Pathol., 68*, 382–384.

McLeisch, M. J., Fraile, A., & García-Arenal, F., (2019). Evolution of plant-virus interactions: Host range and virus emergence. *Curr. Opin. Virol., 34*, 50–55.

Meena, P. N., Kumar, R., Baswaraj, R., & Jeevalatha, A., (2017). Simultaneous detection of potato viruses A and M using CP gene specific primers in an optimized duplex RT-PCR. *Journal of Pharmacognosy and Phytochemistry, 6*(4), 1635–1640.

Mejía-Teniente, L., Torres-Pacheco, I., González-Chavira, M., Ocampo-Velazquez, R., Herrera-Ruiz, G., Chapa-Oliver, A., & Guevara-González, R., (2010). Use of elicitors as an approach for sustainable agriculture. *African Journal of Biotechnology, 9*(54), 9155–9162.

Mondal, S., Wenninger, E. J., Hutchinson, P. J. S., Whitworth, J. L., Shrestha, D., Eigenbrode, S. D., Bosque-Pérez, N. A., & Snyder, W. E., (2017). Responses of aphid vectors of *Potato leaf roll virus* to potato varieties. *Plant Dis., 101*, 1812–1818.

NCBI Virus Portal, (2020). Available online: https://www.ncbi.nlm.nih.gov/labs/virus (accessed on 02 March 2024).

Ng, J. C. K., & Perry, K. L., (2004). Transmission of plant viruses by aphid vectors. *Mol. Plant Pathol., 5*, 505–511.

Nicaise, V., (2014). Crop immunity against viruses: Outcomes and future challenges. *Front. Plant Sci., 5,* 660.

Orfanidou, C. G., Pappi, P. G., Efthimiou, K. E., Katis, N. I., & Maliogka, V. I., (2016). Transmission of *Tomato chlorosis virus* (ToCV) by *Bemisia tabaci* biotype Q and evaluation of four weed species as viral sources. *Plant Dis., 100*, 2043–2049.

Pagán, I., & Holmes, E. C., (2010). Long-term evolution of the *Luteoviridae*: Time scale and mode of virus speciation. *J. Virol., 84*, 6177–6187.

Petrov, N., & Andonova, R., (2012). BION and EXIN as SAR elicitors against *Potato virus Y* infection in tomato. *Science & Technologies, 2*(6), 46–49.

Petrov, N., (2014a). Induction of systemic acquired resistance against *Cucumber mosaic virus* in tomato agro ecosystems. *J. Bal. Ecol., 17*(4), 385–390.

Petrov, N., (2014b). Reduction of the infection with *Tomato mosaic virus* and *Cucumber mosaic virus* in tomatoes trough inducers of resistance. In: *Proceedings of "Seminar of Ecology*" (pp. 129–133). IBER, Sofia, Bulgaria.

Petrov, N., (2015). Distribution and characterization of PVY and PLRV in potato cultivar Agria in different regions of Bulgaria. *Journal of Mountain Agriculture on the Balkans, 18*(3), 612–618.

Petrov, N., Stoyanova, M., & Gaur, R., (2019). Chapter 23 – "post-transcriptional gene silencing as a tool for controlling viruses in plants." In: *Plant Biotechnology: Progress in Genomic Era*. © Springer Nature Singapore Pvt Ltd. ISBN 978-981-13-8498-1 ISBN 978-981-13-8499-8.

Petrov, N., Stoyanova, M., & Valkova, M., (2016b). Antiviral activity of plant extract from *Tanacetum vulgare* against *Cucumber mosaic virus* and *Potato Virus Y*. *J. BioSci. Biotechnol., 5*(2), 189–194. ISSN 1314-6246.

Petrov, N., Stoyanova, M., Andonova, R., & Teneva, A., (2015a). Induction of resistance to *Potato virus Y* strain NTN in potato plants through RNAi. *Biotechnology & Biotechnological Equipment, 29*(1), 21–26. doi: 10.1080/13102818.2014.984968 IF5r. 0.511.

Petrov, N., Stoyanova, M., Stoev, A., & Gaur, R., (2022). Induction of gene silencing of NIb gene region of *Potato virus Y* by dsRNAs and siRNAs and reduction of infection in potato plants cultivar Djeli. *Biotechnology & Biotechnological Equipment, 36*(1), 159–164.

Petrov, N., Teneva, A., Stoyanova, M., Andonova, R., Denev, I., & Tomlekova, N., (2015b). Blocking the systemic spread of *Potato virus Y* in the tissues of potatoes by posttranscriptional gene silencing. *Bul. J. Agri. Sci., 21*(2), 288–294.

Petrov, N., Tishkov, S., Teneva, A., Dimitrova, I., & Bojilova, M., (2016a). Antiviral activity of sukomycin against *Potato Virus Y* and *Tomato mosaic virus*. *J. BioSci. Biotechnol., 5*(3), 203–207. ISSN 1314–6238.

Pospieszny, H., Borodynko-Filas, N., Hasiów-Jaroszewska, B., Rymelska, N., & Elena, S. F., (2019). Transmission rate of two polish *Tomato torrado virus* isolates through tomato seeds. *J. Gen. Plant Pathol., 85*, 109–115.

Riga, E., Larsen, R., Eastwell, K., Guerra, N., Guerra, L., & Crosslin, J. M., (2009). Rapid detection of *Tobacco rattle tobravirus* in viruliferous *Paratrichodorus allius* from greenhouse and field specimens. *J. Nematol., 41*, 60–63.

Rojas, M. R., & Gilbertson, R. L., (2008). Emerging plant viruses: A diversity of mechanisms and opportunities. In: Roossinck, M. J., (ed.), *Plant Virus Evolution* (pp. 27–51). Springer: Berlin/Heidelberg, Germany.

Roossinck, M. J., & García-Arenal, F., (2015). Ecosystem simplification, biodiversity loss and plant virus emergence. *Curr. Opin. Virol., 10*, 56–62.

Roossinck, M. J., (2015). Plants, viruses, and the environment: Ecology and mutualism. *Virology, 479*, 271–277.

Roselló, S., Díez, M. J., & Nuez, F., (1996). Viral diseases causing the greatest economic losses to the tomato crop. I. The *Tomato spotted wilt virus*—A review. *Sci. Hortic., 67*, 117–150.

Rotenberg, D., Jacobson, A. L., Schneweis, D. J., & Whitfield, A. E., (2015). Thrips transmission of *Tospoviruses*. *Curr. Opin. Virol., 15*, 80–89.

Sastry, K. S., Mandal, B., Hammond, J., Scott, S. W., & Briddon, R. W., (2019). *Encyclopedia of Plant Viruses and Viroids* (1st edn.). Springer India: New Delhi, India.

Shates, T. M., Sun, P., Malmstrom, C., Dominguez, C., & Mauck, K., (2019). Addressing research needs in the field of plant virus ecology by defining knowledge gaps and developing wild dicot study systems. *Front. Microbiol., 9*, 3305.

Shew, D., & Lucas, G. B., (1991). *Compendium of Tobacco Diseases*. APS Press, St Paul, USA.

Shteinberg, M., Mishra, R., Anfoka, G., Altaleb, M., Brotman, Y., Moshelion, M., Gorovits, R., & Czosnek, H., (2021). *Tomato yellow leaf curl virus* (TYLCV) promotes plant tolerance to drought. *Cells, 10*(11), 2875. https://doi.org/10.3390/cells10112875.

Singh, B., (2016). Survey and indexing of weeds growing around potato fields for their role as an inoculum source for *Potato leafroll virus* (PLRV). *Br. Biotechnol. J., 16*, 1–8.

Singh, R. P., Valkonen, J. P. T., Gray, S. M., Boonham, N., Jones, R. A., Kerlan, C., & Schubert, J., (2008). Discussion paper: The naming of *Potato virus Y* strains infecting potato. *Archives of Virology, 153*, 1–13.

Smith, K. M., (1931). On the composite nature of certain potato virus diseases of the mosaic group as revealed by the use of plant indicators. *Proceedings of the Royal Society London, B, 109*, 251–267.

Tian, Y. P., Liu, J. L., Zhang, C. L., Liu, Y. Y., Wang, B., Li, X. D., Guo, Z. K., & Valkonen, J. P., (2011). Genetic diversity of *Potato virus Y* infecting tobacco crops in China. *Phytopathology, 101,* 377–387.

Treder, K., Chołuj, J., Zacharzewska, B., Babujee, L., Mielczarek, M., Burzyński, A., & Rakotondrafara, A. M., (2018). Optimization of a magnetic capture RT-LAMP assay for fast and real-time detection of *Potato virus Y* and differentiation of N and O serotypes. *Archives of Virology, 163*(2), 447–458.

Vance, V., & Vaucheret, H., (2001). RNA silencing in plants—Defense and counter defense. *Science, 292*, 2277–2280.

VIDE Database: Plant Viruses, (2020). Available online: http://bio-mirror.im.ac.cn/mirrors/pvo/vide/refs.htm (accessed on 02 March 2024).

Wang, F., Qi, S., Gao, Z., Akinyemi, I. A., Xu, D., & Zhou, B., (2016). Complete genome sequence of *Tobacco virus* 1, a closterovirus from *Nicotiana tabacum*. *Arch. Virol., 161*, 1087–1090.

Xu, P., Chen, F., Mannas, J. P., Feldman, T., Sumner, L. W., & Roossinck, M. J., (2008). Virus infection improves drought tolerance. *New Phytol., 180*, 911–921.

CHAPTER 4

Diversity and Phylogeography of Begomoviruses and DNA Satellites Associated with Weed Hosts

U. PREMCHAND,[1] SHRIDHAR HIREMATH,[2] V. VENKATARAVANAPPA,[3] M. MOHANRAJ,[1] M. MANTESH,[2] C. N. LAKSHMINARAYANA REDDY,[2] V. DEVAPPA,[1] and K. S. SHANKARAPPA[1]

[1]*Department of Plant Pathology, College of Horticulture, Bangalore; University of Horticultural Sciences, Bagalkot, Karnataka, India*

[2]*Department of Plant Pathology, College of Agriculture, University of Agricultural Sciences, GKVK, Bangalore, Karnataka, India*

[3]*Division of Crop Protection, ICAR–Indian Institute of Horticultural Research, Bangalore, Karnataka, India*

ABSTRACT

Begomoviruses are recognized as an important group of plant viruses due to the prevalence and severity of the diseases they cause worldwide. Weeds serve as reservoir hosts for begomoviruses and propagate when the primary host is not present, thanks to their high environmental tolerance. This may allow these viruses to spread to economically significant crop species, ultimately reducing output. Evaluating the genetic diversity of begomoviruses is a crucial element in developing management methods. Based on the DNA-A and DNA-B genome sequences, there are 35 bipartite and 81 monopartite begomovirus species that infect weed hosts globally. Begomoviruses are known to be associated with DNA satellites that have been detected worldwide. Breakpoint studies reveal that distinct begomoviruses that infect weed

New and Emerging Plant Viruses: The Threat to Food Security.
Alireza Golnaraghi and Rajarshi Kumar Gaur (Eds.)

hosts exhibit both intra and interspecies recombination in their DNA-A, DNA-B genomic components, and satellite molecules.

4.1 INTRODUCTION

A plant is considered a weed "if its population grows entirely or predominantly in situations markedly disturbed by humans (without, of course, being deliberately cultivated plants) in any specified geographical area" (Baker, 1965). Weeds have been the subject of research due to their importance in agriculture and horticulture, with the primary focus being on their management or elimination (Adama & Baker, 1962). There are many characteristics of plant species that enable them to perform the role of weeds. Weeds have the capacity to produce abundant seeds, which helps them establish their population rapidly. Moreover, seeds have high dormancy, enabling them to survive longer in the soil. Out of the 2,50,000 plant species known in the world, approximately 8,000 species, or 30%, behave as weeds. In the agricultural ecosystem, weeds pose several problems. They primarily curtail crop yield by competing for space, water, nutrition, and soil (Holzner, 1982). Additionally, they act as reservoirs of plant pathogens and their vectors, significantly influencing disease incidence and spread. The relationship between weeds, plant pathogens, and vectors plays a significant role in determining disease incidence and economic impact (Wisler & Norris, 2005). Compared to their cultivated counterparts, weeds are more genetically diverse and therefore less prone to diseases. Hence, whenever a genetically uniform crop is cultivated in an area, weeds become a reservoir for pathogens that would otherwise exist in low populations in wild plants. Farmers mainly focus on managing crop plants infected with pathogens and often overlook weeds, which act as outcrossing perennials. A plethora of pathogens dwell in weeds, with viruses being the main culprits infecting crop plants. Several weeds serve as natural hosts for many plant viruses. In many cases, viruses transmit from weeds to crops, ultimately resulting in crop loss.

Weeds play a vital role in the persistence and transmission of viruses by serving as a possible source of primary inoculum. Most crops endure in the field for a certain season and are not present during the off-season. The majority of viruses survive in weed hosts over the offseason. Begomoviruses are emerging as a major challenge for crop production due to their increased incidence and severity, aided by their broad host range (Ambrozevicius et al., 2002). These viruses belong to the family *Geminiviridae,* which is divided into 14 genera based on genome structure, vector transmission, and host range, comprising more than 525 species (Walker et al., 2021). Among these, the genus *Begomovirus* is an important group of plant viruses that threaten the cultivation of many crops across the world. Whitefly (*Bemisia tabaci* Genn.) transmits these viruses upon feeding on infected plants (Brown et al., 2012). Begomoviruses have single-stranded circular DNA as their genome, which encapsidates in geminate icosahedral particles (Fiallo-Olive et al., 2021). These viruses are divided into old world (OW) and New World (NW) begomoviruses based on their distribution and genome composition. NW begomoviruses have two genome components and are known as bipartite (two genome components; DNA-A and DNA-B) begomoviruses. On the other hand, OW begomoviruses have a single genome component and are known as monopartite (one genome component homologous to DNA-A bipartite begomoviruses) (Melgarejo et al., 2013). Three types of DNA satellite molecules are known to be associated with begomoviruses, namely betasatellites (Briddon et al., 2003), alphasatellites (Briddon et al., 2004), and deltasatellites (Lozano et al., 2016). Weed hosts play a crucial role in the survival of begomoviruses during off-seasons and serve as major reservoirs of primary inoculum of these viruses upon planting of crop plants (Prajapat et al., 2013). The earliest record of geminivirus symptoms on a weed was on eupatorium, showing yellowing, which is available in a poem written by Japanese Empress Koken during the summer of 752 AD (Saunders et al., 2003). The study and knowledge of begomoviruses diversity in crops and their associated weeds are very important in understanding the population dynamics of the virus and their vectors (Mubin et al., 2009). This chapter attempts to provide information on the diversity of begomoviruses infecting weed hosts.

4.2 IMPORTANCE OF WEED HOSTS BELONGING TO DIFFERENT TAXONOMIC FAMILIES IN SERVING AS RESERVOIRS OF BEGOMOVIRUS INOCULUM

4.2.1 FAMILY: ASTERACEAE

Ageratum, colloquially known as white weed, belongs to the genus *Ageratum* and the family *Asteraceae*. The genus *Ageratum* consists of 40–60 flowering annuals and perennials that are grown in tropical warm temperatures and are native to Central America and Mexico (Nesom, 2006). Among them, *A.*

conyzoides is known to harbor many begomoviruses, such as *Papaya leaf curl China virus* (PaLCuCNV) from China and *Eclipta yellow vein virus* (EYVV) from Pakistan. *Ageratum enation virus* (AEV) has been widely distributed, and *Ageratum leaf curl Cameroon virus* (ALCCMV) has been reported from Cameroon. Similarly, *Ageratum leaf curl Sichuan virus* (ALCScV) has been reported from India and China, and *Ageratum yellow vein China virus* (AYVCNV), *Ageratum yellow vein virus* (AYVV), and *Synedrella leaf curl virus* on *A. conyzoides* have been reported from China. Begomoviruses infecting *A. conyzoides* induce symptoms such as chlorosis, upward leaf curling, yellow veins, reduced leaf size, and stunted growth with poor inflorescence (Khan & Samad, 2014). Recently, ALCScV was found to induce dwarf and downward leaf-curling symptoms in *A. conyzoides, Helianthus annuus,* and *Nicotiana benthamiana* (Li et al., 2020). *N. benthamiana* co-inoculated with ALCScV and a noncognate betasatellite induced more severe symptoms. AEV on *A. conyzoides* has been reported from Nepal, India, and Pakistan, with plants exhibiting foliar vein yellowing that later develops into a bright yellow mosaic. However, in some cases, instead of yellowing, vein swelling, and enation symptoms were reported. Furthermore, AEV is reported to be transmitted by the whitefly (*Bemisia tabaci*) in a circulative and non-propagative manner (Tahir et al., 2015). Apart from *A. conyzoides*, AEV is also known to infect *A. houstonianum* and another *Ageratum* spp. in India (Srivastava et al., 2015). Furthermore, *Ageratum yellow vein Hualian virus* (AYVHuV) infection has been reported in plants of *Ageratum* spp. in Taiwan, exhibiting symptoms of yellow vein mosaic (Tsai et al., 2011).

Emila sonchifolia, commonly known as cupid's shaving brush or tassel flower, is predominantly found in Asian countries such as China, India, and Southeast Asia. It is cultivated as an annual medicinal herb in China and is recognized as one of the 10 sacred flowers of Kerala state in India. In Vietnam, it has been extensively used in traditional medicine for treating fever, sore throat, diarrhea, eczema, and as an antidote for snake bites (Tanaka & Nguyen, 2007). However, in many countries, it is considered a weed that negatively impacts various crops, reducing both their quality and quantity, and serving as a reservoir for crop pathogens. It is classified as an invasive species in India, Brazil, Mexico, Paraguay, and Costa Rica. Yang et al. (2012) identified yellow vein symptoms in *E. sonchifolia*, which were confirmed to be caused by *Emilia yellow vein virus* (EmYVV). During the collection of infected samples near *E. sonchifolia*, yellowing symptoms were observed in thick head plants (*Crassocephalum crepidioides*), a weed belonging to the Asteraceae family. The complete genome sequence obtained from infected

C. crepidioides leaves revealed the presence of EmYVV, suggesting it to be the reservoir of the virus. *Eclipta prostrata*, a native weed of Asia, is widely distributed in warm, tropical, and subtropical regions (Holm et al., 1997). In the Philippines, it is a major contaminant along with the rice seeds (Rao & Moody, 1990) and is thought to be introduced into France along with the contaminated seeds (Jauzein, 1991). These weeds act as reservoirs for several begomoviruses in China and play a significant role in the formation of begomovirus epidemics. Zhang et al. (2015) first reported the incidence of PaLCuCNV and *pepper yellow vein Mali virus* (PYVMIV) on *E. prostrate* from China. Whitefly transmission studies showed that both viruses could individually infect *E. prostrata.* Numerous other identified begomoviruses reported to infect *E. prostrata* include *Alternanthera yellow vein virus* (A1YVV) (Ha et al., 2008), *Eclipta prostrata yellow vein virus* (EPYVV) (Haider et al., 2005), *tomato leaf curl New Delhi virus* (ToLCNDV) (Haider et al., 2006), *Ludwigia yellow vein virus*, *Synedrella leaf curl virus*, *and Erechtites yellow mosaic virus* (ErYMV) reported to infect fireweed (*Erechtites valerianifolia*) from Vietnam (Ha et al., 2008).

Parthenium weed (*Parthenium hysterophorus* L.) is an invasive herbaceous weed that degrades the natural ecosystem and causes serious allergic reactions in domesticated livestock and humans. Additionally, it serves as a habitat for various plant pathogens and insects. Kumar et al. (2016) observed Parthenium plants exhibiting stunting and severe leaf curl symptoms in a farmers' field in Lucknow, India. Sequence analysis revealed the presence of ToLCNDV along with *Papaya leaf curl betasatellite* (PaLCuB) and Ageratum yellow vein India *alphasatellite* (AYVIA). *Sigesbeckia glabrescens*, an annual herb belonging to the *Asteraceae* family (Albu et al., 2016), is traditionally used for managing chronic diseases. Wu & Zhou (2007) collected samples from naturally begomovirus-infected *S. glabrescens* plants showing vein thickening, yellow vein, and stunted symptoms in the Guangzhou region of Guangdong province, China. The virus isolate was transmitted by the whitefly *Bemisia tabaci* (B biotype; currently known as MEAM cryptic species) to *S. glabrescens* and *N. glutinosa*. Sequence analysis confirmed the presence of a begomovirus in the collected samples, which was named *Siegesbeckia yellow vein Guangxi virus* (SgYVGdV) (Wu et al., 2007). *S. glabrescens* has also been reported to be infected with *Synedrella leaf curl virus* in China and Bangladesh (Khatun et al., 2020).

Xanthium strumarium, is a weed speculated to have originated from southern Europe and Asia. This weed is thought to have the toxic chemical carboxyatractyloside and is being used in traditional medicine in South

Asia and China. *X. strumarium* is known to be infected by multiple begomoviruses, as well as satellite molecules that were previously shown to be associated with distinct crops. Mubin et al. (2012) collected five samples from symptomatic *X. strumarium* plants showing leaf curling, vein swelling, and either generalized chlorosis or interveinal yellowing symptoms from the vicinity of Samundri (Punjab province, Pakistan) during the year 2005. Molecular analysis revealed the presence of two begomoviruses, as well as a betasatellite and an alphasatellite, in the samples. *Cotton leaf curl Burewala virus*, associated with cotton varieties resistance breaking in Pakistan, was discovered to be a new recombinant begomovirus species. The second isolate was *tomato leaf curl Gujarat virus* (ToLCGV), a bipartite begomovirus previously reported.

4.2.2 FAMILY: ACANTHACEAE

Andrographis paniculata (Family*: Acanthaceae*), commonly known as 'Kalmegh,' is utilized in traditional medicine due to its anti-inflammatory, antibiotic, anti-malarial, anticancer, anti-hepatotoxic, antiviral, and antioxidant properties (Hiremath et al., 2021). Khan & Samad (2014) observed a disease incidence of 25 to 40% on Kalmegh in Lucknow, India, and nucleotide sequence analysis revealed the presence of the EYVV along with *Ageratum* yellow *vein betasatellite* without alphasatellite.

Blechum pyramidatum (Browne's Blechum plants) is a weed native to tropical America. It has been introduced to southern Taiwan, the Philippines, and islands of the Pacific such as the Mariana Islands, the Caroline Islands, and Western Samoa. The first report of *Blechum interveinal chlorosis virus* (BleICV) was from Mexico (Mauricio-Castillo et al., Unpublished-NC019035). Typically, begomovirus-infected *B. pyramidatum* plants exhibit severe chlorosis symptoms. Similarly, it has also been reported to be infected with *Blechum yellow vein virus* (BleYVV) in the Philippines (Tsai et al., 2011). Infected plants show symptoms of yellow/chlorotic leaf veins and shortened internodes. Sequence analysis (84.8 to 87.6% nucleotide identity) showed a close relationship to *Papaya leaf curl Guangdong virus* isolates.

Dicliptera sexangularis, commonly known as Sixangle foldwing, is a flowering plant belonging to the genus *Dicliptera* in the family *Acanthaceae.* Lotrakul et al. (2000) collected virus-infected leaf samples showing symptoms of severe yellow mottle and leaf distortion from the shoreline of Calusa Island (Lee County, FL). The causal virus was transmitted through grafting and whitefly and identified as a distinct bipartite begomovirus, *Dicliptera yellow mottle virus* (DiYMV). Similarly, on *D. vahliana*, DiYMV was first reported from Cuba by Echemendía et al. (2003) with infected plants showing yellow mottle symptoms during 2000–2001.

4.2.3 FAMILY: EUPHORBIACEAE

Croton bonplandianus, a wild species of croton, due to its resemblance to flower cymes of Tulsi, is popularly known as *Ban Tulsi* (*jungle tulsi*) and is used as fuel and a detergent. *C. bonplandianus* is extensively found in Southern Asia. Hussain et al. (2011) collected leaf samples of *C. bonplandianus* showing yellow vein symptoms. Through PCR and sequence analysis, the complete sequences of begomovirus and associated betasatellite were obtained. The genome sequence showed a maximum nucleotide (nt) identity of 88.9% with *papaya leaf curl virus*, hence it was considered a new species and proposed the name *croton yellow vein virus* (CYVV). The betasatellite associated with CYVV was also identified as a new species and designated as *croton yellow vein mosaic betasatellite* (CroYVMB). In India, *C. bonplandianus* leaves showing severe leaf curl and yellow vein symptoms were collected during a survey conducted at Trichy, Tamil Nadu, India in 2016 (Sravya et al., 2022). Begomovirus sequences obtained from these samples shared a maximum identity of 89.62% with PaLCuV. However, the value was below the species demarcation limit for begomoviruses, so it was considered to be a new species and named *Croton yellow vein leaf curl virus*.

Milkweed (*Euphorbia heterophylla*), a weed native to tropical and subtropical America, subsequently spread to different parts of the world and is difficult to control with many herbicides. In soybean and peanut crops, it is considered a weed of significant economic importance (Wilson, 1981). *E. heterophylla* plants exhibiting bright yellow mosaic symptoms were collected, and the presence of the virus was detected using PCR and sequence analysis. The begomovirus associated with yellow mosaic was found to be bipartite. The DNA-A sequence showed a maximum nucleotide identity of 87.3% with *Euphorbia mosaic Peru virus* (EuMPV). Therefore, it was considered a novel virus and designated as *Euphorbia yellow mosaic virus* (EuYMV). On the other hand, the DNA-B sequence shared a maximum nucleotide identity of 56.2% with DNA-B of *Euphorbia mosaic virus* (EuMV). Furthermore, phylogenetic analysis demonstrated that the virus is an isolate of EuMV reported from Central America (Fernandes et al., 2011).

4.2.4 FAMILY: SOLANACEAE

Datura innoxia (Sacred datura) is flowering plant belonging to the family *Solanaceae* and a familiar hallucinogenic alkaloid containing toxic weed available in India. It is thought to have originated in the Southwestern United States, Central and South America, and was introduced to Africa, Asia, Australia, and Europe (Annapoorani, 2013). Leaf samples showing leaf curling, stunted growth, and decreased leaf size were collected from fields of Rajasthan, India. Molecular characterization of the virus associated with the samples showed a close resemblance to *Chili leaf curl virus*, *tomato leaf curl virus* (ToLCV), and CYVV. This was the first report of begomovirus associated with leaf curl on *D. inoxia* from India (Marwal et al., 2012). *Watermelon chlorotic stunt virus* (WmCSV) was first reported on *D. innoxia* from Sudan by Mohammed et al. (2017). A host range study for ToLCNDV conducted by Juarez et al. (2019) included a total of 683 samples from 13 cultivated species and 203 samples from 24 wild species from three major cucurbits producing areas of Spain. Among the collected weeds, *Datura stromonium* was also found infected with ToLCNDV and it was identified as an alternative host and a potential source of virus inoculum.

European black nightshade (*Solanum nigrum*) is a weed that typically thrives in moist habitats with various types of soils, including stony or deep soils. It poses a significant threat to agriculture when it competes with cultivated crops. Additionally, it serves as a host for several viruses, which can severely impact nearby crops. Nightshade plants exhibiting characteristic symptoms of begomovirus infection, such as leaf curling and puckering, were collected and analyzed at the molecular level. The genomic DNA-A exhibited a maximum nucleotide identity of 94% with the ToLCNDV-ivy gourd isolate, while DNA-B showed 95% identity with the bitter gourd isolate. These findings suggest that *S. nigrum* acts as a reservoir for ToLCNDV and plays a role in the development of epidemics in cucurbit species (Ansar et al., 2021).

4.2.5 FAMILY: SAPINDACEAE

Deinbollia borbonica is a common tropical perennial shrub found from Somalia's coast to northern Mozambique. It is considered a weed in many countries where crops such as beans, cassava, and tomatoes are grown in mixed cropping farming systems. Leaf samples of *D. borbonica* showing yellow mosaic symptoms were collected and the association with begomovirus was molecularly characterized. The genome sequence (DNA-A) of the virus was found to be closely related to *Tomato leaf curl* Mayotte virus with a maximum nucleotide identity of 64.82%, and DNA-B showed a maximum nucleotide identity of 65% with *East African cassava mosaic virus*. Since the nucleotide identity is lower than the ICTV demarcation criteria for begomovirus species, the virus was considered novel and designated as *Deinbollia mosaic virus* (Kyallo et al., 2017).

4.2.6 FAMILY: MALVACEAE

Abutilon is the largest genus of the *Malvaceae* family, and its species are widely distributed throughout the tropics and subtropics of the Americas, Africa, Asia, and Oceania continents. Most of these species are common weeds in agricultural fields. Torres-Herrera et al. (2019) collected leaf samples from *Abutilon permolle*, a weed showing mosaic on leaves from Suytunchen, Yucatán states during the summer of 2007 to assess the diversity of begomoviruses in Southern México. Sequencing results showed that the isolate associated with *A. permolle* is significantly divergent (< 88%) from other begomovirus sequences and considered as a novel, named *Abutilon golden mosaic Yucatan virus* (AbGMYV). Similarly, Ha et al. (2008) identified the novel *Sida leaf curl virus* (SiLCV) infecting abutilon (*Abutilon indicum*) from Vietnam.

Corchoropsis belongs to the family Malvaceae and is native to temperate Asia. *Corchoropsis timentosa* is a common and widespread weed species that thrives on roadsides and fields in regions of southern China, where papaya is extensively cultivated. PaLCuCNV is a distinct begomovirus and a significant threat to the cultivation of papaya in China. Huang & Zhou (2006) reported that *C. timentosa* showing yellow vein symptoms acts as an alternative weed host for PaLCuCNV in China (Huang & Zhou, 2006).

Marshmallow or cheese weed (*Malva parviflora* L., family *Malvaceae*) is native to Southern Europe and Asia (Akbar et al., 2014) and is widespread in major cultivated and non-cultivated agricultural ecosystems (Barros et al., 2010). *M. parviflora* has extensive medicinal and pharmacological uses (Heinrich et al., 2012). Weed plants of the family *Malvaceae* have served as a reservoir for broadening the genetic diversity of many geminiviruses (Paprotka et al., 2010). *M. parviflora* has reportedly been an alternate and/or experimental host for begomoviruses in different parts of Asian countries.

Sattar et al. (2017) recorded *M. parviflora* plants exhibiting vein yellowing and chlorosis in Lahore, Pakistan, and analyzed the presence of begomovirus and associated DNA satellites through PCR and subsequent sequencing. The identified full-length begomovirus genome shared maximum nt sequence identity with *hollyhock leaf curl virus* (HoLCV), representing a new strain. Furthermore, this begomovirus is associated with *Kenaf leaf curl betasatellite* (KLCuB), *Ageratum* conyzoides *symptomless alpha satellite* (ACSLA), and *Ageratum* yellow vein *India alpha satellite* (AYVIA). This was the first report of a complete monopartite begomovirus complex on *M. parviflora* from Pakistan.

Malvastrum coromandelianum, also known as three-lobe false mallow, is an annual or perennial herb or shrub native to North and South America. It has been introduced to many other countries, including Africa, Australia, and southern and eastern Asia. *M. coromandelianum* is likely the reservoir for several begomoviruses. Jiang & Zhou (2005) identified *M. coromandelianum* showing yellow vein symptoms in Yunnan province of China. The complete nucleotide sequence of the begomovirus was determined from this host, which has the maximum sequence identity with *Malvastrum yellow vein virus*. However, it falls below the species demarcation threshold and is reported as a distinct begomovirus species. The proposed name for this virus is *Malvastrum yellow vein Yunnan virus* (MYVYNV). MYVYNV is also found to be associated with a betasatellite molecule that shares sequence identity with a previously reported betasatellite associated with Malvastrum yellow vein virus. Similarly, *Malvastrum yellow vein Lahore virus* (MaYVLaV) was identified in *M. coromandelianum* from Pakistan (Rahman et al., 2019). A monopartite begomoviral infection on *M. coromandelianum* plants exhibiting yellow vein symptoms was identified in Kampot province, Cambodia. The genome sequence of this virus shared maximum sequence identity with AYVV and was considered a novel species, for which the name *Malvastrum yellow vein Cambodia virus* (MaYVCV) was proposed. Furthermore, it was reported that MaYVCV is associated with a distinct betasatellite molecule, Malvastrum yellow vein *Cambodia beta satellite* (MaYVKHB) (Tang et al., 2021). Liu et al. (2009) recorded *M. coromandelianum* plants showing yellow vein symptoms in the provinces of Baoshan, Yunnan, China, and later showed that these plants were associated with a distinct monopartite begomovirus and betasatellite, having maximum sequence identity to TYLCCNV and *tomato leaf curl China betasatellite* (TLCuCNB), respectively. This report concluded that *M. coromandelianum* is an alternate host for TYLCCNV and may contribute to the epidemiology of tomato yellow leaf curl or tobacco leaf curl diseases. Fiallo-Olive et al. (2021) discovered two new strains of *Sida golden mosaic Florida virus* (SiGMFV) in two bipartite begomovirus isolates infecting *M. coromandelianum* in Cuba. The DNA-A component of SiGMFV is found in a monophyletic cluster with begomoviruses that infect malvaceous weeds in the Caribbean, according to phylogenetic analysis. *Sida* spp. is an ubiquitous weed species and is often seen displaying the yellow mosaic symptoms typical of begomovirus infection. Passos et al. (2017) reported two new bipartite begomovirus species from *Sida* sp. plant samples collected from the state of Piaui, Northeastern Brazil. The genomes of these bipartite begomoviruses (*Sida angular mosaic virus* (SiAMV) and *Sida chlorotic vein virus* (SiCVV)) showed genome organization typical of NW begomoviruses. In phylogenetic analysis, these two closely related viruses clustered with other begomoviruses described in tomato and sida plants from Brazil. Three viral isolates, Hn8, Hn40, and Hn41, were obtained from *Sida acuta* plants exhibiting yellow mosaic symptoms in the Hainan province of China. A comparison of partial DNA-A sequences, amplified with degenerate primers, confirmed the presence of a single type of begomovirus in these plants. The complete nucleotide sequence of the DNA-A molecule of Hn8 was determined, showing the highest sequence identity (78%) with that of the AYVCNV. Based on the molecular data, Hn8 is considered a new species, named *Sida yellow mosaic China virus* (SiYMCNV). Betasatellite molecules were also found to be associated with these samples, sharing the highest nucleotide sequence identity with the DNA-β of AYVV (Xiong et al., 2005).

Rodríguez-Negrete et al. (2019) conducted high-throughput sequencing (HTS) to determine viral diversity in seven states of the Northern-Pacific Mexico areas, which have diverse climatic conditions and where different vegetable crops are grown with intensive farming. In total, 132 plant species, including weed species belonging to 34 families, were identified and sampled in the natural ecosystems surrounding cultivated areas during 2012–2016. Sequence analysis identified, for the first time, *Sida mosaic Sinaloa virus* (SiMSiV) infecting *Sida rombifolia*; *Rhynchosia golden mosaic virus* (RhGMV) and *Rhynchosia golden mosaic Sinaloa virus* (RhGMSV) infecting *Rhyncosia minima,* inducing symptoms in their corresponding hosts.

Similarly, Ha et al. (2008) characterized novel begomoviruses, namely *Corchorus golden mosaic virus* (CoGMV), infecting jute mallow (*Corchorus capsularis*), and *Sida yellow vein Vietnam virus* (SiYVVNV), infecting sida (*Sida rhombifolia*), weeds belonging to the Malvaceae family from Vietnam.

A leaf sample exhibiting symptoms of chlorotic mottling was collected from *Sida santaremensis*, a weed host in Monteiro, Manatee County, Florida.

The genome of the associated begomovirus was sequenced, revealing the presence of *Sida golden mottle virus* (SiGMoV), a bipartite begomovirus with a genome organization typical of NW begomoviruses. Sequence analysis of SiGMoV DNA-A showed a maximum identity of 89% and the closest evolutionary relationship with *Sida golden mosaic Buckup virus* (SiGMBuV). On the other hand, SiGMoV DNA-B had a maximum identity of 93% and the closest evolutionary relationship with DNA-B of *Corchorus yellow spot virus* (CoYSV) (Al-Aqeel et al., 2018).

The complete DNA-A sequence of the begomovirus associated with *Sida* spp. plants displaying yellow mosaic symptoms was determined from Jamaica. This revealed the presence of a new begomovirus, SiGMBuV, with the DNA-B component phylogenetically closely related to malvaceous weed-infecting begomoviruses reported from Florida and Mexico. Furthermore, the identified DNA-B was found to be a recombinant molecule with the majority of the genome derived from a virus resembling *Sida yellow mosaic Yucatan virus* (Stewart et al., 2014).

During 2004 and 2008, samples of *Sida* spp., a ubiquitous weed species found in Jamaica, were collected from Braco, Trelawny, and Liguanea, St. Andrew. A DNA-A molecule isolated from a plant in Liguanea, St. Andrew, showed a 90.9% identity with *Sida golden yellow vein virus*, indicating it is a strain of SiGYVV based on sequence analysis. The cognate DNA-B, which was previously unreported, was successfully cloned and found to be similar to that of the *Malvastrum yellow mosaic Jamaica virus* (MaYMJV). Phylogenetic studies suggested that this virus was most directly associated with begomoviruses infecting malvaceous hosts in Jamaica, Cuba, and the United States (Stewart et al., 2011). Fiallo-Olive et al. (2021) conducted surveys on rowing in search of wild reservoirs of begomoviruses in Cuba. They discovered a novel bipartite begomovirus called *Sida yellow mottle virus* (SiYMoV) infecting Sida rhombifolia plants, which exhibited mottle symptoms. The complete genome sequence of DNA-A revealed a close relationship with *Sida golden mosaic virus* (SiGMV), which infects *Phaseolus vulgaris* in Florida. However, the DNA-B sequence showed the highest nucleotide identity with CoYSV. Additionally, a novel strain of Sida yellow vein virus was found infecting *Malvastrum coromandelianum* plants, displaying bright yellow mosaics. Furthermore, an isolate of *Euphorbia mosaic virus* infecting *Euphorbia heterophylla* was reported for the first time in Cuba. Stewart et al. (2011) characterized a *Sida golden yellow vein virus* isolate, SiGYVV-[JM:Lig2:08], from symptomatic *Sida jamaicensis* plant samples collected in Braco, Trelawny, Jamaica. Further analysis using PCR-RFLP identified two genetically distinct DNA-A molecules with a recombinant origin from the Jamaican *Sida jamaicensis* plant. These were considered as two novel begomovirus species, namely *Sida golden mosaic Braco virus* and *Sida golden mosaic Liguanea virus*. Phylogenetically, the DNA-A of these two viruses is closely related to begomoviruses that infect malvaceous weeds in Jamaica, Cuba, Florida, and Mexico. Eight begomovirus isolates from Bolivian weeds, including Sida rhombifolia, Sida micrantha, putative *Solanum* and *Abutilon* spp., were confirmed using RCA and restriction fragment length polymorphism (RFLP). The study revealed a connection between typical NW bipartite begomoviruses. Four out of the eight were new begomovirus species, proposed to be named *Abutilon mosaic Bolivia virus,* Solanum mosaic Bolivia virus, Sida mosaic Bolivia virus 1, and Sida mosaic Bolivia virus 2. SimMV-rho[Bo:CF1:07], SimMV-rho[Bo:CF2:07], and SimMV-rho[BoVi07] were all variants of Sida micrantha mosaic virus (SimMV). However, one of them was a new variant of a previously described SimMV, SimMV-MGS2:07-Bo (Wyant et al., 2011). The diseased Sida plants exhibiting yellow mosaic symptoms were found in agricultural fields of Gujarat, India (Chahwala et al., 2021). Sequence analysis of the associated begomovirus indicated the presence of a distinct monopartite begomovirus (*Sida yellow mosaic Gujarat virus*) along with a betasatellite (*Ludwigia leaf distortion betasatellite)* and an alphasatellite (*Malvastrum yellow mosaic alphasatellite*). Furthermore, it has been confirmed that Sida plants serve as a potential source of virus inoculum, which can be transmitted by whiteflies to other cultivated crops. Paprotka et al. (2010) reported nt sequences of the Brazilian strain of *Abutilon mosaic virus* (AbMV) infecting Sida micrantha, a common weed in Brazil. Sequence comparison studies have demonstrated the association of a complex of at least two begomoviruses (Jovel et al., 2004). These results show that coinfection of begomoviruses can persist over decades, producing a reservoir of partially recombined but distinct geminiviruses. This evidence further supports the hypothesis that weeds, especially those belonging to the genus Sida, strongly impact the evolution of geminiviruses, not only because of the diversity of viruses they harbor but also due to their ubiquitous and perennial nature.

4.2.7 FAMILY: FABACEAE

Ha et al. (2008) characterized monopartite begomoviruses infecting different weed species belonging to the family Fabaceae, namely SiLCV infecting abutilon *(Abutilon indicum)* and *Mimosa yellow leaf curl virus* (MiYLCV) infecting mimosa (Mimosa sp.), as well as a bipartite *Kudzu mosaic virus* (KuMV) infecting Kudzu (*Pueraria montana*) from Vietnam. Silva et al. (2011) conducted a survey of begomoviruses infecting six different leguminous weeds, namely *Macroptilium lathyroides*, *M. atropurpureum*, *Centrosema brasilianum*, *Calopogonium mucunoides*, *Canavalia* sp., and *Macroptilium* spp., from four different states in northeastern Brazil. A total of 26 full-length begomovirus components (19 DNA-A and seven DNA-B, with three pairs of cognate A and B components) were amplified using RCA, cloned, and sequenced. Sequence analysis indicated the presence of four novel species, *Macroptilium yellow net virus* (MaYNV), *Centrosema yellow spot virus* (CenYSV), *Macroptilium yellow spot virus* (MaYSV) and *Macroptilium yellow vein virus* (MaYVV) and three species which were reported earlier, EuYMV, *bean golden mosaic virus* (BGMV) and *tomato crinkle leaf yellows virus* (TCrLYV). Phylogenetic analysis revealed the clustering of four viruses with other *Brazilian begomoviruses*, but one of them (EuYMV) clustered with viruses from other countries in Central and South America. Evidence of recombination was found among isolates of MaYSV. The MaYSV population exhibited a high degree of genetic variability. The results indicated that Macroptilium lathyroides serves as a common host for these viruses and could act as a mixing vessel for viral recombination, leading to the emergence of novel viruses. It was concluded that leguminous weeds in Brazil serve as reservoirs for several begomoviruses and could play a significant role in begomovirus epidemics, both as sources of inoculum and as sources of emerging novel viruses. Sobrinho et al. (2014) conducted a comprehensive survey and collected samples to assess the genetic variability of begomovirus populations infecting cultivated and non-cultivated leguminous host plants (*Macroptilium lathyroides)* from two different regions of Brazil. The full-length DNA-A genome components of the begomoviruses, MaYSV and BGMV, were sequenced.

4.2.8 FAMILY: OXALIDACEAE

Species of the genus *Oxalis*, belonging to the family Oxalidaceae, are widely scattered in subtropical regions of Brazil and are commonly found as weeds in agricultural and urban areas. In Brazil, *Oxalis oxyptera* was reported as one of the hosts of geminivirus in the 1970s (Costa & Betti, 1978). From 2003 to 2015, Fontenele et al. (2018) collected samples from *Oxalis* spp. with golden spots and yellow mosaic symptoms from various parts of Brazil and established transmission *via* the whitefly, *Bemisia tabaci.* The samples were subjected to genome amplification using RCA and were sequenced. Sequence analysis revealed the presence of DNA-A and DNA-B components, indicating an association with the bipartite virus. Furthermore, sequences of both DNA-A and DNA-B components showed maximum nucleotide identity with SiMMV. This is the first report of Oxalis species as a natural host of SiMMV in South America. These *Oxalis* species may serve as a persistent source of SiMMV inoculum for crops such as cotton, okra, tomato, soybean, and snap bean in Brazil (Fernandes-Acioli et al., 2011).

4.2.9 FAMILY: CAPRIFOLIACEAE

Complete nucleotide sequences of two begomoviruses, namely *Nara virus-1* and *Nara virus-2*, a satellite DNA called DNA-Nara, and defective DNAs were isolated and cloned from honeysuckle (*Lonicera japonica*) plants in Japan. These plants exhibited characteristic yellowing of veins with small elliptical-shaped enations along the veins on the underside of the leaves. The genome organization of these viruses is identical to that of other OW monopartite begomoviruses. Comparison of nucleotide sequences with other begomoviruses revealed that *Nara virus-1* and *Nara virus-2* were strains of *honeysuckle yellow vein mosaic virus* (HYVMV). Therefore, it was concluded that these viruses, along with their satellites, are the causative agents of tomato yellow dwarf disease in honeysuckle plants in Japan. Additionally, they may serve as potential reservoir hosts for begomoviruses (Ogawa et al., 2008).

4.2.10 FAMILY: AMARANTHACEAE

Mubin et al. (2009) determined the complete nucleotide sequences of a complex begomovirus disease associated with yellow vein disease of *Digera arvensis*, a common weed. They found that the *Cotton leaf curl Rajasthan virus* had the highest nucleotide sequence identity. Additionally, two betasatellite species were identified. The first was an isolate of *Ageratum yellow leaf curl betasatellite*, and the second was a recombinant of *Tobacco leaf curl* and *Cotton leaf curl Multan betasatellites*. Furthermore, the isolated alphasatellite matched an isolate from potato. Consequently, the presence of multiple and recombinant betasatellites in *D. arvensis* suggests that it serves as an important source of various begomovirus components that impact crop plants.

4.2.11 FAMILY: LAMIACEAE

Leonurus sibiricus (L.) is a subtropical weed that frequently exhibits golden mosaic symptoms. Faria & Maxwell (1999) reported the first begomovirus infection on *L. sibiricus* from Brazil, namely *Leonurus mosaic virus* (LeMV). Later, Fernandes-Acioli et al. (2011); Barbosa et al. (2013); and Colariccio et al. (2007) reported a new bipartite species, *tomato yellow spot virus* (ToYSV) (2007). They hypothesized that under natural conditions, *L. sibiricus* serves as a reservoir for ToYSV. Kwon et al. (2018) conducted field surveys in pepper fields in Korea between 2013 and 2014, collecting samples from *L. sibiricus* plants exhibiting virus-like symptoms. The molecular analysis revealed that affected plants were infected with *beet western yellow virus* (BWYV), *pepper mild mottle virus* (PMMoV), or *tomato spotted wilt virus* (TSWV). It was also proposed that *L. sibiricus* could serve as a reservoir host for viruses that cause damage in pepper fields.

4.2.12 FAMILY: LAMIACEAE AND ONAGRACEAE

Ha et al. (2008) reported *Clerodendrum golden mosaic virus* (ClGMV) infecting glory bower (*Clerodendrum philippinum*) from the family, *Lamiaceae* and *Ludwigia yellow vein Vietnam virus* (LuYVVNV) infecting willow primrose (*Ludwigia octovalvis*), belong to *Onagraceae* family.

4.3 GLOBAL DISTRIBUTION OF BEGOMOVIRUSES ON DIFFERENT WEED HOSTS

4.3.1 ASSOCIATION OF MONOPARTITE AND BIPARTITE BEGOMOVIRUSES WITH WEED HOSTS

Begomoviruses have single-stranded circular DNA as their genome and are broadly categorized into two groups: monopartite and bipartite begomoviruses, based on the number of genome components they possess. Bipartite viruses have two genome components referred to as DNA-A and DNA-B, while monopartite begomoviruses have a single genome component often considered as a homolog of DNA-A in bipartite viruses (Zaidi et al., 2016). These begomoviruses are also referred to as OW and NW begomoviruses, considering their diversity and distribution (Zaidi et al., 2016). In bipartite viruses, both components must be encapsidated and acquired independently for disease symptoms to appear. However, in the past few decades, the association of alphasatellite, betasatellite, and deltasatellite molecules with begomoviruses has been reported (Kumar et al., 2016). Although these molecules are not encoded with CP, their genomes are nearly half or less than half the size of the helper viruses.' These molecules can have an impact on pathogenesis and the accumulation of associated helper viruses. As a result, satellite molecules are considered to be a component of the begomovirus disease complex (Gnanasekaran & Chakraborty, 2018). Due to genetic changes caused by mutation, recombination, and reassortment (pseudo-recombination), begomoviruses have evolved to induce severe symptoms on various crop hosts (Seal et al., 2006). Therefore, it is crucial to investigate the diversity and distribution of begomoviruses infecting weeds in different countries worldwide. Genome sequence data (Tables 4.1–4.6) of all begomoviruses infecting and associated with weed hosts were obtained from the NCBI database and analyzed. Among the numerous begomoviruses reported on different weed hosts globally, 81 were monopartite and 35 were bipartite begomoviruses, accounting for 70% and 30% respectively (Figure 4.1).

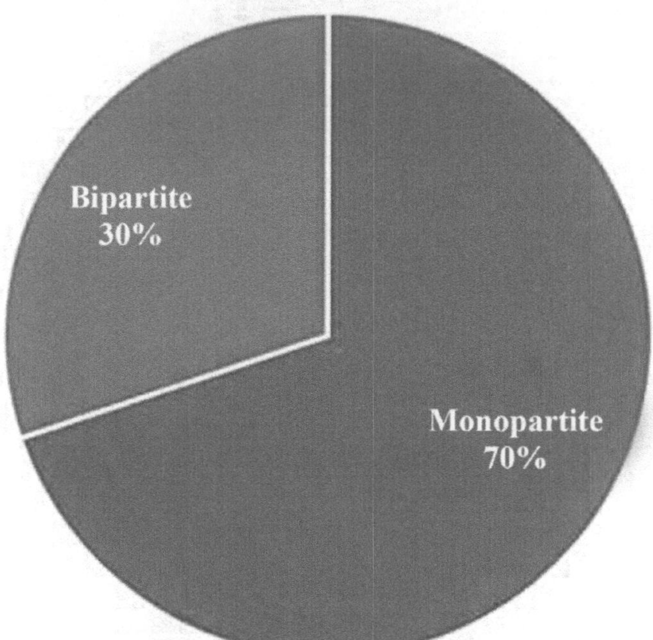

FIGURE 4.1 Percent distribution of old (monopartite) and new world (bipartite) begomoviruses infecting weed hosts worldwide.

This data clearly shows that the percentage of monopartite viruses on weed hosts is higher than that of bipartite begomoviruses, which is a similar trend observed in crop plants as well. The analysis also revealed that the maximum number of monopartite viruses were found to infect weed hosts from the *Malvaceae* (30) family, followed by *Asteraceae* (21), *Fabaceae* (11), *Solanaceae* (9), *Euphorbiaceae* (5), *Acanthaceae* (4), *Lamiaceae* (3), *Sapindaceae* (1), *Amaranthaceae* (1), *Caprifoliaceae* (1), and *Linderniaceae* (1) (Figure 4.2(A) and Table 4.1).

Similarly, a maximum number of bipartite begomoviral species, i.e., 13, have been identified from weed hosts belonging to the *Fabaceae* family. Around 12 were found to infect weed hosts from *Malvaceae*, followed by *Solanaceae* (3). Two were found in each of *Lamiaceae, Sapindaceae*, and *Euphorbaceae,* and one each in *Acanthaceae* and *Sapindaceae* (Figure 4.2(B) and Table 4.2).

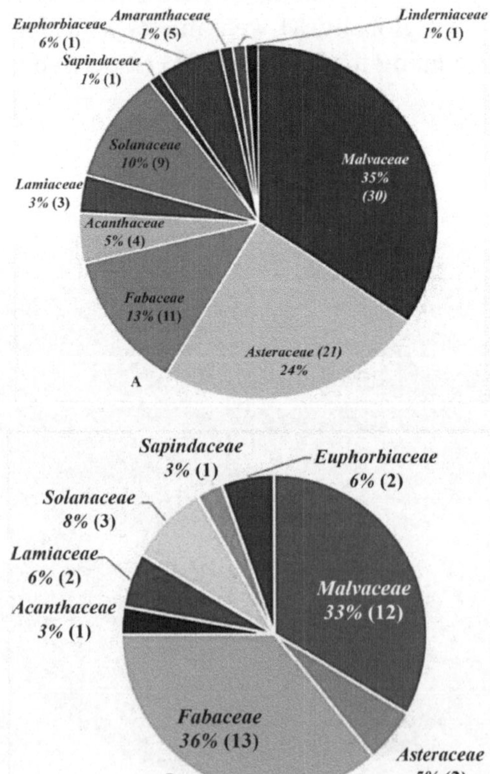

FIGURE 4.2 Distribution of (A) monopartite; and (B) bipartite begmoviruses on weed hosts belonging to different taxonomic families.

TABLE 4.1 Details of Genome Sequences (DNA-A or Homolog of DNA-A) of Begomoviruses Infecting Different Weed Hosts

Sl. No.	Host	Begomovirus	Accession No.	Country	Year	Acronym	Nucleotide Size (nt)
I			**Family: Malvaceae**				
1.	***Abutilon* sp.**						
	Abutilon indicum	*Sida leaf curl virus*	DQ641707	Vietnam	2007	SiLCV-[VI:Abu:07]-DQ641707	2,762
	Abutilon indicum	*Sida leaf curl virus*	DQ641706	Vietnam	2007	SiLCV-[VI:Abu:07]-DQ641706	2,760
	Abutilon megapotamicum	*Abutilon mosaic virus*	LN611623	France	2007	AMV-[FR:Abu:07]-LN611623	2,634
	Abutilon megapotamicum	*Abutilon mosaic virus*	LN611622	Germany	2009	AMV-[FR:Abu:09]-LN611622	2,634
	Abutilon permolle	*Abutilon mosaic Yucatan virus*	NC038438	Mexico	2007	AMYV-[MX:Abu:07]-NC038438	2,629
	Abutilon permolle	*Abutilon Mosaic Yucatan Virus*	KC430935	Mexico	2007	AMYV-[MX:Abu:07]-KC430935	2,629
	Abutilon sellovianum	*Abutilon mosaic virus*	NC001928	West Indian	2018	AMV-[Abu:07]-NC001928	2,632
	Abutilon sellovianum	*Abutilon mosaic virus*	X15983	West Indian	2015	AMV-[Abu:15]-X15983	2,632
	Abutilon sp.	*Abutilon mosaic Bolivia virus*	NC015045	Bolivia	2007	AMBV-[BO:Abu:07]-NC015045	2,717
	Abutilon sp.	*Abutilon mosaic Bolivia virus*	HM585445	Bolivia	2007	AMBV-[BO:Abu:07]-HM585445	2,717
	Abutilon sp.	*Abutilon mosaic virus*	NC014138	Brazil	2007	AMV-[BZ:Abu:07]-NC014138	2,653
	Abutilon sp.	*Abutilon mosaic virus*	FN434438	Brazil	2007	AMV-[BZ:Abu:07]-FN434438	2,653
	Abutilon sp.	*Abutilon mosaic virus*	HQ588901	India	2010	AMV-[IN:Abu:10]-HQ588901	2,637
	Abutilon sp.	*Abutilon mosaic virus*	HQ588900	India	2010	AMV-[IN:Abu:10]-HQ588900	2,635
	Abutilon sp.	*Abutilon mosaic virus*	HQ588899	India	2010	AMV-[IN:Abu:10]-HQ588899	2,633
	Abutilon sp.	*Abutilon mosaic virus*	U51137	Hawaii	2003	AMV-[IN:Abu:03]-U51137	2,634
	Abutilon theophrasti	*Tomato leaf curl virus*	KC852149	China	2012	ToLCV-[CN:Abu:12]-KC852149	2,781
2.	***Corchoropsis timentosa***						
	Corchoropsis timentosa	*Papaya leaf curl China virus*	AJ876548	China	2006	PaLCCNV-[CN:Cor:06]-AJ876548	2,735

TABLE 4.1 *(Continued)*

Sl. No.	Host	Begomovirus	Accession No.	Country	Year	Acronym	Nucleotide Size (nt)
3.	***Malva parviflora***						
	Malva parviflora	*Hollyhock leaf curl virus*	LT716980	Pakistan	2017	HLCV-[PK:Mal:17]-LT716980	2,755
	Malva parviflora	*Malvastrum yellow vein Yunnan virus*	JX679256	China	2011	MYVYNV-[CN:Mal:11]-JX679256	2,747
	Malva parviflora	*Malvastrum yellow vein Yunnan virus*	JX679249	China	2011	MYVYNV-[CN:Mal:11]-JX679249	2,747
	Malva parviflora	*Squash leaf curl virus*	EF532620	Jordan	2008	SLCV-[JO:Mal:08]-EF532620	2,637
	Malva parviflora	*Tomato yellow leaf curl China virus*	JX679252	China	2012	TYLCCNV-[CN:Mal:12]-JX679252	2,738
	Malva parviflora	*Tomato yellow leaf curl China virus*	JX679251	China	2012	TYLCCNV-[CN:Mal:12]-JX679251	2,737
4.	***Malvastrum coromandelianum***						
	Malvastrum coromandelianum	*Malvastrum yellow vein Yunnan virus*	NC006631	China	2018	MaYVYnV-[CN:Mal:18]-NC006631	2,747
	Malvastrum coromandelianum	*Malvastrum yellow vein Lahore virus*	NC055476	Pakistan	2013	MaYVLaV-[PK:Mal:13]-NC055476	2,752
	Malvastrum coromandelianum	*Malvastrum yellow vein Cambodia virus*	NC026763	Cambodia	2014	MaYVCaV-[CA:Mal:14]-NC026763	2,737
	Malvastrum coromandelianum	*Malvastrum yellow vein Cambodia virus*	KP188831	Cambodia	2014	MaYVCaV-[CA:Mal:14]-KP188831	2,737
	Malvastrum coromandelianum	*Tomato yellow leaf curl virus*	KC189894	China	2010	ToYLCV-[CN:Mal:10]-KC189894	2,742
	Malvastrum coromandelianum	*Tomato yellow leaf curl China virus*	KC189893	China	2009	ToYLCCV-[CN:Mal:09]-KC189893	2,737
5.	***Sida* sp.**						
	Sida acuta	*Sida angular mosaic virus*	KX691404	Brazil	2014	SiAMV-[BZ:Sid:14]-KX691404	2,696
	Sida acuta	*Sida angular mosaic virus*	KX691407	Brazil	2014	SiAMV-[BZ:Sid:14]-KX691407	2,692

TABLE 4.1 *(Continued)*

Sl. No.	Host	Begomovirus	Accession No.	Country	Year	Acronym	Nucleotide Size (nt)
	Sida acuta	*Sida angular mosaic virus*	KX691408	Brazil	2014	SiAMV-[BZ:Sid:14]-KX691408	2,692
	Sida acuta	*Sida angular mosaic virus*	NC031456	Brazil	2014	SiAMV-[BZ:Sid:14]-NC031456	2,692
	Sida acuta	*Sida yellow mosaic China virus*	AJ810096	China	2004	SiYMCV-[CN:Sid:04]-AJ810096	2,749
	Sida micrantha	*Sida common mosaic virus*	JX871373	Brazil	2010	SiMCV-[BZ:Sid:10]-JX871373	2,644
	Sida micrantha	*Sida common mosaic virus*	JX871372	Brazil	2010	SiMCV-[BZ:Sid:10]-JX871372	2,644
	Sida micrantha	*Sida common mosaic virus*	JX871371	Brazil	2010	SiMCV-[BZ:Sid:10]-JX871371	2,643
	Sida micrantha	*Sida common mosaic virus*	JX871370	Brazil	2010	SiMCV-[BZ:Sid:10]-JX871370	2,643
	Sida santaremensis	*Sida mottle virus*	JX871378	Brazil	2010	SiMV-[BZ:Sid:10]-JX871378	2,662
	Sida santaremensis	*Sida mottle virus*	JX871377	Brazil	2010	SiMV-[BZ:Sid:10]-JX871377	2,662
	Sida santaremensis	*Sida yellow mosaic virus*	JX871369	Brazil	2010	SiYMV-[BZ:Sid:10]-JX871369	2,657
	Sida sp.	*Euphorbia yellow mosaic virus*	KX348181	Brazil	2010	EuYMV-[BZ:Sid:10]-KX348181	2,813
	Sida sp.	*Euphorbia yellow mosaic virus*	KX348180	Brazil	2010	EuYMV-[BZ:Sid:10]-KX348180	2,618
	Sida sp.	*Sida angular mosaic virus*	KX691406	Brazil	2014	SiAMV-[BZ:Sid:14]-KX691406	2,691
	Sida sp.	*Sida bright yellow mosaic virus*	NC040191	Brazil	2010	SiBYMV-[BZ:Sid:10]-NC040191	2,656
	Sida sp.	*Sida bright yellow mosaic virus*	NC038991	Brazil	2010	SiBYMV-[BZ:Sid:10]-NC038991	2,656
	Sida sp.	*Sida bright yellow mosaic virus*	KX348184	Brazil	2010	SiBYMV-[BZ:Sid:10]-KX348184	2,692
	Sida sp.	*Sida chlorotic leaf virus*	MN013784	Mexico	2018	SiChLV-[MX:Sid:18]-MN013784	2,635
	Sida sp.	*Sida chlorotic leaf virus*	NC055581	Mexico	2018	SiChLV-[MX:Sid:18]-NC055581	2,635
	Sida sp.	*Sida chlorotic mottle virus*	NC038990	Brazil	2010	SiCMOV-[BZ:Sid:10]-NC038990	2,601
	Sida sp.	*Sida chlorotic mottle virus*	KX348183	Brazil	2010	SiCMOV-[BZ:Sid:10]-KX348183	2,601
	Sida sp.	*Sida golden yellow spot virus*	NC038992	Brazil	2010	SiGYSV-[BZ:Sid:10]-NC038992	2,813
	Sida sp.	*Sida golden yellow spot virus*	KX348185	Brazil	2010	SiGYSV-[BZ:Sid:10]-KX348185	2,813

TABLE 4.1 *(Continued)*

Sl. No.	Host	Begomovirus	Accession No.	Country	Year	Acronym	Nucleotide Size (nt)
	Sida sp.	*Sida mottle Alagoas virus*	KX896420	Brazil	2014	SiMAV-[BZ:Sid:10]-KX896420	2,651
	Sida sp.	*Sida mottle Alagoas virus*	KX896421	Brazil	2014	SiMAV-[BZ:Sid:14]-KX896421	2,651
	Sida sp.	*Sida yellow mosaic virus*	KX513859	India	2016	SiYMV-[IN:Sid:16]-KX513859	2,749
	Sida sp.	*Tomato yellow spot virus*	KX348226	Brazil	2010	ToYSV-[BZ:Sid:10]-KX348226	2,595
	Sida spinosa	*Sida angular mosaic virus*	KX691403	Brazil	2014	SiAMV-[BZ:Sid:14]-KX691403	2,689
	Sida urens	*Sida chlorotic yellow vein virus*	NC031455	Brazil	2010	SiChYVV-[BZ:Sid:10]-NC031455	2,654
	Sida urens	*Sida chlorotic vein virus*	KX691405	Brazil	2014	SiChVV-[BZ:Sid:14]-KX691405	2,655
	Sida urens	*Sida chlorotic vein virus*	KX691402	Brazil	2008	SiChVV-[BZ:Sid:08]-KX691402	2,654
	Sida urens	*Sida chlorotic vein virus*	NC_031454	Brazil	2008	SiChVV-[BZ:Sid:08]-NC_031454	2,491
	Sida urens	*Sida yellow mosaic Alagoas virus*	JX871383	Brazil	2010	SiYMAV-[BZ:Sid:10]-JX871383	2,690
	Sida rhombifolia	*Sida common mosaic virus*	NC038457	Brazil	2007	SiMCV-[BZ:Sid:07]-NC038457	2,687
	Sida rhomboidea	*Abutilon mosaic Brazil virus*	NC016574	Brazil	2018	AMBzV-[BZ:Abu:18]-NC016574	2,649
	Sida rhomboidea	*Abutilon mosaic Brazil virus*	JF694482	Brazil	2012	AMBzV-[BZ:Sid:12]-JF694482	2,629
	Sida rhomboidea	*Abutilon mosaic Brazil virus*	JF694481	Brazil	2012	AMBzV-[BZ:Sid:12]-JF694481	2,612
	Sida rhomboidea	*Abutilon mosaic Brazil virus*	JF694480	Brazil	2012	AMBzV-[BZ:Sid:12]-JF694480	2,649
	Sida santaremensis	*Euphorbia yellow mosaic virus*	JX871379	Brazil	2010	EuYMV-[BZ:Sid:10]-JX871379	2,609
	Sida sp.	*Sida mottle Alagoas virus*	KX896424	Brazil	2014	SiMAV-[BZ:Sid:10]-KX896424	2,648
	Sida sp.	*Sida mottle Alagoas virus*	KX896422	Brazil	2014	SiMAV-[BZ:Sid:10]-KX896422	2,647
	Sida sp.	*Sida mottle Alagoas virus*	KX896423	Brazil	2014	SiMAV-[BZ:Sid:10]-KX896423	2,647
	Sida sp.	*Sida mottle Alagoas virus*	KX896415	Brazil	2014	SiMAV-[BZ:Sid:10]-KX896415	2,646
	Sida sp.	*Sida mottle Alagoas virus*	KX896416	Brazil	2014	SiMAV-[BZ:Sid:10]-KX896416	2,646
	Sida sp.	*Sida mottle Alagoas virus*	KX896417	Brazil	2014	SiMAV-[BZ:Sid:10]-KX896417	2,646

TABLE 4.1 *(Continued)*

Sl. No.	Host	Begomovirus	Accession No.	Country	Year	Acronym	Nucleotide Size (nt)
	Sida sp.	*Sida Brazil virus*	NC038887	Brazil	2007	SiAMV-[BZ:Sid:14]-NC038887	2,659
	Sida urens	*Sida mottle Alagoas virus*	JX871385	Brazil	2010	SiMAV-[BZ:Sid:10]-JX871385	2,649
	Sida urens	*Sida mottle Alagoas virus*	NC_020256	Brazil	2010	SiMAV-[BZ:Sid:10]-NC_020256	2,649
	Sida urens	*Sida yellow blotch virus*	NC020254	Brazil	2010	SiYBV-[BZ:Sid:10]-NC020254	2,664
	Sida urens	*Sida yellow blotch virus*	JX871382	Brazil	2010	SiYBV-[BZ:Sid:10]-JX871382	2,664
	Sida urens	*Sida yellow blotch virus*	JX871381	Brazil	2010	SiYBV-[BZ:Sid:10]-JX871381	2,664
	Sida urens	*Sida yellow blotch virus*	JX871380	Brazil	2010	SiYBV-[BZ:Sid:10]-JX871380	2,664
	Sida urens	*Sida yellow mosaic Alagoas virus*	NC020255	Brazil	2010	SiYMAV-[BZ:Sid:10]-NC020255	2,690
	Sida urens	*Tomato mild mosaic virus*	KC706608	Brazil	2010	ToMMV-[BZ:Sid:10]-KC706608	2,707
	Sida urens	*Tomato mild mosaic virus*	KC706607	Brazil	2010	ToMMV-[BZ:Sid:10]-KC706607	2,707
	Sida urens	*Tomato mild mosaic virus*	KC706606	Brazil	2010	ToMMV-[BZ:Sid:10]-KC706606	2,713
	Sida urens	*Tomato mild mosaic virus*	JX871375	Brazil	2010	ToMMV-[BZ:Sid:10]-JX871375	2,720
	Sida sp.	*Sida yellow golden mosaic virus*	MF163258	Brazil	2015	SiYMV-[BZ:Sid:15]-MF163258	2,661
	Sida sp.	*Sida yellow golden mosaic virus*	NC_055548	Brazil	2015	SiYMV-[BZ:Sid:15]-NC055548	2,661
	Sida rhombiforlia	*Sida yellow vein Vietnam virus*	DQ641696	Vietnam	2006	SiYVVV-[VE:Sid:06]-DQ641696	2,753
	Sida sp.	*Sida mottle Alagoas virus*	KX896427	Brazil	2014	SiMAV-[BZ:Sid:14]-KX896427	2,647
II			**Family: Asteraceae**				
6.	***Ageratum* sp.**						
	Ageratum conyzoides	*Papaya leaf curl China virus*	JX294075	China	2012	PLCCV-[CN:Aeg:12]-JX294075	2,754
	Ageratum conyzoides	*Eclipta yellow vein virus*	HG937534	Pakistan	2013	EcYVV-[PK:Ecp:13]-HG937534	2,142
	Ageratum conyzoides	*Ageratum enation virus*	OM719614	India	2021	AEV-[IN:Aeg:21]-OM719614	2,746
	Ageratum conyzoides	*Ageratum enation virus*	JF728866	India	2010	AEV-[IN:Aeg:10]-JF728866	2,748

TABLE 4.1 *(Continued)*

Sl. No.	Host	Begomovirus	Accession No.	Country	Year	Acronym	Nucleotide Size (nt)
	Ageratum conyzoides	*Ageratum enation virus*	JF728864	India	2010	AEV-[IN:Aeg:10]-JF728864	2,748
	Ageratum conyzoides	*Ageratum enation virus*	JF728863	India	2010	AEV-[IN:Aeg:10]-JF728863	2,748
	Ageratum conyzoides	*Ageratum enation virus*	JF728862	India	2010	AEV-[IN:Aeg:10]-JF728862	2,748
	Ageratum conyzoides	*Ageratum enation virus*	JF728861	India	2010	AEV-[IN:Aeg:10]-JF728861	2,748
	Ageratum conyzoides	*Ageratum enation virus*	JF728860	India	2010	AEV-[IN:Aeg:10]-JF728860	2,748
	Ageratum conyzoides	*Ageratum leaf curl Cameroon virus*	FR873230	Cameroon	2009	ALCCV-[CM:Aeg:09]-FR873230	2,794
	Ageratum conyzoides	*Ageratum leaf curl Cameroon virus*	FR873229	Cameroon	2009	ALCCV-[CM:Aeg:09]-FR873229	2,792
	Ageratum conyzoides	*Ageratum leaf curl Cameroon virus*	FR873228	Cameroon	2009	ALCCV-[CM:Aeg:09]-FR873228	2,792
	Ageratum conyzoides	*Ageratum leaf curl Sichuan virus*	NC055492	India	2016	ALCSV-[IN:Aeg:16]-NC055492	2,749
	Ageratum conyzoides	*Ageratum leaf curl Sichuan virus*	MG917698	China	2016	ALCSV-[CN:Aeg:16]-MG917698	2,749
	Ageratum conyzoides	*Ageratum yellow vein China virus*	NC004090	China	2018	AYVCV-[CN:Aeg:18]-NC004090	2,768
	Ageratum conyzoides	*Ageratum yellow vein China virus*	AJ495813	China	2007	AYVCV-[CN:Aeg:07]-AJ495813	2,768
	Ageratum conyzoides	*Ageratum yellow vein China virus*	AJ849916	China	2011	AYVCV-[CN:Aeg:11]-AJ849916	2,740
	Ageratum conyzoides	*Ageratum yellow vein China virus*	MN218666	China	2014	AYVCV-[CN:Aeg:14]-MN218666	2,740
	Ageratum conyzoides	*Ageratum yellow vein China virus*	MN218663	China	2014	AYVCV-[CN:Aeg:14]-MN218663	2,744
	Ageratum conyzoides	*Ageratum yellow vein China virus*	MN218662	China	2014	AYVCV-[CN:Aeg:14]-MN218662	2,745
	Ageratum conyzoides	*Ageratum yellow vein China virus*	MN218661	China	2014	AYVCV-[CN:Aeg:14]-MN218661	2,745
	Ageratum conyzoides	*Ageratum yellow vein China virus*	MN218660	China	2014	AYVCV-[CN:Aeg:14]-MN218660	2,699
	Ageratum conyzoides	*Ageratum yellow vein China virus*	MN218659	China	2014	AYVCV-[CN:Aeg:14]-MN218659	2,699
	Ageratum conyzoides	*Ageratum yellow vein China virus*	MN218658	China	2014	AYVCV-[CN:Aeg:14]-MN218658	2,751
	Ageratum conyzoides	*Ageratum yellow vein virus*	MN233600	China	2012	AYVV-[CN:Aeg:12]-MN233600	2,756
	Ageratum conyzoides	*Synedrella leaf curl virus*	MN210356	China	2017	SyLCV-[CN:Age:17]-MN210356	2,750

TABLE 4.1 *(Continued)*

Sl. No.	Host	Begomovirus	Accession No.	Country	Year	Acronym	Nucleotide Size (nt)
	Ageratum houstonianum	*Ageratum enation virus*	KJ488990	India	2021	AEV-[IN:Aeg:21]-KJ488990	2,753
	Ageratum sp.	*Ageratum enation virus*	JQ911767	India	2012	AEV-[IN:Aeg:12]-JQ911767	2,753
	Ageratum sp.	*Ageratum enation virus*	FN794198	India	2009	AEV-[IN:Aeg:09]-FN794198	2,748
	Ageratum sp.	*Ageratum yellow vein Hualian virus*	NC038913	Taiwan	2000	AYVHV-[TA:Aeg:00]-NC038913	2,753
	Ageratum sp.	*Ageratum yellow vein Hualian virus*	DQ866133	China	2000	AYVHV-[CN:Aeg:00]-DQ866133	2,762
	Ageratum sp.	*Ageratum yellow vein Hualian virus*	DQ866132	China	2000	AYVHV-[CN:Aeg:00]-DQ866132	2,753
	Ageratum sp.	*Ageratum yellow vein virus*	DQ866134	China	2000	AYVV-[CN:Aeg:00]-DQ866134	2,754
7.	***Crassocephalum crepidioides***						
	Crassocephalum crepidioides	*Emilia yellow vein virus*	JQ247188	China	2008	EmYVV-[CN:Cra:08]-JQ247188	2,726
8.	***Emilia sonchifolia***						
	Emilia sonchifolia	*Emilia yellow vein virus*	MN201999	China	2017	EmYVV-[CN:Cra:17]-MN201999	2,725
	Emilia sonchifolia	*Emilia yellow vein virus*	MN201998	China	2017	EmYVV-[CN:Cra:17]-MN201998	2,725
	Emilia sonchifolia	*Emilia yellow vein virus*	MN201997	China	2017	EmYVV-[CN:Cra:17]-MN201997	2,725
9.	***Eclipta prostrate***						
	Eclipta prostrate	*Papaya leaf curl China virus*	AM691552	China	2006	PaLCCNV-[CN:Ecp:06]-AM691552	2,749
	Eclipta prostrata	*Papaya leaf curl China virus*	AM691553	China	2006	PaLCCNV-[CN:Ecp:06]-AM691553	2,749
	Eclipta prostrata	*Ageratum yellow vein China virus*	MN218667	China	2012	AIYVCNV-[CN:Ecp:12]-MN218667	2,759
	Eclipta prostrata	*Alternanthera yellow vein virus*	DQ375456	China	2006	AlYVV-[CN:Ecp:06]-DQ375456	2,745
	Eclipta prostrata	*Eclipta yellow vein virus*	NC016660	Pakistan	2006	EcYVV-[PK:Ecp:06]-NC016660	2,740
	Eclipta prostrata	*Eclipta yellow vein virus*	GQ478343	Pakistan	2006	EcYVV-[PK:Ecp:06]-GQ478343	2,740

TABLE 4.1 *(Continued)*

Sl. No.	Host	Begomovirus	Accession No.	Country	Year	Acronym	Nucleotide Size (nt)
	Eclipta prostrata	*Ludwigia yellow vein virus*	MN210344	China	2017	LuYVV-[CN:Ecp:17]-MN210344	2,755
	Eclipta prostrata	*Ludwigia yellow vein virus*	MN210343	China	2017	LuYVV-[CN:Ecp:17]-MN210343	2,757
	Eclipta prostrata	*Pepper yellow vein Mali virus*	AM691555	China	2006	PepLCYV-[CN:Ecp:06]-AM691555	2,786
	Eclipta prostrata	*Pepper yellow vein Mali virus*	AM691547	China	2006	PepLCYV-[CN:Ecp:06]-AM691547	2,786
	Eclipta prostrata	*Synedrella leaf curl virus*	MN210350	China	2016	SyLCV-[CN:Ecp:16]-MN210350	2,748
	Eclipta prostrata	*Synedrella leaf curl virus*	NC_031306	China	2013	SyLCV-[CN:Ecp:16]-NC_031306	2,749
	Eclipta prostrata	*Synedrella leaf curl virus*	KU933258	China	2013	SyLCV-[CN:Ecp:16]-KU933258	2,749
	Eclipta prostrata	*Alternanthera yellow vein virus*	DQ641704	Vietnam	2006	AYVV-[VE:Ecp:06]-DQ641704	2,745
10.	***Erectites valerianifolia***						
	Erectites valerianifolia	*Erectites yellow mosaic virus*	NC009549	Vietnam	2018	ErYMV-[VI:Ere:18]-NC009549	2,751
	Erectites valerianifolia	*Erectites yellow mosaic virus*	DQ641698	Vietnam	2007	ErYMV-[VI:Ere:07]-DQ641698	2,751
11.	***Parthenium hysterophorus***						
	Parthenium hysterophorus	*Tomato leaf curl Karnataka virus*	MN839537	Pakistan	2018	ToLCKV-[PK:Par:18]-MN839537	2,760
	Parthenium hysterophorus	*Tomato leaf curl New Delhi virus*	KF002409	Pakistan	2013	ToLCNDV-[PK:Par:13]-KF002409	2,743
12.	***Siegesbeckia glabrescens***						
	Siegesbeckia glabrescens	*Siegesbeckia yellow vein Guangxi virus*	NC_008317	China	2005	SBYVGV-[CN: Sig:05]-NC_008317	2,784
	Siegesbeckia orientalis	*Siegesbeckia yellow vein virus*	JF682838	China	2008	SBYVV-[CN: Sig:08]-JF682838	2,770
	Sigesbeckia glabrescens	*Siegesbeckia yellow vein Guangxi virus*	AM238692	China	2005	SBYVGV-[CN: Sig:05]-AM238692	2,784
	Sigesbeckia glabrescens	*Siegesbeckia yellow vein virus*	AM230635	China	2007	SBYVV-[CN: Sig:07]-AM230635	2,768
	Sigesbeckia glabrescens	*Siegesbeckia yellow vein virus*	AM230634	China	2007	SBYVV-[CN: Sig:07]-AM230634	2,768
	Sigesbeckia glabrescens	*Siegesbeckia yellow vein virus*	AM183224	China	2007	SBYVV-[CN: Sig:07]-AM183224	2,768

TABLE 4.1 *(Continued)*

Sl. No.	Host	Begomovirus	Accession No.	Country	Year	Acronym	Nucleotide Size (nt)
	Sigesbeckia glabrescens	*Siegesbeckia yellow vein virus*	KF499589	China	2012	SBYVV-[CN: Sig:12]-KF499589	2,771
	Sigesbeckia glabrescens	*Siegesbeckia yellow vein virus*	JX294076	China	2011	SBYVV-[CN: Sig:11]-JX294076	2,769
	Sigesbeckia glabrescens	*Siegesbeckia yellow vein virus*	NC_008236	China	2018	SBYVV-[CN: Sig:18]-NC_008236	2,768
	Sigesbeckia glabrescens	*Synedrella leaf curl virus*	MN210351	China	2017	SyLCV-[CN:Sig:17]-MN210351	2,751
	Sigesbeckia glabrescens	*Synedrella leaf curl virus*	MN210353	China	2017	SyLCV-[CN: Sig:17]-MN210353	2,751
	Sigesbeckia glabrescens	*Synedrella leaf curl virus*	MN210354	China	2017	SyLCV-[CN: Sig:17]-MN210354	2,751
	Sigesbeckia glabrescens	*Synedrella leaf curl virus*	MN210355	China	2017	SyLCV-[CN: Sig:17]-MN210355	2,751
	Sigesbeckia glabrescens	*Synedrella leaf curl virus*	MN210352	China	2017	SyLCV-[CN: Sig:17]-MN210352	2,750
13.	***Xanthium strumarium***						
	Xanthium strumarium	*Cotton leaf curl Burewala virus*	FR819707	Pakistan	2011	CLCuBuV-[PK:Xan:11]-FR819707	2,756
	Xanthium strumarium	*Tomato leaf curl Gujarat virus*	FR819708	Pakistan	2011	ToLCGuV-[PK:Xan:11]-FR819708	2,754
	Xanthium strumarium	*Tomato leaf curl Patna virus*	KY612435	India	2016	ToLCPaV-[IN:Xan:16]-KY612435	2,747
14.	***Spilanthes paniculata***						
	Spilanthes paniculata	*Spilanthes yellow vein virus*	DQ641694	Vietnam	2006	SYVV-[VE:Spi:06]-DQ641694	2,761
III			**Family: Acanthaceae**				
15.	***Andrographis paniculata***						
	Andrographis paniculata	*Eclipta yellow vein virus*	KC476655	India	2010	EcYVV-[IN:And:10]-KC476655	2,739
16.	***Blechum pyramidatum***						
	Blechum pyramidatum	*Blechum interveinal chlorosis virus*	NC019035	Mexico	2011	BlIChV-[MX:Ble:11]-NC019035	2,640
	Blechum pyramidatum	*Blechum yellow veins virus*	NC055132	Philippines	2012	BlYVV-[MX:Ble:12]-NC055132	2,764
	Blechum pyramidatum	*Papaya leaf curl Guandong virus*	KF446660	Philippines	2012	PaLCGuV-[PH:Ble:12]-KF446660	2,756

TABLE 4.1 *(Continued)*

Sl. No.	Host	Begomovirus	Accession No.	Country	Year	Acronym	Nucleotide Size (nt)
17.	***Dicliptera* sp.**						
	Dicliptera sexangularis	*Dicliptera yellow mottle virus*	AF139168	Florida	2000	DiYMV-[FL:Dic:2000]-AF139168	2,597
	Dicliptera vahliana	*Dicliptera yellow mottle virus*	AJ549960	Cuba	2005	DiYMV-[CU:Dic:05]-AJ549960	2,608
	Dicliptera vahliana	*Dicliptera yellow mottle virus*	NC003856	Florida	2005	DiYMV-[FL:Dic:05]-NC003856	2,607
IV			**Family: Euphorbiaceae**				
18.	***Croton bonplandianus***						
	Croton bonplandianus	*Croton yellow vein mosaic virus*	KX671964	Pakistan	2015	CYVMV-[PK:Cro:15]-KX671964	2,749
	Croton bonplandianus	*Croton yellow vein mosaic virus*	JN817516	India	2008	CYVMV-[IN:Cro:08]-JN817516	2,749
	Croton bonplandianus	*Croton yellow vein virus*	NC014473	Pakistan	2006	CYVV-[PK:Cro:06]-NC014473	2,744
	Croton bonplandianus	*Croton yellow vein virus*	FN543112	Pakistan	2006	CYVV-[PK:Cro:06]-FN543112	2,744
	Croton bonplandianus	*Papaya leaf curl virus*	JN817517	India	2008	PLCV-[IN:Cro:08]-JN817517	2,745
	Croton bonplandianus	*Papaya leaf curl virus*	KM525657	India	2014	PLCV-[IN:Cro:14]-KM525657	2,746
19.	***Euphorbia heterophylla***						
	Euphorbia heterophylla	*Euphorbia mosaic virus*	JQ963888	USA	2011	EuMV-[US:Eup:11]-JQ963888	2,684
	Euphorbia heterophylla	*Euphorbia mosaic virus*	JQ963887	USA	2011	EuMV-[US:Eup:11]-JQ963887	2,609
	Euphorbia heterophylla	*Euphorbia mosaic virus*	NC008304	Mexico	2018	EuMV-[MX:Eup:18]-NC008304	2,613
	Euphorbia heterophylla	*Euphorbia yellow mosaic virus*	NC012553	Brazil	2008	EuYMV-[BR:Eup:08]-NC012553	2,578
	Euphorbia heterophylla	*Euphorbia yellow mosaic virus*	FJ619507	Brazil	2008	EuYMV-[BR:Eup:08]-FJ619507	2,609
V			**Family: Solanaceae**				
20.	***Datura* sp.**						
	Datura inoxia	*Datura leaf curl virus*	NC043531	Sudan	2016	DaLCV-[SU:Dat:16]-NC043531	2,782
	Datura inoxia	*Datura leaf curl virus*	MF402919	Sudan	2016	DaLCV-[SU:Dat:16]-MF402919	2,782

TABLE 4.1 *(Continued)*

Sl. No.	Host	Begomovirus	Accession No.	Country	Year	Acronym	Nucleotide Size (nt)
	Datura inoxia	*Watermelon chlorotic stunt virus*	KY488568	Sudan	2016	WChSV-[SU:Dat:16]-KY488568	2,757
	Datura sp.	*Corchorus yellow vein mosaic virus*	KX513862	India	2016	CoYVMV-[IN:Dat:16]-KX513862	2,742
	Datura sp.	*Datura leaf distortion virus*	JN848773	Venezuela	2007	DaLDV-[VZ:Dat:07]-JN848773	2,599
	Datura sp.	*Tomato leaf curl New Delhi virus*	MH577734	Spain	2015	ToLCNDV-[EX:Dat:15]-MH577734	2,738
	Datura stramonium	*Datura leaf distortion virus*	NC018717	Venezuela	2007	DaLDV-[VZ:Dat:07]-NC018717	2,599
	Datura stramonium	*Tomato leaf curl Sudan virus*	MG868938	Sudan	2016	ToLCSuV-[SU:Dat:16]-MG868938	2,768
	Datura stramonium	*Tomato yellow leaf curl virus*	KC106647	Iran	2011	ToYLCV-[IR:Dat:11]-KC106647	2,781
	Datura stramonium	*Tomato yellow leaf curl China virus*	EF011559	China	2011	ToYLCCNV-[IR:Dat:11]-EF011559	2,739
21.	***Solanum nigrum***						
	Solanum nigrum	*Papaya leaf crumplel virus*	KJ028210	India	2011	PLCrV-[IN:Sol:11]-KJ028210	2,736
	Solanum nigrum	*Tomato leaf curl Axarquia virus*	JN859138	Portugal	2002	ToLCAxV-[PO:Sol:16]-JN859138	2,772
	Solanum nigrum	*Tomato leaf curl Joydebpur virus*	KJ028211	India	2011	ToLCJoV-[IN:Sol:11]-KJ028211	2,768
	Solanum nigrum	*Tomato leaf curl New Delhi virus*	MH465600	India	2016	ToLCNDV-[IN:Sol:16]-MH465600	2,706
	Solanum nigrum	*Tomato leaf curl New Delhi virus*	MH465599	India	2016	ToLCNDV-[IN:Sol:16]-MH465599	2,706
	Solanum nigrum	*Tomato leaf curl New Delhi virus*	AM849548	Pakistan	2004	ToLCNDV-[PK:Sol:04]-AM849548	2,706
	Solanum nigrum	*Tomato leaf curl New Delhi virus*	AJ620187	Pakistan	2006	ToLCNDV-[PK:Sol:06]-AJ620187	2,740
	Solanum nigrum	*Tomato yellow leaf curl Sardinia virus*	LN846597	Morocco	2010	ToLCSV-[MO:Sol:10]-LN846597	2,778
	Solanum nigrum	*Tomato yellow leaf curl virus*	KT099158	Spain	2011	TYLCV-[EX:Sol:11]-KT099158	2,781
VI			**Family: Sapindaceae**				
22.	***Deinbollia borbonica***						
	Deinbollia borbonica	*Deinbollia mosaic virus*	NC029803	Tanzania	2015	DMV-[TZ_Dei_15]-NC029803	2,702

TABLE 4.1 *(Continued)*

Sl. No.	Host	Begomovirus	Accession No.	Country	Year	Acronym	Nucleotide Size (nt)
	Deinbollia borbonica	*Deinbollia mosaic virus*	NC029804	Tanzania	2015	DMV-[TZ:Dei:15]-NC029804	2,759
	Deinbollia borbonica	*Deinbollia mosaic virus*	KT878831	Kenya	2014	DMV-[KY:Dei:14]-KT878831	2,758
	Deinbollia borbonica	*Deinbollia mosaic virus*	KT878829	Tanzania	2015	DMV-[TZ:Dei:14]-KT878829	2,759
	Deinbollia borbonica	*Deinbollia mosaic virus*	KT878826	Tanzania	2015	DMV-[TZ:Dei:14]-KT878826	2,758
	Deinbollia borbonica	*Deinbollia mosaic virus*	KT878824	Tanzania	2015	DMV-[TZ:Dei:14]-KT878824	2,758
VII			**Family: Amaranthaceae**				
23.	***Digeria arvensis***						
	Digeria arvensis	*Cotton leaf curl Rajasthan virus*	FM202328	Pakistan	2009	CLCuRV-[Dig:09]-FM202328	2,751
VIII			**Family: Lamiaceae**				
24.	***Leonurus sibiricus***						
	Leonurus sibiricus	*Euphorbia yellow mosaic virus*	KX348224	Brazil	2010	EuYMV-[BR:Leo:10]-KX348224	2,587
	Leonurus sibiricus	*Euphorbia yellow mosaic virus*	KX348182	Brazil	2011	EuYMV-[BR:Leo:11]-KX348182	2,609
	Leonurus sibiricus	*Leonurus mosaic virus*	NC038450	Brazil	2008	LEMV-[BR:Leo:08]-NC038450	2,652
	Leonurus sibiricus	*Leonurus mosaic virus*	JX863082	Brazil	2010	LEMV-[BR:Leo:10]-JX863082	2,654
	Leonurus sibiricus	*Leonurus mosaic virus*	JX863081	Brazil	2007	LEMV-[BR:Leo:07]-JX863081	2,654
	Leonurus sibiricus	*Leonurus mosaic virus*	KC683374	Paraguay	2010	LEMV-[PR:Leo:07]-KC683374	2,656
	Leonurus sibiricus	*Leonurus mosaic virus*	JQ429791	Brazil	2010	LEMV-[BZ:Leo:10]-JQ429791	2,652
	Leonurus sibiricus	*Tomato yellow spot virus*	KX348223	Brazil	2010	ToYSV-[BR:Leo:10]-KX348223	2,618
	Leonurus sibiricus	*Tomato yellow spot virus*	KX348179	Brazil	2010	ToYSV-[BR:Leo:10]-KX348179	2,618
	Leonurus sibiricus	*Tomato yellow spot virus*	KX348178	Brazil	2010	ToYSV-[BR:Leo:10]-KX348178	2,618
	Leonurus sibiricus	*Tomato yellow spot virus*	KX348177	Brazil	2010	ToYSV-[BR:Leo:10]-KX348177	2,618
	Leonurus sibiricus	*Tomato yellow spot virus*	KX348176	Brazil	2010	ToYSV-[BR:Leo:10]-KX348176	2,618

TABLE 4.1 *(Continued)*

Sl. No.	Host	Begomovirus	Accession No.	Country	Year	Acronym	Nucleotide Size (nt)
	Leonurus sibiricus	*Tomato yellow spot virus*	KX348175	Brazil	2010	ToYSV-[BR:Leo:10]-KX348175	2,618
	Leonurus sibiricus	*Tomato yellow spot virus*	KX348174	Brazil	2010	ToYSV-[BR:Leo:10]-KX348174	2,663
	Leonurus sibiricus	*Tomato yellow spot virus*	KX348173	Brazil	2010	ToYSV-[BR:Leo:10]-KX348173	2,664
	Leonurus sibiricus	*Tomato yellow spot virus*	KX348172	Brazil	2010	ToYSV-[BR:Leo:10]-KX348172	2,661
	Leonurus sibiricus	*Tomato yellow spot virus*	KX348171	Brazil	2010	ToYSV-[BR:Leo:10]-KX348171	2,664
	Leonurus sibiricus	*Tomato yellow spot virus*	KX348170	Brazil	2010	ToYSV-[BR:Leo:10]-KX348170	2,663
	Leonurus sibiricus	*Tomato yellow spot virus*	KX348169	Brazil	2010	ToYSV-[BR:Leo:10]-KX348169	2,661
	Leonurus sibiricus	*Tomato yellow spot virus*	KX348168	Brazil	2010	ToYSV-[BR:Leo:10]-KX348168	2,654
	Leonurus sibiricus	*Tomato yellow spot virus*	KX348167	Brazil	2010	ToYSV-[BR:Leo:10]-KX348167	2,652
	Leonurus sibiricus	*Tomato yellow spot virus*	KX348166	Brazil	2010	ToYSV-[BR:Leo:10]-KX348166	2,660
	Leonurus sibiricus	*Tomato yellow spot virus*	KX348165	Brazil	2011	ToYSV-[BR:Leo:11]-KX348165	2,661
IX.			**Clerodendrum philippinum**				
25.	*Clerodendrum philippinum*	*Clerodendron golden mosaic virus*	DQ641692	Vietnam	2006	ClGMV-[VE:Cle:06]-DQ641692	2,767
X			**Family: Linderniaceae**				
26.	***Lindernia* sp.**						
	Lindernia anagallis	*Lindernia anagallis yellow vein virus*	KC172827	China	2012	LAYVV-[CN:Lin:12]-KC172827	2,739
	Lindernia anagallis	*Lindernia anagallis yellow vein virus*	AY795900	China	2005	LAYVV-[CN:Lin:12]-AY795900	2,739
	Lindernia anagallis	*Lindernia anagallis yellow vein virus*	NC009550	Vietnam	2012	LAYVV-[VI:Lin:12]-NC009550	2,740
	Lindernia procumbens	*Lindernia anagallis yellow vein virus*	DQ641701	Vietnam	2007	LAYVV-[VI:Lin:07]-DQ641701	2,740
XI			**Family: Caprifoliaceae**				
27.	***Lonicera japonica***						
	Lonicera japonica	*Honeysuckle yellow vein virus*	GQ477135	South Korea	2005	HYVV-[SK:Lon:05]-GQ477135	2,763

TABLE 4.1 *(Continued)*

Sl. No.	Host	Begomovirus	Accession No.	Country	Year	Acronym	Nucleotide Size (nt)
	Lonicera japonica	*Honeysuckle yellow vein virus*	FJ817425	New Zealand	2003	HYVV-[NZ:Lon:03]-FJ817425	2,763
	Lonicera japonica	*Honeysuckle yellow vein virus*	AB287441	Japan	2018	HYVV-[JN:Lon:18]-AB287441	2,762
XII			**Family: Fabaceae**				
28.	***Macroptilium lathyroides***						
	Macroptilium lathyroides	*Bean golden mosaic virus*	JN419006	Brazil	2010	BGMV-[BR_Mac_10]-JN419006	2,616
	Macroptilium lathyroides	*Macroptilium common mosaic virus*	NC031453	Brazil	2009	MCMV-[BR:Mac:09]-NC031453	2,598
	Macroptilium lathyroides	*Macroptilium common mosaic virus*	NC031448	Brazil	2009	MCMV-[BR:Mac:09]-NC031448	2,632
	Macroptilium lathyroides	*Macroptilium yellow net virus*	NC017001	Brazil	2009	MYNV-[BR:Mac:09]-NC017001	2,620
	Macroptilium lathyroides	*Macroptilium bright mosaic virus*	KX691400	Brazil	2014	MBMV-[BR:Mac:14]-KX691400	2,636
	Macroptilium lathyroides	*Macroptilium bright mosaic virus*	KX691399	Brazil	2014	MBMV-[BR:Mac:14]-KX691399	2,629
	Macroptilium lathyroides	*Macroptilium bright mosaic virus*	JN419017	Brazil	2010	BGMV-[BR:Mac:10]-JN419017	2,571
	Macroptilium lathyroides	*Macroptilium yellow spot virus*	NC016999	Brazil	2009	MYSV-[BR:Mac:09]-NC016999	2,660
	Macroptilium lathyroides	*Soyabean chlorotic spot virus*	KJ939918	Brazil	2012	SChSV-[BR:Mac:12]-KJ939918	2,623
	Macroptilium lathyroides	*Soyabean chlorotic spot virus*	KJ939917	Brazil	2012	SChSV-[BR:Mac:12]-KJ939917	2,623
	Macroptilium lathyroides	*Macroptilium yellow vein virus*	KJ939913	Brazil	2011	MYVV-[BR:Mac:11]-KJ939913	2,623
	Macroptilium lathyroides	*Macroptilium yellow vein virus*	KJ939912	Brazil	2011	MYVV-[BR:Mac:11]-KJ939912	2,623
	Macroptilium lathyroides	*Macroptilium yellow vein virus*	KJ939911	Brazil	2011	MYVV-[BR:Mac:11]-KJ939911	2,623
	Macroptilium lathyroides	*Macroptilium yellow vein virus*	KJ939910	Brazil	2011	MYVV-[BR:Mac:11]-KJ939910	2,623
	Macroptilium lathyroides	*Macroptilium yellow vein virus*	KJ939909	Brazil	2011	MYVV-[BR:Mac:11]-KJ939909	2,623
	Macroptilium lathyroides	*Macroptilium yellow vein virus*	KJ939908	Brazil	2011	MYVV-[BR:Mac:11]-KJ939908	2,623
	Macroptilium lathyroides	*Macroptilium yellow vein virus*	KJ939907	Brazil	2011	MYVV-[BR:Mac:11]-KJ939907	2,623
	Macroptilium lathyroides	*Macroptilium yellow vein virus*	KJ939906	Brazil	2011	MYVV-[BR:Mac:11]-KJ939906	2,623

TABLE 4.1 *(Continued)*

Sl. No.	Host	Begomovirus	Accession No.	Country	Year	Acronym	Nucleotide Size (nt)
	Macroptilium lathyroides	*Macroptilium yellow vein virus*	KJ939905	Brazil	2011	MYVV-[BR:Mac:11]-KJ939905	2,623
	Macroptilium lathyroides	*Macroptilium yellow vein virus*	KJ939904	Brazil	2011	MYVV-[BR:Mac:11]-KJ939904	2,623
	Macroptilium lathyroides	*Macroptilium yellow vein virus*	KJ939903	Brazil	2011	MYVV-[BR:Mac:11]-KJ939903	2,623
	Macroptilium lathyroides	*Macroptilium yellow vein virus*	KJ939902	Brazil	2011	MYVV-[BR:Mac:11]-KJ939902	2,623
	Macroptilium lathyroides	*Macroptilium yellow vein virus*	KJ939901	Brazil	2011	MYVV-[BR:Mac:11]-KJ939901	2,623
	Macroptilium lathyroides	*Macroptilium yellow vein virus*	KJ939900	Brazil	2011	MYVV-[BR:Mac:11]-KJ939900	2,623
	Macroptilium lathyroides	*Macroptilium yellow vein virus*	KJ939899	Brazil	2011	MYVV-[BR:Mac:11]-KJ939899	2,623
	Macroptilium lathyroides	*Macroptilium yellow vein virus*	KJ939898	Brazil	2011	MYVV-[BR:Mac:11]-KJ939898	2,623
	Macroptilium lathyroides	*Macroptilium yellow spot virus*	KJ939897	Brazil	2012	MYSV-[BR:Mac:11]-KJ939897	2,658
	Macroptilium lathyroides	*Macroptilium yellow spot virus*	KJ939896	Brazil	2012	MYSV-[BR:Mac:11]-KJ939896	2,658
	Macroptilium lathyroides	*Macroptilium yellow spot virus*	KJ939895	Brazil	2012	MYSV-[BR:Mac:11]-KJ939895	2,658
	Macroptilium lathyroides	*Macroptilium yellow vein virus*	KJ939914	Brazil	2012	MYVV-[BR:Mac:12]-KJ939914	2,623
	Macroptilium lathyroides	*Soyabean chlorotic spot virus*	KJ939916	Brazil	2012	SChSV-[BR:Mac:12]-KJ939916	2,623
	Macroptilium lathyroides	*Macroptilium yellow vein virus*	KJ939915	Brazil	2011	MYVV-[BR:Mac:11]-KJ939915	2,623
	Macroptilium lathyroides	*Macroptilium yellow net virus*	JN418998	Brazil	2009	MYNV-[BR:Mac:09]-JN418998	2,620
	Macroptilium lathyroides	*Bean golden mosaic virus*	JN419003	Brazil	2010	BGMV-[BR:Mac:10]-JN419003	2,614
	Macroptilium lathyroides	*Bean golden mosaic virus*	JN419004	Brazil	2010	BGMV-[BR:Mac:10]-JN419004	2,616
	Macroptilium lathyroides	*Macroptilium yellow spot virus*	JN419005	Brazil	2010	MYSV-[BR:Mac:10]-JN419005	2,658
	Macroptilium lathyroides	*Macroptilium yellow spot virus*	JN419007	Brazil	2009	MYSV-[BR:Mac:10]-JN419007	2,658
	Macroptilium lathyroides	*Macroptilium yellow spot virus*	JN419009	Brazil	2010	MYSV-[BR:Mac:10]-JN419009	2,658
	Macroptilium lathyroides	*Macroptilium yellow spot virus*	JN419012	Brazil	2010	MYSV-[BR:Mac:10]-JN419012	2,650
	Macroptilium lathyroides	*Macroptilium yellow spot virus*	JN419013	Brazil	2010	MYSV-[BR:Mac:10]-JN419013	2,660

TABLE 4.1 *(Continued)*

Sl. No.	Host	Begomovirus	Accession No.	Country	Year	Acronym	Nucleotide Size (nt)
	Macroptilium lathyroides	*Macroptilium yellow spot virus*	JN419014	Brazil	2010	MYSV-[BR:Mac:10]-JN419014	2,658
	Macroptilium lathyroides	*Macroptilium yellow spot virus*	JN419018	Brazil	2010	MYSV-[BR:Mac:10]-JN419018	2,649
	Macroptilium lathyroides	*Macroptilium yellow spot virus*	JN419020	Brazil	2010	MYSV-[BR:Mac:10]-JN419020	2,658
	Macroptilium lathyroides	*Macroptilium yellow spot virus*	JN419022	Brazil	2010	MYSV-[BR:Mac:10]-JN419022	2,658
	Macroptilium sp.	*Macroptilium yellow vein virus*	JN419021	Brazil	2010	MYVV-[BR:Mac:10]-JN419021	2,656
	Macroptilium atropurpureum	*Euphorbia yellow mosaic virus*	JN419000	Brazil	2009	EuYMV-[BR:Mac:09]-JN419000	2,609
29.	***Centrosema brasilianum***						
	Centrosema brasilianum	*Centrosema yellow spot virus*	JN419002	Brazil	2009	CYSV-[BR:Mac:09]-JN419002	2,675
30.	***Calopogonium mucunoides***						
	Calopogonium mucunoides	*Macroptilium yellow spot virus*	JN419015	Brazil	2010	MYSV-[BR:Mac:10]-JN419015	2,649
	Calopogonium mucunoides	*Macroptilium yellow spot virus*	JN419016	Brazil	2010	MYSV-[BR:Mac:10]-JN419016	2,658
31.	***Canavalia* sp.**						
	Canavalia sp.	*Macroptilium yellow spot virus*	JN419019	Brazil	2010	MYSV-[BR:Mac:10]-JN419019	2,658
32.	***Mimosa* sp.**						
	Mimosa sp.	*Mimosa yellow leaf curl virus*	DQ641695	Vietnam	2007	MiYLCV-[BR:Mac:07]-DQ641695	2,757
33.	***Rhynchosia minima***						
	Rhynchosia minima	*Cabbage leaf curl virus*	OK044492	Venezuela	2017	CLCV-[VZ:Rhy:17]-OK044492	2,575
	Rhynchosia minima	*Rhynchosia golden mosaic Colombia virus*	KX216480	Colombia	2014	RhGMCV-[CO:Rhy:14]-KX216480	2,620
	Rhynchosia minima	*Rhynchosia yellow mosaic virus*	NC038885	Pakistan	2007	RhYMV-[PK:Rhy:07]-NC038885	2,740
	Rhynchosia minima	*Rhynchosia golden mosaic Yucatan virus*	NC012482	Mexico	2018	RhGMYuV-[MX:Rhy:18]-NC012482	2,542

TABLE 4.1 *(Continued)*

Sl. No.	Host	Begomovirus	Accession No.	Country	Year	Acronym	Nucleotide Size (nt)
	Rhynchosia minima	*Rhynchosia mottle virus*	OK044488	Venezuela	2017	RhMoV-[VZ:Rhy:17]-OK044488	2,592
	Rhynchosia minima	*Rhynchosia golden mosaic Havana virus*	HM236368	Cuba	2007	RhGMHaV-[CU:Rhy:07]-HM236368	2,525
	Rhynchosia minima	*Rhynchosia rugose golden mosaic virus*	NC038805	Cuba	2009	RhRuGMV-[CU:Rhy:09]-NC038805	2,631
	Rhynchosia minima	*Rhynchosia mild mosaic virus*	NC015488	Puerto Rico	1997	RhMMV-[PR:Rhy:1997]-NC015488	2,611
XIII			**Onagraceae**				
34.	***Ludwigia octovalvis***						
	Ludwigia octovalvis	*Ludwigia yellow vein virus*	DQ641708	Vietnam	2006	LuYVV-[VE:Lud:06]-DQ641708	2,757
	Ludwigia octovalvis	*Ludwigia yellow vein virus*	DQ641699	Vietnam	2007	LuYVV-[VE:Lud:07]-DQ641699	2,751

Source: Retrieved from: NCBI and used for analysis.

TABLE 4.2 Details of DNA-B Genome Sequences of Bipartite Begomoviruses Infecting Different Weed Hosts

Sl. No.	Host	Begomovirus	Accession No.	Country	Year	Acronym	Nucleotide Size (nt)
I			***Family: Malvaceae***				
1.	***Abutilon* sp.**						
	Abutilon megapotamicum	*Abutilon mosaic virus*	LN611625	France	2007	AMV-[FR:Abu:07]-LN611625	2,606
	Abutilon megapotamicum	*Abutilon mosaic virus*	LN611624	Germany	2009	AMV-[GR:Abu:09]-LN611624	2,590
	Abutilon pannosum	*Abutilon mosaic virus*	AM886130	South Africa	2007	AMV-[SA:Abu:07]-AM886130	2,609
	Abutilon pictum	*Abutilon mosaic virus*	JN236209	India	2010	AMV-[IN:Abu:10]-JN236209	2,609
	Abutilon sellovianum	*Abutilon mosaic virus*	NC001929	West Indian	2018	AMV-[Abu:18]-NC001929	2,585
	Abutilon sellovianum	*Abutilon mosaic virus*	X15984	West Indian	2015	AMV-[Abu:15]-X15984	2,585
	Abutilon sp.	*Abutilon mosaic Bolivia virus*	HM585446	Bolivia	2007	AMBV-[BO:Abu:07]-HM585446	2,682
	Abutilon sp.	*Abutilon mosaic Bolivia virus*	NC_015048	Bolivia	2007	AMBV-[BO:Abu:07]-NC015048	2,682
	Abutilon sp.	*Abutilon mosaic virus*	NC014139	Brazil	2007	AMV-[BZ:Abu:07]-NC014139	2,618
	Abutilon sp.	*Abutilon mosaic virus*	FN434439	Brazil	2007	AMV-[BZ:Abu:07]-FN434439	2,618
	Abutilon sp.	*Abutilon mosaic virus*	U51138	Hawaii	2003	AMV-[HW:Abu:03]-U51138	2,571
2.	***Malva parviflora***						
	Malva parviflora	*Squash leaf curl virus*	EF532621	Jordan	2008	SLCV-[JO:Mal:08]-EF532621	2,608
3.	***Sida* sp.**						
	Sida acuta	*Sida angular mosaic virus*	KX691415	Brazil	2014	SiAMV-[BZ:Sid:14]-KX691415	2,657
	Sida acuta	*Sida angular mosaic virus*	NC_031457	Brazil	2014	SiAMV-[BZ:Sid:14]-NC031457	2,657
	Sida rhomboidea	*Abutilon mosaic Brazil virus*	JF694483	Brazil	2012	AMBzV-[BZ:Sid:12]-JF694483	2,629
	Sida sp.	*Sida Brazil virus*	FN436002	Brazil	2007	SiBzV-[BZ:Sid:14]-FN436002	2,610
	Sida sp.	*Sida bright yellow mosaic virus*	KX348225	Brazil	2010	SiBYMV-[BZ:Sid:10]-KX348225	2,656
	Sida sp.	*Sida chlorotic leaf virus*	MN013785	Mexico	2018	SiChLV-[MX:Sid:18]-MN013785	2,580

TABLE 4.2 *(Continued)*

Sl. No.	Host	Begomovirus	Accession No.	Country	Year	Acronym	Nucleotide Size (nt)
	Sida sp.	*Sida chlorotic leaf virus*	NC_060102	Mexico	2018	SiChLV-[MX:Sid:18]-NC_060102	2,580
	Sida sp.	*Sida micrantha mosaic virus*	KX348204	Brazil	2011	SiMIMV-[BZ:Sid:11]-KX348204	2,656
	Sida sp.	*Sida micrantha mosaic virus*	KX348202	Brazil	2011	SiMIMV-[BZ:Sid:11]-KX348202	2,647
	Sida sp.	*Sida micrantha mosaic virus*	KX348201	Brazil	2011	SiMIMV-[BZ:Sid:11]-KX348201	2,637
	Sida sp.	*Sida micrantha mosaic virus*	KX348200	Brazil	2011	SiMIMV-[BZ:Sid:11]-KX348200	2,646
	Sida sp.	*Sida micrantha mosaic virus*	KX348199	Brazil	2011	SiMIMV-[BZ:Sid:11]-KX348199	2,655
	Sida urens	*Sida chlorotic yellow vein virus*	NC031454	Brazil	2008	SiChYVV-[BZ:Sid:08]-NC031454	2,491
	Sida urens	*Sida chlorotic vein virus*	KX691414	Brazil	2014	SiChVV-[BZ:Sid:14]-KX691414	2,499
	Sida urens	*Sida chlorotic vein virus*	KX691413	Brazil	2008	SiChVV-[BZ:Sid:08]-KX691413	2,491
	Sida urens	*Sida chlorotic vein virus*	NC_031454	Brazil	2008	SiChVV-[BZ:Sid:08]-NC_031454	2,491
	Sida urens	*Tomato mild mosaic virus*	KC706609	Brazil	2010	ToMMV-[BZ:Sid:10]-KC706609	2,672
	Sida sp.	*Sida yellow golden mosaic virus*	MF163259	Brazil	2015	SiYGMV-[BZ:Sid:15]-MF163259	2,589
	Sida sp.	*Sida yellow golden mosaic virus*	NC_060100	Brazil	2015	SiYGMV-[BZ:Sid:15]-NC_060100	2,589
	Sida spinosa	*Sida chlorotic vein virus*	KX691413	Brazil	2008	SiChVV-[BZ:Sid:08]-KX691413	2,491
	Sida spinosa	*Sida chlorotic vein virus*	KX691415	Brazil	2008	SiChVV-[BZ:Sid:08]-KX691415	2,491
	Sida acuta	*Sid mosaic Sinola virus*	MK643154	Mexico	2015	SiMSiV-[MX:Sid:15]-MK643154	2,583
II			**Family: Solanaceae**				
4.			***Solanum nigrum***				
	Solanum nigrum	*Tomato leaf curl New Delhi virus*	MH465600	India	2016	ToLCNDV-[IN:Sol:16]-MH465600	2,706
	Solanum nigrum	*Tomato leaf curl New Delhi virus*	AJ620188	Pakistan	2006	ToLCNDV-[PK:Sol:06]-AJ620188	2,702

TABLE 4.2 *(Continued)*

Sl. No.	Host	Begomovirus	Accession No.	Country	Year	Acronym	Nucleotide Size (nt)
5.	***Datura* sp.**						
	Datura inoxia	*Watermelon chlorotic stunt virus*	KY488569	Sudan	2016	WChSV-[SU:Dat:16]-KY488569	2,757
	Datura sp.	*Datura leaf distortion virus*	JN848774	Venezuela	2007	DaLDV-[VZ:Dat:07]-JN848774	2,564
	Datura sp.	*Tomato leaf curl New Delhi virus*	MH577654	Spain	2015	ToLCNDV-[EX:Dat:15]-MH577654	2,684
	Datura stramonium	*Datura leaf distortion virus*	NC018715	Venezuela	2007	DaLDV-[VZ:Dat:07]-NC018715	2,564
III			***Family: Asteraceae***				
6.	***Blainvillea rhomboidea***						
	Blainvillea rhomboidea	*Abutilon mosaic Brazil virus*	NC016577	Brazil	2018	AMBzV-[BZ:Abu:18]-NC016577	2,633
	Blainvillea rhomboidea	*Abutilon mosaic Brazil virus*	JF694479	Brazil	2012	AMBzV-[BZ:Bla:12]-JF694479	2,633
7.	***Parthenium hysterophorus***						
	Parthenium hysterophorus	*Tomato leaf curl New Delhi virus*	KF002410	Pakistan	2013	ToLCNDV-[PK:Par:13]-KF002410	2,689
IV			**Family: Acanthaceae**				
8.	***Blechum pyramidatum***						
	Blechum pyramidatum	*Blechum interveinal chlorosis virus*	NC019036	Mexico	2011	BlChV-[MX:Ble:11]-NC019036	2,640
9.	***Dicliptera* sp.**						
	Dicliptera sexangularis	*Dicliptera yellow mottle virus*	AF170101	Calusa Island	2000	DiYMV-[Dic:05]-AF170101	2,597
	Dicliptera vahliana	*Dicliptera yellow mottle virus*	NC003857	Calusa Island	2018	DiYMV-[Dic:05]-NC003857	2,597
V			**Family: Sapindaceae**				
10.	***Deinbollia borboica***						
	Deinbollia borboica	*Deinbollia mosaic virus*	NC029803	Tanzania	2015	DMV-[TZ:Dei:15]-NC029803	2,702

TABLE 4.2 *(Continued)*

Sl. No.	Host	Begomovirus	Accession No.	Country	Year	Acronym	Nucleotide Size (nt)
	Deinbollia borboica	*Deinbollia mosaic virus*	KT878830	Kenya	2014	DMV-[KY:Dei:14]-KT878830	2,706
	Deinbollia borboica	*Deinbollia mosaic virus*	KT878828	Kenya	2014	DMV-[KY:Dei:14]-KT878828	2,702
	Deinbollia borboica	*Deinbollia mosaic virus*	KT878827	Tanzania	2014	DMV-[TZ:Dei:14]-KT878827	2,702
	Deinbollia borboica	*Deinbollia mosaic virus*	KT878825	Tanzania	2014	DMV-[TZ:Dei:14]-KT878825	2,702
VI			***Family: Euphorbiaceae***				
11.	***Euphorbia heterophylla***						
	Euphorbia heterophylla	*Euphorbia mosaic virus*	NC008305	Mexico	2018	EuMV-[MX:Eup:18]-NC008305	2,602
	Euphorbia heterophylla	*Euphorbia mosaic virus*	HQ896202	Cuba	2007	EuMV-[CU:Eup:07]-HQ896202	2,573
	Euphorbia heterophylla	*Euphorbia yellow mosaic virus*	NC012554	Brazil	2008	EuYMV-[BR:Eup:08]-NC012554	2,578
	Euphorbia heterophylla	*Euphorbia yellow mosaic virus*	FJ619508	Brazil	2008	EuYMV-[BR:Eup:08]-FJ619508	2,578
VII			**Family: Lamiaceae**				
12.	***Leonurus sibiricus***						
	Leonurus sibiricus	*Tomato yellow spot virus*	KX348222	Brazil	2010	ToYSV-[BR:Leo:10]-KX348222	2,608
	Leonurus sibiricus	*Tomato yellow spot virus*	KX348221	Brazil	2010	ToYSV-[BR:Leo:10]-KX348221	2,605
	Leonurus sibiricus	*Tomato yellow spot virus*	KX348220	Brazil	2010	ToYSV-[BR:Leo:10]-KX348220	2,605
	Leonurus sibiricus	*Tomato yellow spot virus*	KX348219	Brazil	2010	ToYSV-[BR:Leo:10]-KX348219	2,605
	Leonurus sibiricus	*Tomato yellow spot virus*	KX348218	Brazil	2010	ToYSV-[BR:Leo:10]-KX348218	2,605
	Leonurus sibiricus	*Tomato yellow spot virus*	KX348217	Brazil	2010	ToYSV-[BR:Leo:10]-KX348217	2,605
	Leonurus sibiricus	*Tomato yellow spot virus*	KX348216	Brazil	2010	ToYSV-[BR:Leo:10]-KX348216	2,605
	Leonurus sibiricus	*Tomato yellow spot virus*	KX348215	Brazil	2010	ToYSV-[BR:Leo:10]-KX348215	2,605
	Leonurus sibiricus	*Tomato yellow spot virus*	KX348214	Brazil	2010	ToYSV-[BR:Leo:10]-KX348214	2,605
	Leonurus sibiricus	*Tomato yellow spot virus*	KX348213	Brazil	2010	ToYSV-[BR:Leo:10]-KX348213	2,605

TABLE 4.2 *(Continued)*

Sl. No.	Host	Begomovirus	Accession No.	Country	Year	Acronym	Nucleotide Size (nt)
	Leonurus sibiricus	*Tomato yellow spot virus*	KX348212	Brazil	2010	ToYSV-[BR:Leo:10]-KX348212	2,605
	Leonurus sibiricus	*Tomato yellow spot virus*	KX348211	Brazil	2010	ToYSV-[BR:Leo:10]-KX348211	2,605
	Leonurus sibiricus	*Tomato yellow spot virus*	KX348210	Brazil	2010	ToYSV-[BR:Leo:10]-KX348210	2,605
	Leonurus sibiricus	*Tomato yellow spot virus*	KX348209	Brazil	2010	ToYSV-[BR:Leo:10]-KX348209	2,605
	Leonurus sibiricus	*Tomato yellow spot virus*	KX348208	Brazil	2010	ToYSV-[BR:Leo:10]-KX348208	2,605
	Leonurus sibiricus	*Tomato yellow spot virus*	KX348207	Brazil	2010	ToYSV-[BR:Leo:10]-KX348207	2,605
	Leonurus sibiricus	*Tomato yellow spot virus*	KX348206	Brazil	2010	ToYSV-[BR:Leo:10]-KX348206	2,605
	Leonurus sibiricus	*Tomato yellow spot virus*	KX348205	Brazil	2010	ToYSV-[BR:Leo:10]-KX348205	2,605
VIII			**Family: Fabaceae**				
13.	***Macroptilium lathyroides***						
	Macroptilium lathyroides	*Macroptilium yellow net virus*	NC017002	Brazil	2009	MYNVV-[BR:Mac:09]-NC017002	2,577
	Macroptilium lathyroides	*Macroptilium bright mosaic virus*	JN419006	Brazil	2010	BGMV-[BR:Mac:10]-JN419006	2,571
	Macroptilium lathyroides	*Macroptilium common mosaic virus*	NC031453	Brazil	2009	MCMV-[BR:Mac:09]-NC031453	2,598
	Macroptilium lathyroides	*Macroptilium bright mosaic virus*	JN419017	Brazil	2010	MBMV-[BR:Mac:10]-JN419017	2,571
	Macroptilium lathyroides	*Macroptilium yellow net virus*	JN418999	Brazil	2009	MYNV-[BR:Mac:09]-JN418999	2,577
	Macroptilium lathyroides	*Bean golden mosaic virus*	JN419008	Brazil	2010	BGMV-[BR:Mac:10]-JN419008	2,556
	Macroptilium atropurpureum	*Tomato crinkle leaf yellows virus*	JN419010	Brazil	2010	ToCrLYV-[BR:Mac:10]-JN419010	2,586
	Macroptilium atropurpureum	*Tomato crinkle leaf yellows virus*	JN419011	Brazil	2010	ToCrLYV-[BR:Mac:10]-JN419011	2,582

TABLE 4.2 *(Continued)*

Sl. No.	Host	Begomovirus	Accession No.	Country	Year	Acronym	Nucleotide Size (nt)
	Macroptilium atropurpureum	*Euphorbia yellow mosaic virus*	JN419001	Brazil	2009	EuYMV-[BR:Mac:09]-JN419001	2,579
15.	***Rhynchosia minima***						
	Rhynchosia minima	*Cabbage leaf curl virus*	OK044493	Venezuela	2017	CLCV-[VZ:Rhy:17]-OK044493	2,535
	Rhynchosia minima	*Rhynchosia golden mosaic Colombia virus*	KX216481	Colombia	2014	RhGMCV-[CO:Rhy:14]-KX216481	2,620
	Rhynchosia minima	*Rhynchosia mottle virus*	OK044489	Venezuela	2017	RhMoV-[VZ:Rhy:17]-OK044489	2,592
	Rhynchosia minima	*Rhynchosia golden mosaic Sinaloa virus*	MK618663	Mexico	2008	RhGMSiV-[MX:Rhy:08]-MK618663	2,525
	Rhynchosia minima	*Rhynchosia rugose golden mosaic virus*	NC038804	Cuba	2009	RhRuGMV-[CU:Rhy:09]-NC038804	2,555
	Rhynchosia minima	*Rhynchosia mild mosaic virus*	NC015489	Puerto Rico	1997	RhMMV-[PR:Rhy:1997]-NC015489	2,551
	Rhynchosia minima	*Rhynchosia golden mosaic virus*	MK634539	Mexico	2019	RhGMV-[MX:Rhy:19]-MK634539	2,568
XI			Family: Lamiaceae				
16.	Clerodendrum philippinum						
	Clerodendrum philippinum	*Clerodendron golden mosaic virus*	DQ641693	Vietnam	2006	ClGMV-[VE:Cle:06]-DQ641693	2,757

Source: Retrieved from: NCBI and used for analysis.

4.3.2 ASSOCIATION OF SATELLITE MOLECULES OF BEGOMOVIRUSES WITH WEED HOSTS

Satellites associated with begomoviruses enhance their ability to produce severe symptoms on the host. Among these satellites, the association of betasatellites with begomovirus poses a heavy threat to crops and weeds. Among the total (33) satellites reported on weed hosts, betasatellites dominated with 72%, followed by alphasatellites (19%), associated satellite DNA (6%), and deltasatellite (3%) (Figure 4.3). Similarly, among the total alphasatellites (6) reported on weed hosts, accounting for 45%, were identified from weed hosts belonging to *Malvaceae*, followed by two alphasatellites from *Asteraceae* and one each alphasatellite from *Amaranthaceae* and *Lamiaceae* (Figure 4.4(A) and Table 4.4). Around 32% of betasatellites were identified from weed hosts belonging to the families *Asteraceae* and *Malvaceae* (Figure 4.4(B) and Table 4.3). Furthermore, the association of deltasatellite, i.e., *Deltasatellite sat*-603 associated with *Sidastrum micranthum*, was reported only in Cuba (Table 4.5).

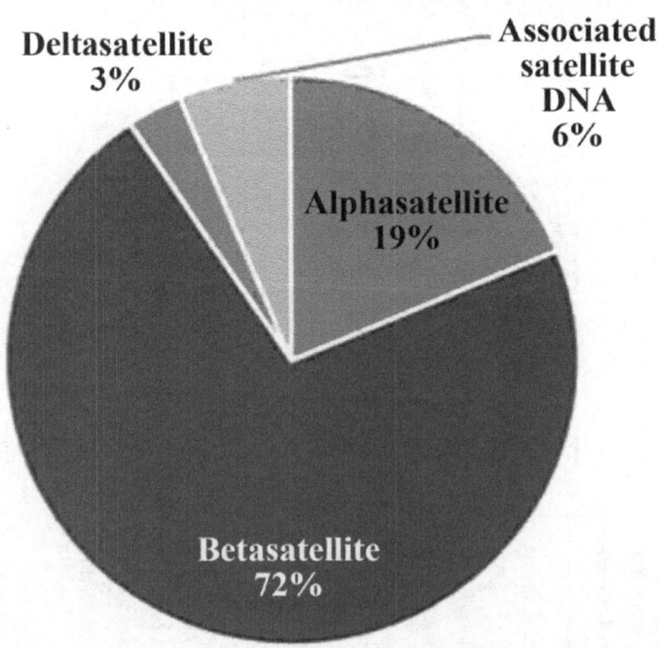

FIGURE 4.3 Percent distribution of begomoviruses associated satellites worldwide on weed hosts.

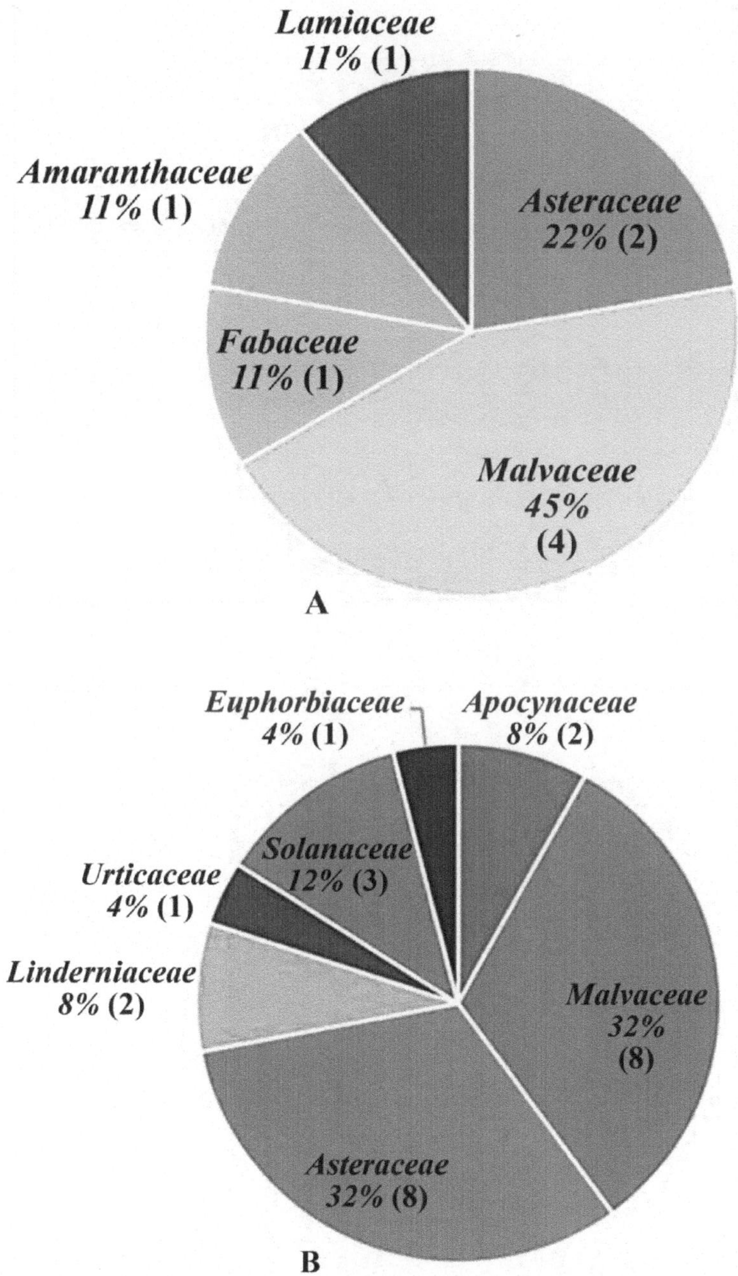

FIGURE 4.4 Distribution of (A) alphasatellite; and (B) betasatellite associated with begmoviruses on weed hosts belong to different taxonomic families worldwide.

TABLE 4.3 Details of Betasatellite Genome Sequences of Begomoviruses Infecting Different Weed Hosts

Sl. No.	Host	Begomovirus Betasatellite	Accession No.	Country	Year	Acronym	Nucleotide Size (nt)
I			**Family: Asteraceae**				
1.	***Ageratum conyzoides***						
	Ageratum conyzoides	*Ageratum yellow leaf curl betasatellite*	JF728868	India	2010	AgYLCB-[IN:Age:10]-JF728868	1,359
	Ageratum conyzoides	*Ageratum yellow leaf curl betasatellite*	AM698010	Pakistan	2006	AgYLCB-[PK:Age:06]-AM698010	1,355
	Ageratum conyzoides	*Ageratum yellow leaf curl betasatellite*	FM164738	Cameroon	2009	AgYLCB-[CM:Age:06]-FM164738	1,355
	Ageratum conyzoides	*Ageratum yellow leaf curl betasatellite*	AJ316027	Pakistan	2003	AgYLCB-[PK:Age:03]-AJ316027	1,352
	Ageratum conyzoides	*Ageratum yellow leaf curl betasatellite*	AJ316026	Pakistan	2003	AgYLCB-[PK:Age:03]-AJ316026	1,352
	Ageratum conyzoides	*Ageratum yellow vein India betasatellite*	NC043428	India	2019	AgYVInB-[IN:Age:19]-NC043428	1,353
	Ageratum conyzoides	*Ageratum yellow vein India betasatellite*	AJ557441	India	2018	AgYVInB-[IN:Age:18]-AJ557441	1,353
	Ageratum conyzoides	*Tomato leaf curl betasatellite*	AM691551	China	2005	ToLCB-[CN:Age:05]-AM691551	1,350
	Ageratum conyzoides	*Tobacco curly shoot betasatellite*	AM260736	China	2006	ToCSB-[CN:Age:06]-AM260736	1,355
	Ageratum conyzoides	*Malvastrum yellow vein betasatellite*	AJ744882	China	2004	MaYVB-[CN:Age:04]-AJ744882	1,351
2.	***Erectites valerianfolia***						
	Erectites valerianfolia	*Erectites yellow mosaic betasatellite*	DQ641713	Vietnam	2006	ErYMB-[VE:Ere:06]-DQ641713	1,342
3.	***Parthenium hysterophorus***						
	Parthenium hysterophorus	*Papaya leaf curl betasatellite*	LN906595	Pakistan	2016	PaLCuB-[PK:Par:16]-LN906595	1,362
	Parthenium hysterophorus	*Papaya leaf curl betasatellite*	MF683837	Pakistan	2012	PaLCuB-[PK:Par:12]-MF683837	1,360
	Parthenium hysterophorus	*Papaya leaf curl betasatellite*	MF683836	Pakistan	2012	PaLCuB-[PK:Par:12]-MF683836	1,362
	Parthenium hysterophorus	*Papaya leaf curl betasatellite*	JX987089	India	2012	PaLCuB-[IN:Par:12]-JX987089	1,367
4.	***Sigesbeckia glabrescens***						
	Sigesbeckia glabrescens	*Siegesbeckia yellow vein betasatellite*	AM230645	China	2007	SiYVB-[CN:Sig;07]-AM230645	1,359
	Sigesbeckia glabrescens	*Siegesbeckia yellow vein betasatellite*	AM230644	China	2007	SiYVB-[CN:Sig;07]-AM230644	1,358

TABLE 4.3 *(Continued)*

Sl. No.	Host	Begomovirus Betasatellite	Accession No.	Country	Year	Acronym	Nucleotide Size (nt)
	Sigesbeckia glabrescens	*Siegesbeckia yellow vein betasatellite*	NC_038682	China	2012	SiYVB-[CN:Sig;07]-NC_038682	1,359
	Sigesbeckia glabrescens	*Siegesbeckia yellow vein betasatellite*	KF499590	China	2012	SiYVB-[CN:Sig;07]-KF499590	1,359
	Sigesbeckia glabrescens	*Siegesbeckia yellow vein betasatellite*	AM230643	China	2007	SiYVB-[CN:Sig;07]-AM230643	1,358
	Sigesbeckia glabrescens	*Siegesbeckia yellow vein betasatellite*	NC_008237	China	2018	SiYVB-[CN:Sig;07]-NC_008237	1,358
	Sigesbeckia glabrescens	*Siegesbeckia yellow vein betasatellite*	AM238695	China	2007	SiYVB-[CN:Sig;07]-AM238695	1,355
5.	Xanthium strumarium						
	Xanthium strumarium	*Tomato leaf curl betasatellite*	FR819710	Pakistan	2011	ToLCB-[PK:Xan:11]-FR819710	1,348
II			**Family: Apocynaceae**				
6.	***Calotropis* sp.**						
	Calotropis procera	*Ageratum leaf curl betasatellite*	KU376492	India	2015	AgLCB-[IN:Cal:15]-KU376492	1,362
	Calotropis sp.	*Ludwigia leaf distortion betasatellite*	KX513857	India	2016	LuLDB-[IN:cal:16]-KX513857	1,351
III			**Family: Euphorbiaceae**				
7.	***Croton bonplandianus***						
	Croton bonplandianus	*Croton yellow vein mosaic betasatellite*	LT674485	Pakistan	2014	CYVMB-[PK:Cro:14]-LT674485	1,346
	Croton bonplandianus	*Croton yellow vein mosaic betasatellite*	LT674484	Pakistan	2014	CYVMB-[PK:Cro:14]-LT674484	1,346
	Croton bonplandianus	*Croton yellow vein mosaic betasatellite*	LT674483	Pakistan	2014	CYVMB-[PK:Cro:14]-LT674483	1,346
	Croton bonplandianus	*Croton yellow vein mosaic betasatellite*	LT674482	Pakistan	2014	CYVMB-[PK:Cro:14]-LT674482	1,346
	Croton bonplandianus	*Croton yellow vein mosaic betasatellite*	LT674481	Pakistan	2014	CYVMB-[PK:Cro:14]-LT674481	1,346
	Croton bonplandianus	*Croton yellow vein mosaic betasatellite*	LT674480	Pakistan	2014	CYVMB-[PK:Cro:14]-LT674480	1,346
	Croton bonplandianus	*Croton yellow vein mosaic betasatellite*	LT674479	Pakistan	2014	CYVMB-[PK:Cro:14]-LT674479	1,346
	Croton bonplandianus	*Croton yellow vein mosaic betasatellite*	LT674478	Pakistan	2014	CYVMB-[PK:Cro:14]-LT674478	1,346
	Croton bonplandianus	*Croton yellow vein mosaic betasatellite*	LT674477	Pakistan	2014	CYVMB-[PK:Cro:14]-LT674477	1,346

TABLE 4.3 *(Continued)*

Sl. No.	Host	Begomovirus Betasatellite	Accession No.	Country	Year	Acronym	Nucleotide Size (nt)
	Croton bonplandianus	*Croton yellow vein mosaic betasatellite*	LT674476	Pakistan	2014	CYVMB-[PK:Cro:14]-LT674476	1,346
	Croton bonplandianus	*Croton yellow vein mosaic betasatellite*	LT674475	Pakistan	2014	CYVMB-[PK:Cro:14]-LT674475	1,346
	Croton bonplandianus	*Croton yellow vein mosaic betasatellite*	LT674474	Pakistan	2014	CYVMB-[PK:Cro:14]-LT674474	1,346
	Croton bonplandianus	*Croton yellow vein mosaic betasatellite*	LT674473	Pakistan	2014	CYVMB-[PK:Cro:14]-LT674473	1,346
	Croton bonplandianus	*Croton yellow vein mosaic betasatellite*	LT674472	Pakistan	2014	CYVMB-[PK:Cro:14]-LT674472	1,346
	Croton bonplandianus	*Croton yellow vein mosaic betasatellite*	LT674471	Pakistan	2014	CYVMB-[PK:Cro:14]-LT674471	1,346
	Croton bonplandianus	*Croton yellow vein mosaic betasatellite*	LT674470	Pakistan	2014	CYVMB-[PK:Cro:14]-LT674470	1,346
	Croton bonplandianus	*Croton yellow vein mosaic betasatellite*	LT674469	Pakistan	2014	CYVMB-[PK:Cro:14]-LT674469	1,346
	Croton bonplandianus	*Croton yellow vein mosaic betasatellite*	LT674468	Pakistan	2014	CYVMB-[PK:Cro:14]-LT674468	1,346
	Croton bonplandianus	*Croton yellow vein mosaic betasatellite*	LT674467	Pakistan	2014	CYVMB-[PK:Cro:14]-LT674467	1,346
	Croton bonplandianus	*Croton yellow vein mosaic betasatellite*	JQ354987	India	2007	CYVMB-[IN:Cro:07]-JQ354987	1,347
	Croton bonplandianus	*Croton yellow vein mosaic betasatellite*	JN831448	India	2014	CYVMB-[PK:Cro:14]-JN831448	1,366
	Croton bonplandianus	*Croton yellow vein mosaic betasatellite*	JN831447	India	2014	CYVMB-[PK:Cro:14]-JN831447	1,346
	Croton bonplandianus	*Croton yellow vein mosaic betasatellite*	NC008579	Pakistan	2005	CYVMB-[PK:Cro:05]-NC008579	1,346
	Croton bonplandianus	*Croton yellow vein mosaic betasatellite*	AM410551	Pakistan	2005	CYVMB-[PK:Cro:05]-AM410551	1,346
IV			**Family: Solanaceae**				
8.	***Datura* sp.**						
	Datura sp.	*Corchorus yellow vein mosaic betasatellite*	KX513863	India	2016	CoYVMB-[IN:Dat:16]-KX513863	1,361
	Datura sp.	*Cotton leaf curl Gezira betasatellite*	DQ644564	Sudan	2009	CoLCuGB-[SU:Dat:09]-DQ644564	1,349
9.	***Solanum nigrum***						
	Solanum nigrum	*Ageratum yellow leaf curl betasatellite*	KJ028213	India	2011	AgYLCB-[IN:Sol:11]-KJ028213	1,364

TABLE 4.3 *(Continued)*

Sl. No.	Host	Begomovirus Betasatellite	Accession No.	Country	Year	Acronym	Nucleotide Size (nt)
V			**Family: Linderniaceae**				
10.	***Lindernia anagallis***						
	Lindernia anagallis	*Lindernia anagallis yellow vein betasatellite*	DQ641715	Vietnam	2006	LaYVVB-[VE:Lin:02]-DQ641715	1,346
	Lindernia anagallis	*Lindernia anagallis yellow vein betasatellite*	NC009561	Vietnam	2002	LaYVVB-[VE:Lin:02]-NC009561	1,346
	Ludwigia hyssopifolia	*Ludwigia yellow vein mosaic betasatellite*	NC007212	China	2018	LuYVMB-[CN:Lud:18]-NC007212	1,347
	Ludwigia hyssopifolia	*Ludwigia yellow vein mosaic betasatellite*	AJ965541	China	2006	LuYVMB-[CN:Lud:06]-AJ965541	1,347
VI			**Family: Malvaceae**				
11.	***Sida*** **sp.**						
	Sida acuta	*Chayote yellow mosaic Benin betasatellite*	KT454830	Nigeria	2014	ChYMBB-[NG:Sid:14]-KT454830	1,367
	Sida acuta	*Sida yellow vein Vietnam betasatellite*	KF990602	China	2011	SiYVVB-[CN:Lin:22]-KF990602	1,333
	Sida acuta	*Malvastrum yellow vein Yunnan betasatellite*	FN806781	China	2008	MaYVYuB-[CN:Lin:08]-FN806781	1,341
	Sida acuta	*Malvastrum yellow vein Yunnan betasatellite*	FN806780	China	2008	MaYVYuB-[CN:Lin:08]-FN806780	1,341
	Sida acuta	*Sida yellow mosaic China betasatellite*	AM048833	China	2006	SiYMCNB-[CN:Lin:06]-AM048833	1,355
	Sida sp.	*Ludwigia leaf distortion betasatellite*	KX513860	India	2016	LuLDB-[IN:Sid:16]-KX513860	1,368
	Sida rhombiforlia	*Sida yellow vein Vietnam betasatellite 1*	DQ641712	Vietnam	2006	SiYVB-[VE:Sid:06]-DQ641712	1,340

TABLE 4.3 *(Continued)*

Sl. No.	Host	Begomovirus Betasatellite	Accession No.	Country	Year	Acronym	Nucleotide Size (nt)
12.	***Abutilon***						
	Abutilon indicum	*Sida leaf curl virus satellite DNA beta*	DQ641711	Vietnam	2006	SiYVB-[VE:Sid:06]-DQ641711	1,367
13.	***Malvastrum coromandelianum***						
	Malvastrum coromandelianum	*Malvastrum yellow vein Cambodia betasatellite*	KP188832	Cambodia	2014	MaYVKHB-[CA:Mal:14]-KP188832	1,346
	Malvastrum coromandelianum	*Tomato leaf curl China betasatellite*	AM980510	China	2009	TYLCCNB-[CN:Mal:09]-AM980510	1,334
	Malvastrum coromandelianum	*Tomato leaf curl China betasatellite*	AM980512	China	2009	TYLCCNB-[CN:Mal:09]-AM980512	1,335
	Malvastrum coromandelianum	*Tomato leaf curl China betasatellite*	AM980511	China	2009	TYLCCNB-[CN:Mal:09]-AM980511	1,342
VII			**Family: Fabaceae**				
14.	***Mimosa* sp.**						
	Mimosa sp.	*Mimosa yellow leaf curl virus satellite DNA beta*	DQ641710	Vietnam	2006	MiYLCCB-[VE:Mim:06]-DQ641710	1,358
VIII			**Family: Urticaceae**				
15.	***Urtica diocia***						
	Urtica diocia	*Ageratum yellow leaf curl betasatellite*	KT716083	Pakistan	2013	AgYLCB-[PK:Urt:13]-KT716083	1,354
	Urtica diocia	*Ageratum yellow leaf curl betasatellite*	LN880825	Pakistan	2013	AgYLCB-[PK:Urt:13]-LN880825	1,354

Source: Retrieved from: NCBI used for analysis.

TABLE 4.4 Details of Alphasatellite Genome Sequences of Begomoviruses Infecting Different Weed Hosts

Sl. No.	Host	Begomovirus	Accession No.	Country	Year	Acronym	Nucleotide Size (nt)
I	**Family: Asteraceae**						
1.	***Ageratum* sp.**						
	Ageratum sp.	*Ageratum yellow vein India alphasatellite*	HG515070	Pakistan	2012	AYVIA-[PK:Age:12]-HG515070	1,384
	Ageratum sp.	*Ageratum yellow vein India alphasatellite*	HG515071	Pakistan	2012	AYVIA-[PK: Age:12]-HG515071	1,384
2.	***Parthenium hysterophorus***						
	Parthenium hysterophorus	*Ageratum yellow vein India alphasatellite*	JX570736	India	2012	AYVIA-[IN:Par:12]-JX570736	1,389
	Parthenium hysterophorus	*Ageratum yellow vein India alphasatellite*	NC_019546	India	2012	AYVIA-[IN:Par:12]-NC_019546	1,389
	Parthenium hysterophorus	*Ageratum yellow vein India alphasatellite*	MF683835	Pakistan	2012	AYVIA-[PK: Par:12]-MF683835	1,389
	Parthenium hysterophorus	*Parthenium leaf curl alphasatellite*	LN906596	Pakistan	2015	PLCA-[PK:Par:15]-LN906596	1,386
	Parthenium hysterophorus	*Parthenium leaf curl alphasatellite*	MF683834	Pakistan	2012	PLCA-[PK:Par:12]-MF683834	1,389
	Parthenium hysterophorus	*Parthenium leaf curl alphasatellite*	MN191632	Pakistan	2015	PLCA-[PK:Par:15]-MN191632	1,386
	Parthenium hysterophorus	*Parthenium leaf curl alphasatellite*	NC_030226	Pakistan	2016	PLCA-[PK:Par:16]-NC_030226	1,386
II	**Family: Fabaceae**						
3.	***Cassia occidentalis***						
	Cassia occidentalis	*Ageratum yellow vein India alphasatellite*	KU852743	India	2013	AYVIA-[IN:Par:13]-KU852743	1,389
III	**Family: Amaranthaceae**						
4.	***Digera muricata***						
	Digera muricata	*Ageratum yellow vein India alphasatellite*	MH973694	Pakistan	2017	AYVIA-[IN:Dig:17]-MH973694	1,385
IV	**Family: Lamiaceae**						
5.	***Leonurus sibiricus***						
	Leonurus sibiricus	*Tomato yellow spot alphasatellite*	KX348228	Brazil	2011	ToYSA-[BZ:Leo:11]-KX348228	1,367
	Leonurus sibiricus	*Tomato yellow spot alphasatellite*	NC038993	Brazil	2011	ToYSA-[BZ:Leo:11]-NC_038993	1,367

TABLE 4.4 *(Continued)*

Sl. No.	Host	Begomovirus	Accession No.	Country	Year	Acronym	Nucleotide Size (nt)
V			Family: Malvaceae				
6.	Malva parviflora						
	Malva parviflora	*Ageratum yellow vein India alphasatellite*	LT716983	Pakistan	2017	AYVIA-[IN:Mal:17]-LT716983	1,388
7.	Sida sp.						
	Sida spp.	*Cuban alphasatellite 1*	HE806451	Cuba	2007	CA-[CU:Sid:07]-HE806451	1,292
	Sida spp.	*Cuban alphasatellite 1*	NC021708	Cuba	2007	CA-[CU:Sid:07]-NC_021708	1,292
	Sida rhombiforlia	*Sida yellow vein Vietnam alphasatellite*	DQ641718	Vietnam	2006	SiYVVA-[VE:Sid06]-DQ641718	1,372
8.	Abutilon sp.						
	Abutilon indicum	*Sida leaf curl alphasatellite*	DQ641717	Vietnam	2006	SiLCA-[VE:Sid:06]-DQ641717	1,379
9.	Weed						
	Weed	*Ageratum yellow vein India alphasatellite*	MN191626	Pakistan	2015	AYVIA-[PK:weed:15]-MN191626	1,389
	Weed	*Ageratum yellow vein India alphasatellite*	MN191627	Pakistan	2015	AYVIA-[PK: weed:15]-MN191627	1,389

Source: Retrieved from: NCBI used for analysis.

Furthermore, the association of Deltasatellite, specifically *Deltasatellite sat*-603, with *Sidastrum micranthum* was only reported in Cuba (Table 4.5).

TABLE 4.5 Details of Deltasatellite Genome Sequences of Begomoviruses Infecting Weeds

Sl. No.	Host	Deltasatellite	Accession No.	Country	Year	Acronym	Nucleotide Size
I				Family: Malvaceae			
1.	***Sidastrum micranthum***						
	Sidastrum micranthum	*Deltasatellite sat*-603	KU232893	Cuba	2011	DELTA-[CU:Sid:11]-KU232893	694
	Sidastrum micranthum	*Deltasatellite sat*-603	NC043205	Cuba	2011	DELTA-[CU:Sid:11]-NC043205	694

Source: Retrieved from: NCBI used for analysis.

Whereas, two associated satellite DNAs from Vietnam and Cuba, *namely Mimosa yellow leaf curl virus-associated DNA 1* on *Mimosa* sp. and *New World begomovirus-associated satellite DNA* infecting *Malvastrum coromandelianum* and *Sidastrum micranthum*, respectively (Table 4.6).

4.4 PHYLOGENETIC ANALYSIS OF GENOME COMPONENTS SEQUENCES OF WEED INFECTING BEGOMOVIRUSES

Nucleotide sequences of 301 DNA-A, including both mono and bipartite begomoviruses infecting weeds reported from different parts of the world, were retrieved from the GenBank. These sequences were used for phylogenetic analysis using MEGA X software (Kumar et al., 2018) with the Maximum-Likelihood method to understand the evolutionary relatedness among the begomovirus population.

The analysis revealed that the weeds infecting begomoviruses reported from Asia, America, Europe, and the Southwestern Pacific Ocean were grouped into four distinct clades (I, II, III, and IV). The majority of isolates in clade I were from Asia, infecting different weed hosts and containing both mono and bipartite viruses, which formed separate clusters. Clade II contained both mono and bipartite virus isolates belonging to American origin, which also formed separate clusters. Clade III contained both mono and bipartite virus isolates from mixed origins of Asian, American, and southwestern Pacific Ocean. Clade IV consisted of mixed isolates from India, Africa, America, and Europe, which contained both mono and bipartite begomoviruses and formed separate clusters. Between Clades III and I, there were mixed origins of both mono and bipartite begomoviruses, which may be recombinants of both clade III and clade I groups. These isolates belonged

TABLE 4.6 Details of Associated Satellite DNA Genome Sequences of Begomoviruses Infecting Weeds

Sl. No.	Host	Associated DNA	Accession No.	Country	Year	Acronym	Nucleotide Size
I	**Family: Malvaceae**						
1.	***Malvastrum coromandelianum***						
	Malvastrum coromandelianum	*New World begomovirus associated satellite DNA*	JN819486	Cuba	2009	NWB-[CU:Mal:09]-JN819486	689
	Malvastrum coromandelianum	*New World begomovirus associated satellite DNA*	NC016963	Cuba	2009	NWB-[CU:Mal:09]-NC_016963	689
	Malvastrum coromandelianum	*New World begomovirus associated satellite DNA*	JN819487	Cuba	2009	NWB-[CU:Mal:09]-JN819487	687
	Malvastrum coromandelianum	*New World begomovirus associated satellite DNA*	JN819488	Cuba	2009	NWB-[CU:Mal:09]-JN819488	687
	Malvastrum coromandelianum	*New World begomovirus associated satellite DNA*	JN819489	Cuba	2009	NWB-[CU:Mal:09]-JN819489	687
	Malvastrum coromandelianum	*New World begomovirus associated satellite DNA*	JN986808	Cuba	2009	NWB-[CU:Mal:09]-JN986808	687
	Malvastrum coromandelianum	*New World begomovirus associated satellite DNA*	NC_043200	Cuba	2009	NWB-[CU:Mal:09]-NC_043200	687
2.	***Sidastrum micranthum***						
	Sidastrum micranthum	*New World begomovirus associated satellite DNA*	JN819495	Cuba	2010	NWB-[CU:Sid:09]-JN819495	685
	Sidastrum micranthum	*New World begomovirus associated satellite DNA*	NC043204	Cuba	2010	NWB-[CU:Sid:09]-NC_043204	685
	Sidastrum micranthum	*New World begomovirus associated satellite DNA*	JN819496	Cuba	2010	NWB-[CU:Sid:09]-JN819496	685
	Sidastrum micranthum	*New World begomovirus associated satellite DNA*	JN819497	Cuba	2010	NWB-[CU:Sid:09]-JN819497	685
II	**Family: Fabaceae**						
4.	***Mimosa* sp.**						
	Mimosa sp.	*Mimosa yellow leaf curl virus-associated DNA 1*	DQ641719	Vietnam	2007	MiYLCA-[VE:Mim:07]-DQ641719	1,378

Source: Retrieved from: NCBI and used for analysis.

to Asian, American, and southwestern Pacific Ocean origins. However, the analysis also showed that a few begomoviruses reported from India closely clustered with those of American origin (Figure 4.5).

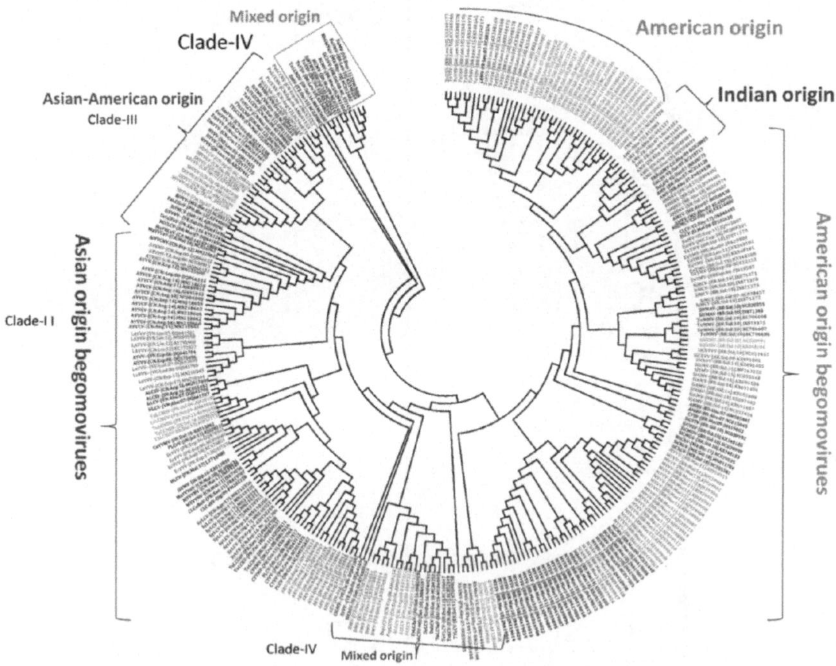

FIGURE 4.5 Phylogenetic tree based on the alignment of the DNA-A and homologous sequences of begomoviruses infecting weed hosts using the maximum-likelihood method by employing the MEGA X tool with 1,000 bootstrap replicates. Details of viruses with accession numbers are given in Table 4.1.

Similarly, 90 complete DNA-B components associated with bipartite begomoviruses infecting weeds reported from different parts of the world were retrieved from the GenBank and used for phylogenetic analysis using MEGA X software (Kumar et al., 2018) with the Maximum-Likelihood method to understand the evolutionary relatedness among DNA-B sequences. The analysis indicated that the DNA-B sequences associated with weed-infecting begomoviruses reported from different parts of the world were grouped into four distinct clades (I, II, III, and Clade IV). The majority of sequences grouped into clade I were from America, whereas clade II contains sequences associated with begomoviruses from Asia-Europe origin and Clade III from African origin. However, a few of them belong to clade IV, which originated from different countries infecting diverse weeds (Figure 4.6). The bipartite begomoviruses associated with different weeds are well segregated based on their geographical origin, similar to DNA-A of both mono and bipartite begomoviruses.

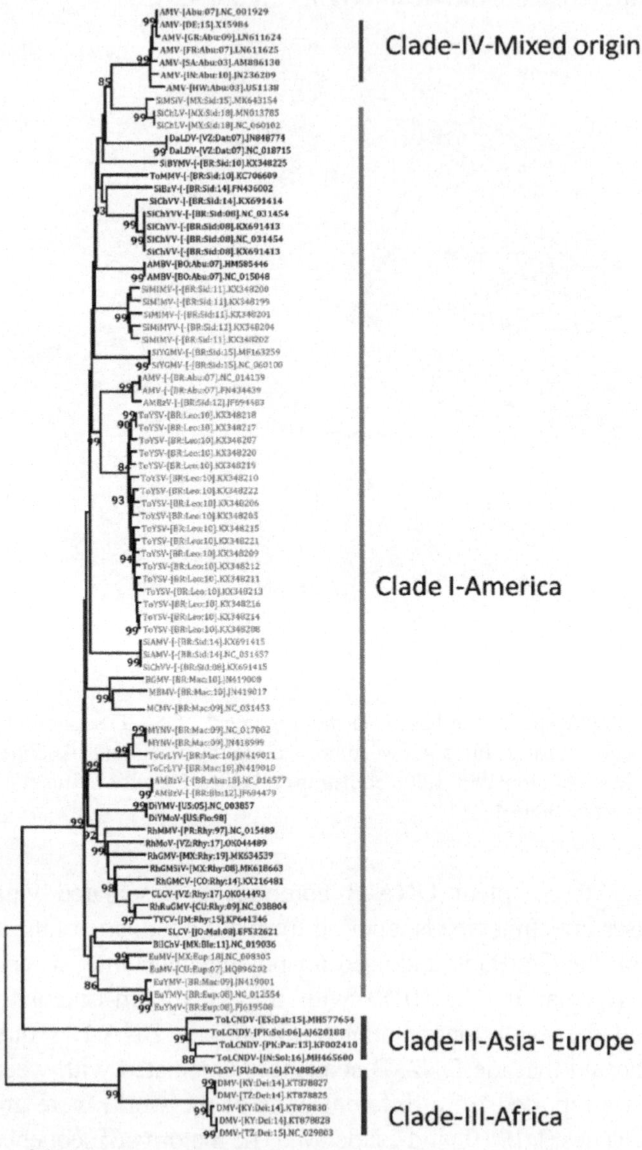

FIGURE 4.6 Phylogenetic tree based on the alignment of the DNA-B sequences of bipartite begomoviruses infecting weed hosts using the maximum-likelihood method by employing the MEGA X tool with 1,000 bootstrap replicates. Detailed sequence accession numbers are given in Table 4.2.

4.4.1 *PHYLOGENETIC ANALYSIS OF DNA SATELLITES*

A dataset of 23 complete nucleotide sequences of alphasatellite molecules associated with weed-infecting begomoviruses was retrieved from the database for phylogenetic analysis. The analysis was conducted using MEGA X software (Kumar et al., 2018) with the Maximum-Likelihood method to determine the evolutionary relationship among the alphasatellites.

The analysis clearly showed that the 23 alphasatellites associated with different weeds cluster into two groups, Clade I and II. The majority of alphasatellites associated with weed-infecting begomoviruses belong to Clade I, which originated from the Indian subcontinent. On the other hand, the weed-infecting begomoviruses in Clade II originated from the American continent. Furthermore, the 23 alphasatellites associated with weed-infecting begomoviruses belong to the genus – *Colecusatellite*, subfamily – *Gemini alphasatellitinae*, and family – *Alphasatellitidae* (Figure 4.7). Similar to DNA-A of both mono and bipartite begomoviruses, the alphasatellites associated with different begomoviruses infecting weeds are segregated based on their geographical origin.

A dataset of 72 complete sequences of betasatellites associated with weed-infecting begomoviruses, available in the GenBank, was retrieved and utilized for phylogenetic analysis using MEGA X software (Kumar et al., 2018) with the Maximum-Likelihood method to understand the evolutionary relatedness among the betasatellites. The phylogenetic analysis revealed that the 72 betasatellites associated with weeds formed two groups. The majority of the betasatellites associated with weed-infecting begomoviruses belong to Southeast Asian origin, while a few betasatellites belong to Asia-American origin. The betasatellites associated with weed-infecting begomoviruses were grouped based on their geographical region (Figure 4.8).

4.5 RECOMBINATION ANALYSIS OF WEED INFECTING BEGOMOVIRUSES GENOMES

Recombination is a major factor in geminivirus evolution and may be responsible for the emergence of new strains or species. The increased frequency of mixed infections of begomoviruses in different hosts allows for the emergence of new viruses resulting from species recombination. Numerous studies have found evidence of recombination in members of the *Geminiviridae* family due to advancements in computational recombination detection tools and an increasing number of available genome sequences. The datasets of complete sequences of weed-infecting begomoviruses (301 DNA-A, including both mono and bipartite begomoviruses DNA-A homolog

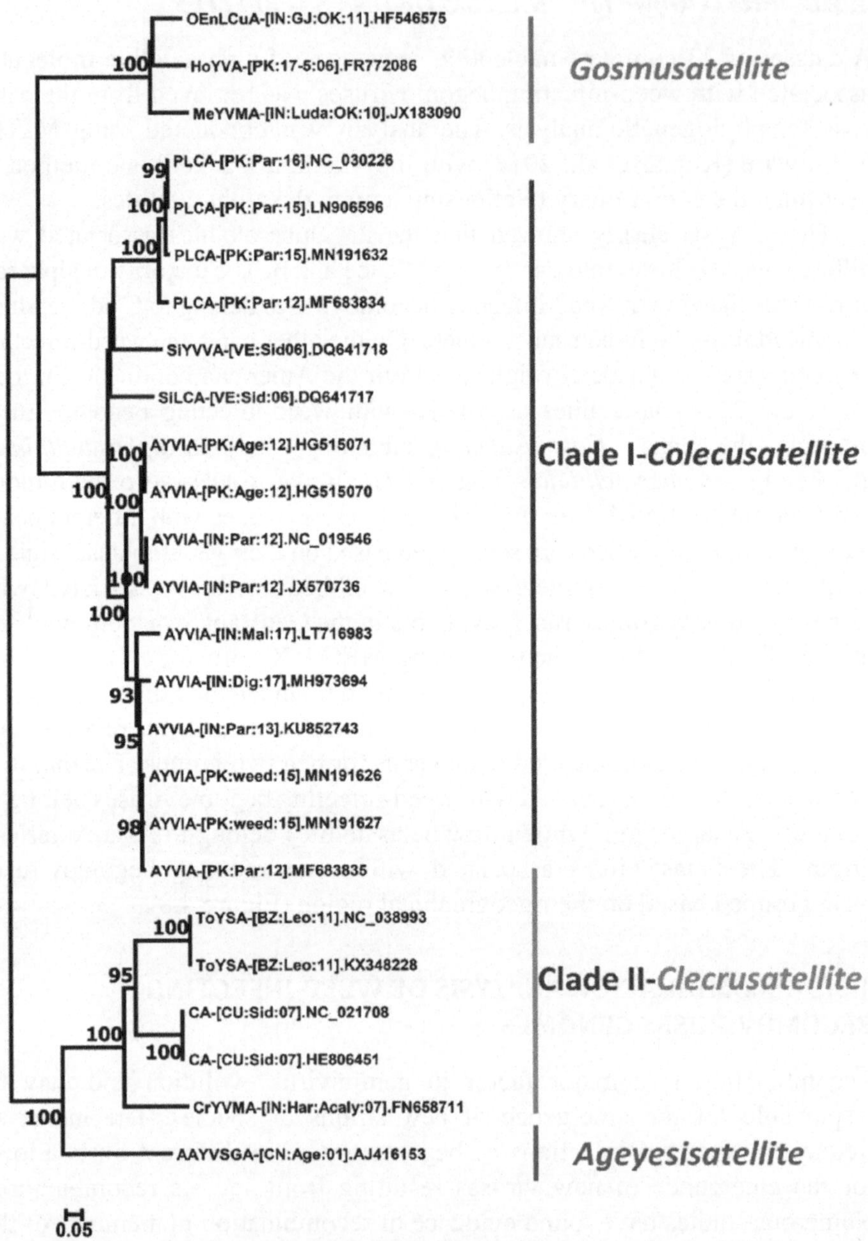

FIGURE 4.7 A phylogenetic tree was constructed based on the alignment of alphasatellite sequences of begomoviruses infecting weed hosts using the maximum-likelihood method. The MEGA X tool was utilized with 1,000 bootstrap replicates. Details of alphasatellites with accession numbers can be found in Table 4.4.

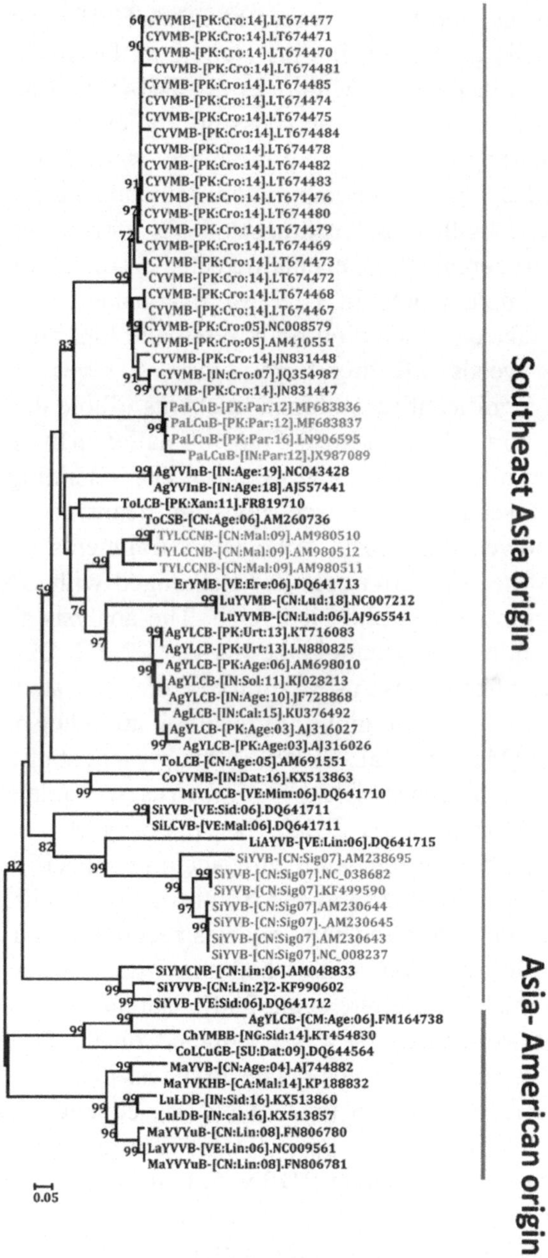

FIGURE 4.8 Phylogenetic tree based on the alignment of the betasatellite sequences of begomoviruses infecting weed hosts using the maximum-likelihood method by employing the MEGA X tool with 1,000 bootstrap replicates. Details of betasatellite with accession numbers are given in the Table 4.3.

monopartite viruses, and 90 sequences of DNA-B of bipartite begomoviruses) were initially assessed using the Neighbor-Net method implemented in the program SplitsTree4 v (Muhire et al., 2014). Our analysis revealed evidence of non-tree-like evolution (possibly indicative of recombination) of weed-infecting begomoviruses. Furthermore, recombination breakpoints were determined using the recombination detection program (RDP) v. 5 (Martin et al., 2021) with default settings and a Bonferroni-corrected p-value cutoff of 0.05. Recombination events detected by more than three of the seven methods implemented in RDP were considered significant. RDP analysis of 301 DNA-A, including both mono and bipartite begomoviruses of the different weeds infecting begomoviruses, revealed that only 120 sequences have significant recombination events (Table 4.7). The PHI test also strongly supported the presence of recombination in these 120 sequences in begomoviruses infecting weeds ($p < 0.001$). The remaining 181 sequences of begomovirus sequences showed non-significant recombination. The analysis also showed that most of the DNA fragments derived from the DNA-A of bipartite begomoviruses are exchanged with DNA-A homolog monopartite begomoviruses and vice versa. The analysis also showed that most of the recombination occurs in the IR, V1, V2, C1, C2, and C4 regions of 120 sequences of begomoviruses. Similarly, previous studies also showed that the IR, V1, V2, and C1 regions are recombination hotspots in the begomovirus genome (Mondal et al., 2019). In the present study, recombination events identified in the genes of 120 begomoviruses correlated well with the recent analysis (Mondal et al., 2019). The majority of the weed-infecting begomoviruses have reported both intra and inter-specific recombination from Asia, America, Europe, and the Southwestern Pacific Ocean origin begomoviruses. Such inter and intra-species recombination are one of the key factors for the evolution of new begomoviruses in many crops in the agricultural ecosystem (Lefeuvre et al., 2007; Garcia-Andres et al., 2007).

Similarly, network analysis was conducted on 90 DNA-B sequences of begomoviruses infecting weeds. The PHI test also strongly supported the presence of recombination in 22 DNA-B sequences associated with begomoviruses infecting weeds ($p < 0.001$). Furthermore, recombination breakpoints were identified using RDP v. 5. Out of the 90 DNA-B component sequences of begomoviruses, 22 showed significant recombination, which were detected by more than three methods. The majority of DNA-B sequences associated with begomoviruses infecting weeds were reported from Asia, Europe, and Africa, and exhibited both intra- and inter-specific recombination (Table 4.8).

TABLE 4.7 Details of Recombination Analysis of Genome Sequences (DNA-A and Homolog of DNA-A) of Begomoviruses Infecting Weed Hosts Using RDP5

Recombinant	Recombination		Minor Parents	Major Parents	Detection Methods					
	Begin	End			RDP	GENECONV	Maxchi	Chimaera	SiSscan	3Seq
ToYLCCNV-[IR:Dat:11]-EF011559	15	494	MYVYNV-[CN:Mal:11]-JX679249	TYLCCNV-[CN:Mal:12]-JX679252	1.59E-56	8.10E-59	1.46E-18	1.15E-18	8.82E-28	3.75E-57
ToLCSuV-[SU:Dat:16]-MG868938	2097	2669	DaLCV-[SU:Dat:16]-MF402919	ToYLCV-[IR:Dat:11]-KC106647	1.13E-52	1.26E-57	3.45E-24	3.77E-24	1.70E-29	3.46E-69
ToYLCV-[CN:Mal:10]-KC189894	2062	2681	ToLCKV-[PK:Par:18]-MN839537	TYLCCNV-[CN:Mal:12]-JX679252	6.59E-36	3.76E-26	6.82E-24	3.52E-20	1.02E-14	5.19E-53
ToLCSuV-[SU:Dat:16]-MG868938	413	1049	DaLCV-[SU:Dat:16]-NC043531	ToYLCV-[IR:Dat:11]-KC106647	2.99E-32	3.10E-31	1.61E-24	9.95E-24	9.52E-31	9.00E-52
SyLCV-[CN:Ecp:16]-MN210350	2130	2683	AYVCV-[CN:Aeg:14]-MN218666	AYVV-[CN:Aeg:12]-MN233600	3.39E-46	1.82E-51	8.14E-20	8.73E-20	1.72E-21	1.82E-52
ToLCKV-[PK:Par:18]-MN839537	536	2140	SyLCV-[CN:Ecp:16]-MN210350	ToLCGuV-[PK:Xan:11]-FR819708	5.47E-08	1.91E-09	8.55E-09	2.90E-10	9.51E-15	5.39E-50
PLCV-[IN:Cro:14]-KM525657	2147	2747	TYLCCNV-[CN:Mal:12]-JX679252	ToLCKV-[PK:Par:18]-MN839537	4.37E-03	3.27E-12	3.41E-11	1.71E-11	1.17E-11	2.24E-68
SLCV-[JO:Mal:08]-EF532620	366	2456	MYNV-[BR:Mac:09]-NC017001	ToLCKV-[PK:Par:18]-MN839537	1.48E-16	9.86E-22	6.67E-14	8.36E-11	4.22E-17	5.61E-64
MYVYNV-[CN:Mal:11]-JX679249	1	478	EcYVV-[PK:Ecp:06]-GQ478343	ToLCKV-[PK:Par:18]-MN839537	2.83E-11	3.84E-19	6.74E-11	2.30E-11	3.62E-21	7.95E-61
PepLCYV-[CN:Ecp:06]-AM691547	1067	1473	AlYVV-[CN:Ecp:06]-DQ375456	AlYVCNV-[CN:Ecp:12]-MN218667	8.62E-33	1.92E-46	1.04E-15	6.09E-13	5.75E-18	7.96E-15
MaYVYnV-[CN:Mal:18]-NC006631	1075	2138	EcYVV-[PK:Ecp:06]-GQ478343	SyLCV-[CN:Sig:17]-MN210352	7.81E-27	3.69E-17	2.28E-19	3.37E-20	1.27E-19	2.52E-43
ToLCSV-[MO:Sol:10]-LN846597	18	1745	ToLCAxV-[PO:Sol:16]-JN859138	AYVV-[CN:Aeg:12]-MN233600	1.05E-38	2.76E-40	4.01E-28	8.25E-14	2.00E-60	1.32E-07
PaLCCNV-[CN:Cor:06]-AJ876548	970	2087	AYVCV-[CN:Aeg:14]-MN218658	ToYLCCV-[CN:Mal:09]-KC189893	1.38E-36	7.35E-41	1.12E-24	6.71E-24	4.05E-31	6.88E-41
AYVCV-[CN:Aeg:14]-MN218666	2097	2756	PLCCV-[CN:Aeg:12]-JX294075	AYVCV-[CN:Aeg:11]-AJ849916	1.45E-42	1.92E-37	4.58E-23	2.42E-22	9.94E-28	1.12E-64
AYVCV-[CN:Aeg:18]-NC004090	2129	2720	PaLCCNV-[CN:Cor:06]-AJ876548	AYVCV-[CN:Aeg:11]-AJ849916	6.63E-20	2.83E-19	2.72E-21	8.24E-20	8.85E-22	3.97E-44
SiAMV-[BZ:Sid:14]-KX691404	2039	2696	SiChVV-[BZ:Sid:14]-KX691405	SiAMV-[BZ:Sid:14]-KX691406	2.98E-36	1.53E-29	8.72E-23	1.45E-17	4.71E-21	9.55E-64
AMBzV-[BZ:Sid:12]-JF694480	1208	2649	AMV-[BZ:Abu:07]-FN434438	SiYBV-[BZ:Sid:10]-NC020254	3.48E-09	5.48E-19	1.49E-21	6.95E-15	8.09E-34	2.14E-35
DaLCV-[SU:Dat:16]-NC043531	62	1061	ToYLCV-[IR:Dat:11]-KC106647	BlYVV-[MX:Ble:12]-NC055132	1.61E-28	3.44E-35	2.68E-20	6.84E-14	3.16E-31	2.79E-10
LuYVV-[VE:Lud:06]-DQ641708	2118	2578	ClGMV-[VE:Cle:06]-DQ641692	LuYVV-[VE:Lud:07]-DQ641699	5.11E-35	8.51E-10	1.39E-24	2.52E-15	2.39E-07	1.26E-38
AYVCV-[CN:Aeg:14]-MN218660	2100	2699	PLCCV-[CN:Aeg:12]-JX294075	AYVV-[CN:Aeg:00]-DQ866134	9.85E-34	9.58E-31	6.43E-21	2.60E-21	1.62E-25	1.03E-50
RhGMYuV-[MX:Rhy:18]-NC012482	80	418	MYVV-[BR:Mac:11]-KJ939904	EuMV-[US:Eup:11]-JQ963887	NS	3.47E-04	NS	NS	1.24E-09	3.92E-33
EuYMV-[BR:Eup:08]-FJ619507	1987	2461	AlYVCNV-[CN:Ecp:12]-MN218667	SiCMOV-[BZ:Sid:10]-NC038990	5.09E-26	5.48E-18	2.38E-14	1.21E-16	3.11E-22	5.27E-34

TABLE 4.7 *(Continued)*

Recombinant	Recombination		Minor Parents	Major Parents	Detection Methods					
	Begin	End			RDP	GENECONV	Maxchi	Chimaera	SiSscan	3Seq
AMV-[BZ:Abu:07]-FN434438	1993	2583	AEV-[IN:Aeg:10]-JF728866	ToYSV-[BR:Leo:10]-KX348176	5.92E-27	8.69E-15	1.07E-15	1.57E-15	4.28E-19	2.00E-34
ToLCKV-[PK:Par:18]-MN839537	519	2589	PaLCCNV-[CN:Ecp:06]-AM691553	CYVMV-[PK:Cro:15]-KX671964	2.88E-25	3.07E-26	3.22E-12	1.40E-13	2.32E-17	3.16E-31
CLCuBuV-[PK:Xan:11]-FR819707	144	504	SyLCV-[CN:Ecp:16]-KU933258	EcYVV-[PK:Ecp:13]-HG937534	1.14E-03	3.50E-03	1.77E-08	NS	4.75E-11	2.52E-31
CYVMV-[IN:Cro:08]-JN817516	536	1350	CLCuBuV-[PK:Xan:11]-FR819707	CYVV-[PK:Cro:06]-FN543112	2.69E-29	5.39E-23	6.56E-20	7.89E-16	7.45E-19	2.71E-31
AlYVCNV-[CN:Ecp:12]-MN218667	40	300	SyLCV-[CN:Ecp:16]-MN210350	AYVCV-[CN:Aeg:14]-MN218658	5.76E-29	3.36E-28	1.66E-15	3.83E-13	2.63E-17	8.62E-31
EuYMV-[BR:Leo:10]-KX348224	222	486	ToYSV-[BR:Leo:10]-KX348176	EuMV-[US:Eup:11]-JQ963887	NS	NS	NS	NS	NS	NS
SyLCV-[CN:Sig:17]-MN210352	521	2101	CYVMV-[PK:Cro:15]-KX671964	EmYVV-[CN:Cra:17]-MN201999	3.50E-14	4.42E-14	4.28E-14	1.41E-14	5.13E-20	2.21E-29
ToLCGuV-[PK:Xan:11]-FR819708	2119	2665	TYLCCNV-[CN:Mal:12]-JX679252	PLCV-[IN:Cro:08]-JN817517	7.11E-04	1.38E-06	1.57E-09	2.76E-06	5.33E-15	3.28E-65
AMYV-[MX:Abu:07]-KC430935	1980	2564	SiLCV-[VI:Abu:07]-DQ641706	AMV-[IN:Abu:10]-HQ588900	1.95E-28	6.66E-21	4.55E-16	1.72E-14	1.27E-18	5.75E-29
ToMMV-[BZ:Sid:10]-KC706607	1	260	ToMMV-[BZ:Sid:10]-JX871375	ToMMV-[BZ:Sid:10]-KC706606	NS	NS	NS	NS	NS	NS
CYVMV-[PK:Cro:15]-KX671964	536	1118	CLCuBuV-[PK:Xan:11]-FR819707	CYVV-[PK:Cro:06]-FN543112	5.37E-22	6.38E-07	1.96E-17	7.52E-15	2.21E-37	1.33E-27
EuYMV-[BR:Eup:08]-FJ619507	180	373	SiCMOV-[BZ:Sid:10]-NC038990	LuYVV-[CN:Ecp:17]-MN210344	NS	5.21E-04	1.70E-04	0.047622	1.26E-08	1.37E-27
MYSV-[BR:Mac:10]-JN419015	2118	2570	ToLCNDV-[EX:Dat:15]-MH577734	MYSV-[BR:Mac:11]-KJ939897	1.79E-27	6.53E-12	4.36E-19	7.65E-13	2.53E-09	7.71E-25
SiYBV-[BZ:Sid:10]-NC020254	1997	20	MYSV-[BR:Mac:10]-JN419015	SiYBV-[BZ:Sid:10]-JX871382	2.44E-05	1.78E-04	5.48E-08	2.72E-06	9.47E-52	1.05E-33
SLCV-[JO:Mal:08]-EF532620	412	1977	SiMAV-[BZ:Sid:10]-KX896427	AEV-[IN:Aeg:10]-JF728864	5.06E-09	8.05E-09	1.38E-11	5.62E-12	8.20E-21	1.05E-27
SiGYSV-[BZ:Sid:10]-NC038992	1142	2179	SiCMOV-[BZ:Sid:10]-NC038990	LuYVV-[CN:Ecp:17]-MN210343	2.19E-09	2.76E-11	6.20E-13	2.98E-13	1.60E-23	2.51E-26
SiGYSV-[BZ:Sid:10]-NC038992	219	1141	ToYSV-[BR:Leo:10]-KX348167	PaLCCNV-[CN:Ecp:06]-AM691553	7.99E-09	1.94E-09	2.41E-10	6.07E-09	4.04E-17	8.09E-102
ToLCNDV-[IN:Sol:16]-MH465599	2056	2709	ToLCJoV-[IN:Sol:11]-KJ028211	CYVMV-[IN:Cro:08]-JN817516	1.32E-17	5.48E-14	1.13E-15	1.71E-16	5.70E-19	3.40E-25
MaYVCaV-[CA:Mal:14]-KP188831	182	1217	AYVCV-[CN:Aeg:14]-MN218658	SiYVVV-[VE:Sid:06]-DQ641696	2.56E-19	2.09E-23	5.61E-18	2.00E-16	1.07E-25	1.97E-18
BGMV-[BR:Mac:10]-JN419017	2356	2571	BGMV-[BR:Mac:10]-JN419004	SiMAV-[BZ:Sid:10]-KX896422	1.09E-22	1.30E-19	1.63E-06	1.10E-03	4.05E-11	NS
ToLCKV-[PK:Par:18]-MN839537	1057	2109	PLCV-[IN:Cro:08]-JN817517	SiYMV-[IN:Sid:16]-KX513859	7.58E-06	1.13E-05	1.33E-11	5.49E-09	6.89E-22	2.81E-21
ToLCGuV-[PK:Xan:11]-FR819708	592	1063	PLCrV-[IN:Sol:11]-KJ028210	ToLCJoV-[IN:Sol:11]-KJ028211	4.87E-12	4.07E-07	6.17E-11	9.81E-12	3.28E-20	7.60E-17
ToLCNDV-[PK:Sol:06]-AJ620187	1101	2086	PLCV-[IN:Cro:08]-JN817517	ToLCJoV-[IN:Sol:11]-KJ028211	1.82E-02	NS	1.16E-10	5.41E-06	4.18E-16	2.18E-47

TABLE 4.7 *(Continued)*

Recombinant	Recombination		Minor Parents	Major Parents	Detection Methods					
	Begin	End			RDP	GENECONV	Maxchi	Chimaera	SiSscan	3Seq
ToLCNDV-[EX:Dat:15]-MH577734	1	398	CYVMV-[IN:Cro:08]-JN817516	SiLCV-[VI:Abu:07]-DQ641706	9.12E-11	8.30E-12	3.18E-12	8.09E-07	4.43E-40	1.16E-48
PLCV-[IN:Cro:08]-JN817517	2251	2673	AMV-[FR:Abu:09]-LN611622	AEV-[IN:Aeg:21]-KJ488990	1.43E-30	1.21E-20	1.96E-15	4.67E-13	2.77E-18	8.24E-29
ToLCAxV-[PO:Sol:16]-JN859138	1739	2772	TYLCV-[EX:Sol:11]-KT099158	ToLCSuV-[SU:Dat:16]-MG868938	1.63E-18	7.93E-16	3.53E-09	3.53E-08	7.03E-18	5.33E-21
MaYVLaV-[PK:Mal:13]-NC055476	523	1077	CLCuBuV-[PK:Xan:11]-FR819707	HLCV-[PK:Mal:17]-LT716980	2.46E-19	1.69E-16	6.43E-14	1.39E-14	9.34E-21	2.65E-17
DaLCV-[SU:Dat:16]-NC043531	2268	2782	AMV-[Abu:07]-NC001928	AEV-[IN:Aeg:21]-KJ488990	6.46E-16	4.89E-19	2.28E-11	9.24E-08	6.55E-20	3.18E-12
SiYMV-[BZ:Sid:10]-JX871369	2014	2513	AYVCV-[CN:Aeg:14]-MN218658	SiMAV-[BZ:Sid:10]-KX896417	1.64E-19	5.91E-11	9.71E-13	2.64E-12	1.86E-06	4.58E-11
PaLCCNV-[CN:Cor:06]-AJ876548	351	958	MiYLCV-[BR:Mac:07]-DQ641695	ToYLCCV-[CN:Mal:09]-KC189893	2.34E-07	4.15E-06	1.54E-14	3.04E-14	2.03E-11	1.82E-18
AMBzV-[BZ:Sid:12]-JF694481	19	106	AMV-[FR:Abu:07]-LN611623	AMBzV-[BZ:Abu:18]-NC016574	NS	6.45E-12	9.86E-05	2.29E-04	7.73E-07	8.25E-18
MBMV-[BR:Mac:14]-KX691400	1896	2501	SChSV-[BR:Mac:12]-KJ939918	MYSV-[BR:Mac:10]-JN419022	5.89E-12	4.52E-09	6.44E-12	2.85E-13	2.22E-17	6.95E-19
MYSV-[BR:Mac:10]-JN419005	136	1080	MYSV-[BR:Mac:10]-JN419015	MYSV-[BR:Mac:10]-JN419014	1.36E-06	2.98E-04	1.59E-12	3.02E-12	2.33E-10	9.26E-20
MCMV-[BR:Mac:09]-NC031448	2535	2632	MCMV-[BR:Mac:09]-NC031453	MBMV-[BR:Mac:14]-KX691400	5.37E-17	5.75E-17	3.25E-03	2.02E-03	6.46E-10	8.36E-10
SiChVV-[BZ:Sid:14]-KX691405	1987	2585	SiAMV-[BZ:Sid:14]-KX691406	SiAMV-[BZ:Sid:14]-NC038887	1.11E-16	7.98E-17	5.77E-14	1.91E-09	1.14E-22	2.91E-03
CLCuBuV-[PK:Xan:11]-FR819707	1309	2756	EcYVV-[PK:Ecp:06]-NC016660	AEV-[IN:Aeg:21]-OM719614	2.42E-04	4.04E-08	8.24E-11	7.17E-10	1.05E-07	1.06E-16
CLCuBuV-[PK:Xan:11]-FR819707	871	1308	PLCV-[IN:Cro:14]-KM525657	CLCuRV-[Dig:09]-FM202328	0.01652	NS	1.84E-10	1.52E-08	2.76E-12	3.63E-17
ToYSV-[BR:Leo:10]-KX348223	65	617	CYVV-[PK:Cro:06]-FN543112	ToYSV-[BZ:Sid:10]-KX348226	NS	NS	NS	NS	NS	NS
AYVCV-[CN:Aeg:14]-MN218658	2102	2715	EcYVV-[IN:And:10]-KC476655	AYVCV-[CN:Aeg:11]-AJ849916	8.71E-10	NS	1.64E-16	5.71E-10	NS	1.47E-04
ToMMV-[BZ:Sid:10]-KC706607	2508	2656	RhMoV-[VZ:Rhy:17]-OK044488	ToMMV-[BZ:Sid:10]-KC706608	4.55E-16	1.25E-09	8.31E-08	3.81E-03	NS	3.35E-09
ToLCPaV-[IN:Xan:116]-KY612435	2076	2748	EcYVV-[PK:Ecp:06]-NC016660	PaLCCNV-[CN:Ecp:06]-AM691553	9.66E-16	2.47E-10	6.42E-10	1.15E-09	1.39E-17	4.66E-14
SBYVV-[CN:Sig:07]-AM230634	410	1063	MiYLCV-[BR:Mac:07]-DQ641695	SBYVGV-[CN:Sig:05]-NC_008317	7.27E-17	2.42E-08	8.75E-15	8.64E-15	7.65E-09	1.87E-19
DMV-[KY:Dei:14]-KT878831	2653	2758	DMV-[TZ:Dei:15]-NC029803	PaLCGuV-[PH:Ble:12]-KF446660	1.62E-13	2.15E-18	9.13E-08	2.15E-03	4.05E-11	1.12E-09
AlYVV-[CN:Ecp:06]-DQ375456	2311	2651	RhMMV-[PR:Rhy:1997]-NC015488	EcYVV-[IN:And:10]-KC476655	8.48E-08	1.79E-08	9.96E-07	7.00E-08	4.01E-15	1.25E-05
SiLCV-[VI:Abu:07]-DQ641706	2002	2726	SiYVVV-[VE:Sid:06]-DQ641696	SiLCV-[VI:Abu:07]-DQ641707	1.92E-14	NS	3.70E-14	1.81E-07	NS	4.49E-08
PaLCCNV-[CN:Ecp:06]-AM691553	1001	2065	AYVV-[CN:Aeg:00]-DQ866134	EcYVV-[IN:And:10]-KC476655	1.92E-13	9.65E-14	8.01E-16	9.70E-13	1.81E-33	3.02E-13

TABLE 4.7 *(Continued)*

Recombinant	Recombination		Minor Parents	Major Parents	Detection Methods					
	Begin	End			RDP	GENECONV	Maxchi	Chimaera	SiSscan	3Seq
SiBYMV-[BZ:Sid:10]-NC040191	2525	2647	SiBYMV-[BZ:Sid:10]-KX348184	ToLCNDV-[EX:Dat:15]-MH577734	1.71E-09	1.10E-13	6.44E-07	4.81E-06	6.85E-11	5.27E-08
WChSV-[SU:Dat:16]-KY488568	2281	2600	AMV-[Abu:07]-NC001928	MaYVCaV-[CA:Mal:14]-KP188831	1.20E-07	6.56E-06	7.26E-06	7.76E-04	1.62E-13	2.85E-06
RhYMV-[PK:Rhy:07]-NC038885	2260	2596	RhRuGMV-[CU:Rhy:09]-NC038805	MaYVCaV-[CA:Mal:14]-KP188831	4.63E-06	3.07E-05	4.11E-07	2.11E-02	2.30E-13	6.73E-06
MYNV-[BR:Mac:09]-NC017001	2054	2456	EmYVV-[CN:Cra:17]-MN201997	MYSV-[BR:Mac:09]-NC016999	2.44E-13	3.30E-04	2.51E-09	8.18E-10	6.52E-08	2.79E-10
MYSV-[BR:Mac:11]-KJ939896	1964	2440	AEV-[IN:Aeg:10]-JF728864	SiMAV-[BZ:Sid:10]-KX896423	5.20E-13	6.07E-11	3.18E-07	4.43E-08	4.69E-10	1.49E-10
SiAMV-[BZ:Sid:14]-NC038887	2172	2526	EcYVV-[IN:And:10]-KC476655	SiMAV-[BZ:Sid:10]-KX896427	4.85E-12	1.52E-09	4.89E-08	8.61E-07	1.02E-06	5.58E-10
PLCrV-[IN:Sol:11]-KJ028210	192	1059	ToLCJoV-[IN:Sol:11]-KJ028211	AYVV-[CN:Aeg:00]-DQ866134	1.23E-11	1.67E-06	2.10E-08	1.29E-08	5.57E-08	1.34E-08
SYVV-[VE:Spi:06]-DQ641694	1985	2247	RhYMV-[PK:Rhy:07]-NC038885	EmYVV-[CN:Cra:08]-JQ247188	1.41E-06	4.31E-11	4.08E-04	4.17E-05	2.20E-11	NS
DiYMV-[CU:Dic:05]-AJ549960	1959	23	MYSV-[BR:Mac:10]-JN419018	DiYMV-[FL:Dic:2000]-AF139168	NS	NS	NS	NS	7.57E-40	2.93E-11
HYVV-[JN:Lon:18]-AB287441	2432	2710	PaLCGuV-[PH:Ble:12]-KF446660	HYVV-[SK:Lon:05]-GQ477135	4.29E-11	NS	6.36E-05	1.52E-05	3.72E-04	8.06E-08
SBYVV-[CN:Sig:11]-JX294076	2352	2672	EmYVV-[CN:Cra:17]-MN201997	SBYVV-[CN:Sig:07]-AM230634	4.32E-11	NS	8.55E-10	5.28E-08	4.74E-04	1.39E-09
MCMV-[BR:Mac:09]-NC031448	2100	2490	LAYVV-[VI:Lin:12]-NC009550	SiAMV-[BZ:Sid:14]-KX691408	3.89E-12	5.41E-09	2.25E-08	8.66E-08	1.77E-22	2.79E-10
TYLCV-[EX:Sol:11]-KT099158	2371	2532	AMV-[Abu:07]-NC001928	ToYLCV-[IR:Dat:11]-KC106647	1.61E-11	2.61E-02	1.79E-06	0.016978	2.95E-04	6.11E-04
PepLCYV-[CN:Ecp:06]-AM691547	52	1066	ToLCV-[CN:Abu:12]-KC852149	TYLCCNV-[CN:Mal:12]-JX679252	1.35E-04	3.41E-04	8.66E-08	4.28E-08	5.52E-15	1.64E-44
ToLCJoV-[IN:Sol:11]-KJ028211	1787	2359	TYLCV-[EX:Sol:11]-KT099158	PepLCYV-[CN:Ecp:06]-AM691547	NS	NS	3.06E-03	1.04E-03	1.83E-06	2.25E-12
PLCV-[IN:Cro:14]-KM525657	429	679	SiYMV-[IN:Sid:16]-KX513859	CYVV-[PK:Cro:06]-FN543112	8.35E-11	5.52E-04	1.15E-08	2.59E-08	1.91E-06	1.20E-08
AEV-[IN:Aeg:21]-OM719614	92	1981	AEV-[IN:Aeg:10]-JF728861	AEV-[IN:Aeg:21]-KJ488990	2.77E-02	NS	1.64E-10	3.50E-05	NS	1.38E-03
ToLCNDV-[IN:Sol:16]-MH465600	2596	2679	ToLCNDV-[EX:Dat:15]-MH577734	AYVV-[CN:Aeg:12]-MN233600	6.44E-09	2.58E-10	2.08E-02	3.81E-04	1.40E-08	2.57E-08
DMV-[TZ:Dei:14]-KT878826	1941	2091	RhYMV-[PK:Rhy:07]-NC038885	ToYLCV-[IR:Dat:11]-KC106647	1.04E-09	2.29E-05	8.10E-10	2.56E-08	2.52E-09	2.79E-10
DMV-[TZ:Dei:14]-KT878829	89	1941	AEV-[IN:Aeg:10]-JF728864	DaLDV-[VZ:Dat:07]-JN848773	NS	1.46E-02	5.45E-03	1.59E-02	1.22E-07	9.57E-80
ToYSV-[BR:Leo:10]-KX348223	628	2074	ToYSV-[BR:Leo:10]-KX348179	MBMV-[BR:Mac:14]-KX691399	NS	0.019393	NS	NS	3.37E-45	2.79E-10
BGMV-[BR:Mac:10]-JN419003	19	55	MYVV-[BR:Mac:11]-KJ939904	BGMV-[BR:Mac:10]-JN419004	NS	NS	NS	NS	NS	NS
SiLCV-[VI:Abu:07]-DQ641707	2038	2762	CLCuRV-[Dig:09]-FM202328	EcYVV-[IN:And:10]-KC476655	1.28E-03	NS	1.05E-06	2.91E-07	4.86E-04	8.36E-10

TABLE 4.7 *(Continued)*

Recombinant	Recombination		Minor Parents	Major Parents	Detection Methods					
	Begin	End			RDP	GENECONV	Maxchi	Chimaera	SiSscan	3Seq
ToYSV-[BR:Leo:10]-KX348176	15	748	LEMV-[BR:Leo:07]-JX863081	ToYSV-[BR:Leo:10]-KX348171	4.16E-07	8.39E-07	9.42E-10	4.72E-08	4.03E-07	7.88E-13
ToYSV-[BR:Leo:10]-KX348178	2144	2659	SiMV-[BZ:Sid:10]-JX871378	ToYSV-[BR:Leo:10]-KX348179	6.42E-07	6.71E-06	5.09E-12	2.52E-10	1.13E-08	4.27E-15
PLCrV-[IN:Sol:11]-KJ028210	2344	2664	AEV-[IN:Aeg:21]-KJ488990	ToLCJoV-[IN:Sol:11]-KJ028211	2.05E-09	8.79E-06	1.24E-06	3.25E-07	8.42E-10	9.68E-10
AEV-[IN:Aeg:10]-JF728864	2111	2483	BlIChV-[MX:Ble:11]-NC019035	ErYMV-[VI:Ere:07]-DQ641698	8.11E-08	1.36E-06	7.09E-07	3.00E-06	1.91E-19	1.12E-09
CoYVMV-[IN:Dat:16]-KX513862	2198	2531	RhRuGMV-[CU:Rhy:09]-NC038805	EmYVV-[CN:Cra:08]-JQ247188	3.34E-09	5.81E-07	6.37E-07	2.15E-05	1.31E-09	3.08E-06
ALCSV-[IN:Aeg:16]-NC055492	27	485	ToYLCCV-[CN:Mal:09]-KC189893	CoYVMV-[IN:Dat:16]-KX513862	5.22E-06	NS	6.96E-06	7.95E-03	1.43E-11	NS
ErYMV-[VI:Ere:07]-DQ641698	1122	1481	SiYMCV-[Sid:14]-AJ810096	EmYVV-[CN:Cra:17]-MN201997	3.00E-05	8.37E-03	2.72E-06	4.61E-07	1.93E-09	6.41E-08
ToYSV-[BR:Leo:10]-KX348223	2486	2618	ToYSV-[BR:Leo:10]-KX348172	AEV-[IN:Aeg:21]-OM719614	1.42E-02	5.60E-03	1.99E-03	2.05E-03	3.06E-09	6.44E-05
LuYVV-[VE:Lud:07]-DQ641699	1364	2318	AYVV-[CN:Aeg:00]-DQ866134	ToYLCCNV-[IR:Dat:11]-EF011559	1.43E-11	3.29E-09	2.73E-10	NS	8.14E-10	1.76E-13
MiYLCV-[BR:Mac:07]-DQ641695	2145	2486	ToMMV-[BZ:Sid:10]-KC706606	PaLCCNV-[CN:Ecp:06]-AM691553	7.49E-09	3.38E-09	7.45E-10	4.09E-08	5.24E-06	1.17E-08
ToYSV-[BZ:Sid:10]-KX348226	2474	2583	ToYSV-[BR:Leo:10]-KX348179	LAYVV-[VI:Lin:12]-NC009550	1.61E-06	1.54E-07	9.10E-04	1.63E-03	8.22E-09	NS
BGMV-[BR:Mac:10]-JN419017	1170	1585	ToYLCV-[CN:Mal:10]-KC189894	MCMV-[BR:Mac:09]-NC031453	NS	NS	NS	NS	NS	NS
BGMV-[BR:Mac:10]-JN419017	130	707	ToLCNDV-[IN:Sol:16]-MH465600	MCMV-[BR:Mac:09]-NC031453	NS	NS	NS	NS	NS	NS
PLCCV-[CN:Aeg:12]-JX294075	2357	2754	EmYVV-[CN:Cra:17]-MN201999	PaLCCNV-[CN:Cor:06]-AJ876548	NS	NS	1.34E-08	8.58E-07	1.26E-02	6.97E-08
RhMMV-[PR:Rhy:1997]-NC015488	1300	2094	SiAMV-[BZ:Sid:14]-KX691408	BlIChV-[MX:Ble:11]-NC019035	5.83E-09	1.35E-02	1.04E-08	7.13E-08	8.15E-17	6.01E-06
SiMCV-[BZ:Sid:07]-NC038457	1252	1723	ToMMV-[BZ:Sid:10]-KC706607	AMV-[BZ:Abu:07]-FN434438	6.60E-08	4.07E-06	3.17E-06	6.12E-03	3.38E-08	8.24E-05
AMBV-[BO:Abu:07]-HM585445	2228	2577	MaYVCaV-[CA:Mal:14]-KP188831	SiAMV-[BZ:Sid:14]-NC031456	NS	NS	9.78E-07	4.71E-05	2.63E-27	4.07E-08
ToYLCCNV-[IR:Dat:11]-EF011559	2085	2699	EmYVV-[CN:Cra:17]-MN201997	AYVHV-[CN:Aeg:00]-DQ866133	1.89E-10	1.12E-09	4.17E-08	4.47E-07	5.95E-10	3.21E-05
SiYMCV-[Sid:14]-AJ810096	1573	2060	TYLCCNV-[CN:Mal:12]-JX679251	SiYVVV-[VE:Sid:06]-DQ641696	2.94E-04	NS	2.41E-08	4.70E-07	4.31E-04	5.30E-09
SiChVV-[BZ:Sid:08]-NC_031454	1957	2091	MCMV-[BR:Mac:09]-NC031453	EuYMV-[BR:Leo:10]-KX348224	NS	NS	NS	NS	NS	NS
MaYVCaV-[CA:Mal:14]-KP188831	2245	2507	BlIChV-[MX:Ble:11]-NC019035	ToYLCV-[IR:Dat:11]-KC106647	5.57E-07	3.21E-08	8.43E-10	8.94E-04	2.44E-12	2.13E-04
BlYVV-[MX:Ble:12]-NC055132	415	2007	AYVCV-[CN:Aeg:11]-AJ849916	AYVHV-[CN:Aeg:00]-DQ866132	5.42E-04	3.64E-04	4.01E-12	3.59E-10	2.33E-27	3.49E-11
AYVV-[CN:Aeg:00]-DQ866134	2563	2689	HLCV-[PK:Mal:17]-LT716980	AYVV-[CN:Aeg:12]-MN233600	1.00E-10	NS	0.014617	NS	NS	3.79E-07

TABLE 4.7 *(Continued)*

Recombinant	Recombination		Minor Parents	Major Parents	Detection Methods					
	Begin	End			RDP	GENECONV	Maxchi	Chimaera	SiSscan	3Seq
AYVV-[CN:Aeg:00]-DQ866134	2094	2244	EmYVV-[CN:Cra:17]-MN201999	SBYVGV-[CN:Sig:05]-NC_008317	2.31E-04	6.75E-12	4.66E-07	2.19E-03	1.20E-11	5.58E-10
HLCV-[PK:Mal:17]-LT716980	417	1136	EcYVV-[PK:Ecp:06]-NC016660	EcYVV-[IN:And:10]-KC476655	2.01E-02	NS	2.09E-09	1.91E-06	4.56E-03	6.94E-08
ToMMV-[BZ:Sid:10]-KC706608	50	1150	SiMCV-[BZ:Sid:10]-JX871373	SiMAV-[BZ:Sid:10]-NC_020256	NS	NS	2.79E-04	1.65E-03	1.42E-06	8.06E-08
MaYVLaV-[PK:Mal:13]-NC055476	2549	2735	HLCV-[PK:Mal:17]-LT716980	EcYVV-[PK:Ecp:06]-NC016660	1.14E-04	1.49E-02	3.45E-03	NS	1.20E-07	NS
CYVV-[PK:Cro:06]-FN543112	1854	2039	SiYMV-[IN:Sid:16]-KX513859	AEV-[IN:Aeg:09]-FN794198	8.28E-07	NS	4.70E-05	3.02E-04	4.21E-17	2.37E-07
ToLCNDV-[EX:Dat:15]-MH577734	579	2086	LEMV-[BR:Leo:10]-JX863082	AMV-[Abu:07]-NC001928	6.43E-03	NS	2.63E-07	5.40E-06	2.11E-36	5.72E-06
EuMV-[US:Eup:11]-JQ963888	1127	1358	ALCSV-[CN:Aeg:16]-MG917698	SiChVV-[BZ:Sid:08]-NC_031454	NS	NS	NS	NS	NS	NS
SiYMCV-[Sid:14]-AJ810096	2554	2749	EmYVV-[CN:Cra:08]-JQ247188	SiYVVV-[VE:Sid:06]-DQ641696	9.47E-04	3.35E-02	1.10E-06	NS	1.70E-16	8.36E-10
RhYMV-[PK:Rhy:07]-NC038885	234	523	WChSV-[SU:Dat:16]-KY488568	EmYVV-[CN:Cra:17]-MN201997	1.99E-04	NS	6.39E-06	1.15E-07	9.56E-08	1.62E-08
PLCV-[IN:Cro:14]-KM525657	1818	2036	SiYMV-[IN:Sid:16]-KX513859	AEV-[IN:Aeg:21]-KJ488990	7.81E-05	8.74E-03	3.67E-05	8.38E-03	1.79E-25	2.79E-07
LuYVV-[VE:Lud:07]-DQ641699	826	1116	AMBzV-[BZ:Sid:12]-JF694482	TYLCCNV-[CN:Mal:12]-JX679251	NS	NS	1.51E-07	2.25E-05	3.77E-02	5.19E-07
CLCuRV-[Dig:09]-FM202328	2693	2734	SyLCV-[CN:Sig:17]-MN210352	SyLCV-[CN:Ecp:16]-KU933258	NS	6.00E-07	NS	NS	NS	3.95E-02
MYNV-[BR:Mac:09]-NC017001	2572	2584	SiChVV-[BZ:Sid:08]-NC_031454	MYNV-[BR:Mac:09]-JN418998	NS	NS	NS	NS	NS	NS
CLCuRV-[Dig:09]-FM202328	871	1595	EcYVV-[PK:Ecp:06]-NC016660	ToLCJoV-[IN:Sol:11]-KJ028211	1.17E-03	NS	1.02E-05	7.26E-05	1.90E-09	1.06E-06
SBYVV-[CN:Sig:07]-AM230634	1	405	ALCCV-[CM:Aeg:09]-FR873230	EmYVV-[CN:Cra:17]-MN201999	NS	NS	9.66E-10	1.01E-03	NS	1.27E-02
TYLCCNV-[CN:Mal:12]-JX679251	2057	47	AYVCV-[CN:Aeg:14]-MN218662	AYVHV-[CN:Aeg:00]-DQ866133	NS	NS	NS	NS	NS	NS
PaLCGuV-[PH:Ble:12]-KF446660	1041	1993	AlYVV-[CN:Ecp:06]-DQ375456	DiYMV-[FL:Dic:2000]-AF139168	NS	NS	3.59E-04	1.30E-04	6.90E-10	1.23E-06
SiYMV-[BZ:Sid:15]-NC055548	2313	2622	SiChLV-[MX:Sid:18]-NC055581	SiAMV-[BZ:Sid:14]-KX691407	5.99E-05	7.17E-04	7.62E-05	NS	2.02E-06	7.20E-05
MaYVLaV-[PK:Mal:13]-NC055476	2061	2098	SBYVGV-[CN:Sig:05]-NC_008317	EcYVV-[PK:Ecp:06]-GQ478343	4.18E-04	2.30E-06	2.40E-03	NS	NS	2.62E-03
MCMV-[BR:Mac:09]-NC031448	440	842	MYSV-[BR:Mac:09]-NC016999	MYVV-[BR:Mac:11]-KJ939904	NS	NS	NS	NS	NS	NS
AMBV-[BO:Abu:07]-HM585445	24	34	SiAMV-[BZ:Sid:14]-KX691403	AMBV-[BO:Abu:07]-NC015045	NS	NS	NS	NS	NS	NS
AYVCV-[CN:Aeg:14]-MN218660	779	2094	AYVCV-[CN:Aeg:14]-MN218666	AYVCV-[CN:Aeg:11]-AJ849916	NS	NS	1.71E-05	1.19E-02	1.14E-09	4.16E-06
SiAMV-[BZ:Sid:14]-KX691406	1	113	SiAMV-[BZ:Sid:14]-KX691403	SiAMV-[BZ:Sid:14]-KX691408	NS	NS	NS	NS	NS	NS

TABLE 4.7 *(Continued)*

Recombinant	Recombination		Minor Parents	Major Parents	Detection Methods					
	Begin	End			RDP	GENECONV	Maxchi	Chimaera	SiSscan	3Seq
ToYSV-[BZ:Sid:10]-KX348226	268	2181	SiCMOV-[BZ:Sid:10]-NC038990	MYVV-[BR:Mac:10]-JN419021	NS	NS	NS	NS	NS	NS
SiMCV-[BZ:Sid:10]-JX871372	853	974	LEMV-[BR:Leo:10]-JX863082	SiMCV-[BZ:Sid:10]-JX871370	7.67E-06	1.18E-02	1.52E-07	1.88E-02	NS	5.11E-05
SiAMV-[BZ:Sid:14]-KX691404	2039	102	SiYMV-[BZ:Sid:15]-NC055548	SiAMV-[BZ:Sid:14]-KX691403	NS	NS	NS	NS	4.79E-32	8.62E-06
AYVV-[CN:Aeg:12]-MN233600	2725	376	TYLCCNV-[CN:Mal:12]-JX679252	AYVHV-[CN:Aeg:00]-DQ866133	2.43E-03	NS	8.69E-06	3.97E-03	NS	2.33E-02
ToMMV-[BZ:Sid:10]-KC706608	2592	2665	SiYBV-[BZ:Sid:10]-JX871382	ToMMV-[BZ:Sid:10]-JX871375	NS	NS	NS	NS	NS	NS
AYVCV-[CN:Aeg:11]-AJ849916	2515	373	AEV-[IN:Aeg:09]-FN794198	AYVHV-[CN:Aeg:00]-DQ866133	NS	NS	1.21E-05	3.27E-05	NS	NS
ToYLCV-[IR:Dat:11]-KC106647	2215	2505	RhRuGMV-[CU:Rhy:09]-NC038805	ALCCV-[CM:Aeg:09]-FR873230	NS	NS	NS	NS	NS	NS
SiYMV-[IN:Sid:16]-KX513859	2064	2101	PLCV-[IN:Cro:08]-JN817517	EcYVV-[PK:Ecp:06]-GQ478343	NS	NS	NS	NS	NS	NS
SiChVV-[BZ:Sid:08]-NC_031454	980	1178	HYVV-[NZ:Lon:03]-FJ817425	MCMV-[BR:Mac:09]-NC031453	NS	NS	NS	NS	NS	NS
SyLCV-[CN:Ecp:16]-NC_031306	2695	2724	EcYVV-[PK:Ecp:06]-NC016660	SyLCV-[CN:Sig:17]-MN210352	NS	NS	NS	NS	NS	NS
SyLCV-[CN:Sig:17]-MN210354	1075	1758	SyLCV-[CN:Ecp:16]-MN210350	ToLCSuV-[SU:Dat:16]-MG868938	NS	6.75E-03	4.42E-07	8.69E-03	1.14E-27	3.43E-05
AEV-[IN:Aeg:09]-FN794198	24	513	SyLCV-[CN:Sig:17]-MN210352	EmYVV-[CN:Cra:08]-JQ247188	NS	NS	1.03E-06	7.42E-03	2.71E-22	3.46E-05
AEV-[IN:Aeg:10]-JF728862	2660	2674	AYVV-[CN:Aeg:12]-MN233600	AEV-[IN:Aeg:10]-JF728861	NS	NS	NS	NS	NS	NS
ErYMV-[VI:Ere:07]-DQ641698	101	327	MiYLCV-[BR:Mac:07]-DQ641695	LAYVV-[CN:Lin:12]-AY795900	3.33E-04	2.38E-02	2.06E-05	3.33E-06	1.49E-16	3.10E-03
PepLCYV-[CN:Ecp:06]-AM691547	2147	2496	SiMAV-[BZ:Sid:10]-KX896420	ErYMV-[VI:Ere:07]-DQ641698	NS	NS	9.15E-04	6.05E-03	6.46E-08	2.27E-03
TYLCV-[EX:Sol:11]-KT099158	1133	2127	LAYVV-[CN:Lin:12]-AY795900	BGMV-[BR:Mac:10]-JN419003	1.07E-04	NS	2.74E-07	2.56E-05	5.19E-17	4.39E-05
EmYVV-[CN:Cra:08]-JQ247188	1	545	ALCCV-[CM:Aeg:09]-FR873230	SBYVGV-[CN:Sig:05]-NC_008317	NS	NS	8.63E-07	1.25E-03	1.23E-23	NS
MYSV-[BR:Mac:10]-JN419015	1	72	MBMV-[BR:Mac:14]-KX691400	MYSV-[BR:Mac:10]-JN419012	NS	NS	NS	NS	NS	NS
MYVV-[BR:Mac:10]-JN419021	48	130	BGMV-[BR:Mac:10]-JN419004	MYVV-[BR:Mac:11]-KJ939904	NS	NS	NS	NS	NS	NS
CYSV-[BR:Mac:09]-JN419002	2554	2653	BGMV-[BR:Mac:10]-JN419003	MBMV-[BR:Mac:14]-KX691400	NS	NS	NS	NS	NS	NS
SChSV-[BR:Mac:12]-KJ939918	1971	2084	SiAMV-[BZ:Sid:14]-NC038887	MCMV-[BR:Mac:09]-NC031448	4.56E-07	8.41E-03	6.09E-03	2.48E-03	5.07E-07	NS
PLCV-[IN:Cro:08]-JN817517	522	596	CLCuRV-[Dig:09]-FM202328	CYVV-[PK:Cro:06]-FN543112	NS	1.94E-02	3.03E-03	2.53E-02	0.016985	1.08E-04
CYVMV-[PK:Cro:15]-KX671964	1351	1786	PLCV-[IN:Cro:08]-JN817517	CYVMV-[IN:Cro:08]-JN817516	NS	NS	1.90E-04	NS	4.61E-05	4.33E-04

TABLE 4.7 *(Continued)*

Recombinant	Recombination		Minor Parents	Major Parents	Detection Methods					
	Begin	End			RDP	GENECONV	Maxchi	Chimaera	SiSscan	3Seq
AYVV-[VE:Ecp:06]-DQ641704	1740	1957	HLCV-[PK:Mal:17]-LT716980	TYLCCNV-[CN:Mal:12]-JX679252	NS	NS	NS	NS	NS	NS
PaLCCNV-[CN:Ecp:06]-AM691553	45	104	AYVCV-[CN:Aeg:14]-MN218666	AYVHV-[CN:Aeg:00]-DQ866132	NS	NS	NS	NS	NS	NS
AMV-[IN:Abu:03]-U51137	2123	2283	DaLCV-[SU:Dat:16]-MF402919	SiBYMV-[BZ:Sid:10]-KX348184	NS	NS	1.17E-04	8.49E-04	9.53E-11	7.37E-05
RhGMHaV-[CU:Rhy:07]-HM236368	2136	2370	RhRuGMV-[CU:Rhy:09]-NC038805	SiAMV-[BZ:Sid:14]-KX691406	NS	NS	NS	NS	NS	NS
SiYBV-[BZ:Sid:10]-JX871382	2348	2603	SiYMAV-[BZ:Sid:10]-JX871383	SiMV-[BZ:Sid:10]-JX871378	6.11E-04	1.86E-11	3.68E-06	4.08E-06	1.39E-04	3.09E-05
SiYBV-[BZ:Sid:10]-JX871382	1898	2344	SiMAV-[BZ:Sid:10]-KX896427	SiMV-[BZ:Sid:10]-JX871378	2.39E-06	4.64E-04	3.01E-07	2.15E-05	3.55E-14	NS
SiYBV-[BZ:Sid:10]-NC020254	184	465	AMBzV-[BZ:Sid:12]-JF694482	SiMAV-[BZ:Sid:14]-KX896421	NS	NS	NS	NS	NS	NS
CYVV-[PK:Cro:06]-FN543112	2121	2245	SiMAV-[BZ:Sid:10]-KX896427	WChSV-[SU:Dat:16]-KY488568	NS	NS	NS	NS	NS	NS
AYVV-[VE:Ecp:06]-DQ641704	1115	1443	MiYLCV-[BR:Mac:07]-DQ641695	LAYVV-[CN:Lin:12]-AY795900	NS	NS	NS	NS	NS	NS
ClGMV-[VE:Cle:06]-DQ641692	606	1102	SiMCV-[BZ:Sid:10]-JX871371	HYVV-[NZ:Lon:03]-FJ817425	NS	NS	1.62E-04	3.66E-04	NS	1.53E-04
ToYLCV-[CN:Mal:10]-KC189894	145	328	ToLCPaV-[IN:Xan:116]-KY612435	ToYLCCV-[CN:Mal:09]-KC189893	2.65E-03	NS	1.48E-03	4.84E-03	1.47E-04	1.53E-04
ToYSV-[BR:Leo:10]-KX348223	1	61	ToYSV-[BR:Leo:10]-KX348179	SiYMV-[BZ:Sid:15]-MF163258	NS	NS	NS	NS	NS	NS
SiBYMV-[BZ:Sid:10]-NC040191	1	204	AMV-[FR:Abu:09]-LN611622	EuMV-[US:Eup:11]-JQ963888	NS	NS	NS	NS	NS	NS
SiMAV-[BZ:Sid:10]-KX896424	207	214	SiBYMV-[BZ:Sid:10]-NC040191	SiMAV-[BZ:Sid:10]-KX896423	NS	NS	NS	NS	NS	NS
RhRuGMV-[CU:Rhy:09]-NC038805	2374	2483	LAYVV-[VI:Lin:12]-NC009550	SiMCV-[BZ:Sid:10]-JX871370	NS	NS	NS	NS	NS	NS
RhRuGMV-[CU:Rhy:09]-NC038805	2484	2567	RhMMV-[PR:Rhy:1997]-NC015488	MYSV-[BR:Mac:10]-JN419016	NS	NS	NS	NS	NS	NS
SiMAV-[BZ:Sid:10]-NC_020256	111	124	BGMV-[BR:Mac:10]-JN419006	SiMAV-[BZ:Sid:10]-JX871385	NS	NS	NS	NS	NS	NS
MYSV-[BR:Mac:11]-KJ939896	127	1717	MYSV-[BR:Mac:10]-JN419007	MYSV-[BR:Mac:11]-KJ939895	NS	NS	NS	NS	NS	NS
AMV-[BZ:Abu:07]-FN434438	28	70	SiYBV-[BZ:Sid:10]-JX871381	AMBzV-[BZ:Sid:12]-JF694482	NS	NS	NS	NS	NS	NS
RhYMV-[PK:Rhy:07]-NC038885	2129	2259	TYLCV-[EX:Sol:11]-KT099158	SiMAV-[BZ:Sid:10]-KX896415	NS	NS	NS	NS	NS	NS
HYVV-[NZ:Lon:03]-FJ817425	2400	2717	ClGMV-[VE:Cle:06]-DQ641692	EmYVV-[CN:Cra:17]-MN201999	NS	NS	2.58E-06	1.74E-03	2.20E-08	8.47E-04
TYLCCNV-[CN:Mal:12]-JX679251	1256	1448	MiYLCV-[BR:Mac:07]-DQ641695	AYVHV-[CN:Aeg:00]-DQ866133	NS	NS	5.11E-04	4.85E-02	2.10E-11	NS
MYSV-[BR:Mac:10]-JN419012	1985	2074	HLCV-[PK:Mal:17]-LT716980	SiMAV-[BZ:Sid:10]-KX896423	NS	NS	NS	NS	NS	NS

TABLE 4.7 *(Continued)*

Recombinant	Recombination		Minor Parents	Major Parents	Detection Methods					
	Begin	End			RDP	GENECONV	Maxchi	Chimaera	SiSscan	3Seq
SiYMAV-[BZ:Sid:10]-JX871383	325	941	AMBzV-[BZ:Sid:12]-JF694482	AMYV-[MX:Abu:07]-KC430935	NS	NS	NS	NS	NS	NS
ToYLCCV-[CN:Mal:09]-KC189893	1	918	MiYLCV-[BR:Mac:07]-DQ641695	AYVHV-[CN:Aeg:00]-DQ866132	NS	NS	NS	NS	NS	NS
SiMCV-[BZ:Sid:10]-JX871373	932	1003	ToYSV-[BR:Leo:10]-KX348177	SiMCV-[BZ:Sid:10]-JX871370	NS	NS	NS	NS	NS	NS
MYNV-[BR:Mac:09]-JN418998	2522	2530	AYVHV-[CN:Aeg:00]-DQ866133	MYNV-[BR:Mac:09]-NC017001	NS	NS	NS	NS	NS	NS
DMV-[TZ:Dei:15]-NC029803	108	424	AYVV-[CN:Aeg:00]-DQ866134	SChSV-[BR:Mac:12]-KJ939918	NS	NS	NS	NS	NS	NS
PLCrV-[IN:Sol:11]-KJ028210	19	113	ToLCNDV-[PK:Sol:06]-AJ620187	ToLCNDV-[PK:Par:13]-KF002409	NS	NS	NS	NS	7.26E-07	1.16E-03
SiCMOV-[BZ:Sid:10]-NC038990	2558	2571	SiMAV-[BZ:Sid:10]-KX896420	SiCMOV-[BZ:Sid:10]-KX348183	NS	NS	NS	NS	NS	NS
EuMV-[US:Eup:11]-JQ963887	65	138	RhMoV-[VZ:Rhy:17]-OK044488	BIYVV-[MX:Ble:12]-NC055132	NS	NS	NS	NS	NS	NS
AYVV-[CN:Aeg:00]-DQ866134	105	1524	AYVCV-[CN:Aeg:14]-MN218663	AYVHV-[CN:Aeg:00]-DQ866132	NS	NS	NS	9.70E-03	7.09E-04	1.61E-03
SiMAV-[BZ:Sid:14]-KX896421	113	127	MCMV-[BR:Mac:09]-NC031453	SiMAV-[BZ:Sid:10]-KX896420	NS	NS	NS	NS	NS	NS
SiMCV-[BZ:Sid:10]-JX871371	2573	794	LEMV-[BR:Leo:07]-JX863081	SiMCV-[BZ:Sid:10]-JX871372	NS	NS	NS	NS	NS	NS
BGMV-[BR:Mac:10]-JN419004	238	823	RhMMV-[PR:Rhy:1997]-NC015488	MYVV-[BR:Mac:10]-JN419021	NS	NS	1.39E-04	NS	NS	NS
SiMAV-[BZ:Sid:10]-JX871385	1222	1636	AMBzV-[BZ:Sid:12]-JF694482	SiYBV-[BZ:Sid:10]-JX871380	NS	NS	NS	NS	NS	NS
SiMCV-[BZ:Sid:10]-JX871370	160	297	AMBzV-[BZ:Sid:12]-JF694482	ToYSV-[BR:Leo:10]-KX348173	NS	NS	NS	NS	NS	NS
AMV-[Abu:15]-X15983	2531	2633	LAYVV-[VI:Lin:12]-NC009550	SiMCV-[BZ:Sid:10]-JX871373	NS	NS	NS	NS	NS	NS
WChSV-[SU:Dat:16]-KY488568	1077	2013	LAYVV-[CN:Lin:12]-AY795900	DiYMV-[FL:Dic:2000]-AF139168	1.56E-02	NS	3.99E-04	1.85E-03	2.43E-05	0.003128
RhYMV-[PK:Rhy:07]-NC038885	608	1261	DiYMV-[FL:Dic:2000]-AF139168	LAYVV-[CN:Lin:12]-KC172827	NS	1.87E-02	3.34E-06	NS	3.22E-03	0.006829
SiChLV-[MX:Sid:18]-NC055581	1	42	SiAMV-[BZ:Sid:14]-KX691408	SiChLV-[MX:Sid:18]-MN013784	NS	NS	NS	NS	NS	NS
MiYLCV-[BR:Mac:07]-DQ641695	1268	1367	LAYVV-[VI:Lin:12]-NC009550	AYVHV-[CN:Aeg:00]-DQ866132	1.38E-02	4.38E-03	4.06E-03	NS	2.42E-12	NS
MiYLCV-[BR:Mac:07]-DQ641695	929	1268	AYVCV-[CN:Aeg:14]-MN218663	ToLCPaV-[IN:Xan:116]-KY612435	NS	1.67E-04	6.40E-05	3.45E-04	5.14E-06	NS
CoYVMV-[IN:Dat:16]-KX513862	1464	1648	SiLCV-[VI:Abu:07]-DQ641707	CYVV-[PK:Cro:06]-FN543112	NS	NS	NS	NS	NS	NS
SiMAV-[BZ:Sid:10]-KX896427	109	119	SiMCV-[BZ:Sid:10]-JX871373	SiMAV-[BZ:Sid:10]-KX896417	NS	NS	NS	NS	NS	NS
AEV-[IN:Aeg:10]-JF728864	1093	2098	SBYVGV-[CN:Sig:05]-NC_008317	DiYMV-[FL:Dic:2000]-AF139168	NS	1.33E-02	3.54E-03	NS	1.53E-14	6.36E-03

TABLE 4.7 *(Continued)*

Recombinant	Recombination		Minor Parents	Major Parents	Detection Methods					
	Begin	End			RDP	GENECONV	Maxchi	Chimaera	SiSscan	3Seq
SiAMV-[BZ:Sid:14]-KX691408	2469	2629	SiLCV-[VI:Abu:07]-DQ641706	BlIChV-[MX:Ble:11]-NC019035	NS	NS	NS	NS	NS	NS
SiAMV-[BZ:Sid:14]-NC031456	43	63	EuMV-[US:Eup:11]-JQ963888	SiAMV-[BZ:Sid:14]-KX691408	NS	NS	NS	NS	NS	NS
SiAMV-[BZ:Sid:14]-KX691408	32	66	EuMV-[US:Eup:11]-JQ963888	SiAMV-[BZ:Sid:14]-KX691407	NS	NS	NS	NS	NS	NS
SiAMV-[BZ:Sid:14]-KX691406	1069	1660	AMBzV-[BZ:Sid:12]-JF694482	SiBYMV-[BZ:Sid:10]-KX348184	NS	NS	NS	NS	NS	NS
ToYSV-[BR:Leo:10]-KX348176	67	107	SiYMV-[BZ:Sid:10]-JX871369	ToYSV-[BR:Leo:10]-KX348167	NS	NS	NS	NS	NS	NS
ToYSV-[BR:Leo:10]-KX348178	67	107	SiYMV-[BZ:Sid:10]-JX871369	ToYSV-[BR:Leo:10]-KX348167	NS	NS	NS	NS	NS	NS
SBYVV-[CN:Sig:12]-KF499589	1	157	SiLCV-[VI:Abu:07]-DQ641706	MaYVYnV-[CN:Mal:18]-NC006631	NS	NS	NS	NS	NS	NS
ALCCV-[CM:Aeg:09]-FR873230	17	625	ToLCV-[CN:Abu:12]-KC852149	HYVV-[JN:Lon:18]-AB287441	NS	NS	NS	NS	2.44E-14	8.49E-03
SBYVGV-[CN:Sig:05]-AM238692	28	141	MYVYNV-[CN:Mal:11]-JX679256	SiLCV-[VI:Abu:07]-DQ641706	NS	NS	NS	NS	NS	NS
PLCV-[IN:Cro:08]-JN817517	21	516	SyLCV-[CN:Sig:17]-MN210354	EcYVV-[PK:Ecp:13]-HG937534	NS	NS	5.30E-04	NS	2.36E-15	1.13E-02
AYVCV-[CN:Aeg:18]-AJ495813	96	106	SiLCV-[VI:Abu:07]-DQ641707	AYVCV-[CN:Aeg:18]-NC004090	NS	NS	NS	NS	NS	NS
AMYV-[MX:Abu:07]-KC430935	23	52	AMBzV-[BZ:Abu:18]-NC016574	SiChVV-[BZ:Sid:14]-KX691405	NS	NS	NS	NS	NS	NS
MYSV-[BR:Mac:10]-JN419018	166	1863	MYSV-[BR:Mac:10]-JN419014	MYSV-[BR:Mac:11]-KJ939895	NS	NS	3.15E-06	NS	NS	1.51E-02
ClGMV-[VE:Cle:06]-DQ641692	2135	2309	SiMV-[BZ:Sid:10]-JX871378	PLCCV-[CN:Aeg:12]-JX294075	NS	NS	NS	NS	NS	NS
EcYVV-[PK:Ecp:13]-HG937534	25	123	HLCV-[PK:Mal:17]-LT716980	EcYVV-[PK:Ecp:06]-GQ478343	NS	NS	NS	NS	NS	NS
MYVV-[BR:Mac:10]-JN419021	207	979	RhMoV-[VZ:Rhy:17]-OK044488	MBMV-[BR:Mac:14]-KX691399	NS	NS	1.65E-03	0.012384	1.50E-08	NS
ToYSV-[BR:Leo:10]-KX348175	67	175	SiYMV-[BZ:Sid:10]-JX871369	ToYSV-[BR:Leo:10]-KX348167	NS	NS	NS	NS	NS	NS
MiYLCV-[BR:Mac:07]-DQ641695	419	929	AYVCV-[CN:Aeg:14]-MN218663	AYVCV-[CN:Aeg:14]-MN218666	NS	NS	NS	NS	NS	NS
SiAMV-[BZ:Sid:14]-NC038887	187	215	LEMV-[PR:Leo:07]-KC683374	ToYSV-[BR:Leo:10]-KX348170	NS	NS	NS	NS	NS	NS
MCMV-[BR:Mac:09]-NC031453	893	1034	MYVV-[BR:Mac:11]-KJ939908	SiChVV-[BZ:Sid:08]-NC_031454	NS	NS	NS	NS	NS	NS
ToYSV-[BZ:Sid:10]-KX348226	2584	262	SiMAV-[BZ:Sid:10]-KX896415	SiCMOV-[BZ:Sid:10]-KX348183	NS	NS	NS	NS	NS	NS
MCMV-[BR:Mac:09]-NC031453	37	52	MYNV-[BR:Mac:09]-NC017001	AMV-[Abu:07]-NC001928	NS	NS	NS	NS	NS	NS
MCMV-[BR:Mac:09]-NC031453	424	677	ToLCPaV-[IN:Xan:116]-KY612435	EuMV-[US:Eup:11]-JQ963888	NS	NS	NS	NS	NS	NS

TABLE 4.7 *(Continued)*

Recombinant	Recombination		Minor Parents	Major Parents	Detection Methods					
	Begin	End			RDP	GENECONV	Maxchi	Chimaera	SiSscan	3Seq
EuMV-[US:Eup:11]-JQ963888	2132	2299	SiMAV-[BZ:Sid:10]-KX896423	MCMV-[BR:Mac:09]-NC031453	NS	NS	NS	NS	NS	NS
SBYVV-[CN:Sig:11]-JX294076	38	154	SiLCV-[VI:Abu:07]-DQ641706	SyLCV-[CN:Sig:17]-MN210353	NS	NS	NS	NS	NS	NS
AYVHV-[CN:Aeg:00]-DQ866133	2075	52	HLCV-[PK:Mal:17]-LT716980	AYVHV-[CN:Aeg:00]-DQ866132	NS	NS	NS	NS	NS	NS
EcYVV-[PK:Ecp:13]-HG937534	617	1088	ToYSV-[BR:Leo:10]-KX348167	ALCSV-[CN:Aeg:16]-MG917698	NS	NS	4.84E-02	NS	NS	NS
LuYVV-[VE:Lud:07]-DQ641699	2675	2729	EcYVV-[PK:Ecp:13]-HG937534	LuYVV-[VE:Lud:06]-DQ641708	NS	NS	4.85E-21	NS	NS	NS
SiYMV-[BZ:Sid:10]-JX871369	2000	2657	LEMV-[BR:Leo:08]-NC038450	SiMV-[BZ:Sid:10]-JX871378	5.97E-10	2.36E-09	1.11E-02	1.63E-03	2.97E-09	1.37E-05
ToYSV-[BZ:Sid:10]-KX348226	2269	2459	SiMAV-[BZ:Sid:10]-KX896420	ToYSV-[BR:Leo:10]-KX348223	NS	NS	NS	NS	NS	NS
PaLCCNV-[CN:Ecp:06]-AM691553	331	564	EcYVV-[PK:Ecp:06]-GQ478343	SBYVGV-[CN:Sig:05]-AM238692	NS	NS	NS	NS	NS	NS
SiBYMV-[BZ:Sid:10]-NC040191	2350	2486	ToMMV-[BZ:Sid:10]-KC706606	SiChVV-[BZ:Sid:08]-NC_031454	NS	NS	NS	NS	NS	NS
SiMCV-[BZ:Sid:07]-NC038457	2047	2687	SiChLV-[MX:Sid:18]-MN013784	LEMV-[BR:Leo:10]-JX863082	NS	NS	4.97E-06	4.23E-08	4.86E-06	6.25E-08
BlChV-[MX:Ble:11]-NC019035	173	1443	MYSV-[BR:Mac:11]-KJ939895	SiMAV-[BZ:Sid:10]-KX896417	NS	NS	NS	NS	NS	NS
BGMV-[BR:Mac:10]-JN419017	707	2039	MCMV-[BR:Mac:09]-NC031453	SiCMOV-[BZ:Sid:10]-NC038990	NS	NS	2.77E-05	2.06E-04	1.16E-10	NS
SiCMOV-[BZ:Sid:10]-NC038990	2026	2577	SiMAV-[BZ:Sid:14]-KX896421	SiMV-[BZ:Sid:10]-JX871378	NS	NS	NS	NS	NS	NS
SiMCV-[BZ:Sid:10]-JX871370	1761	2594	SiChLV-[MX:Sid:18]-NC055581	ToYSV-[BR:Leo:10]-KX348167	NS	NS	8.04E-05	6.65E-06	3.77E-07	2.91E-05
LuYVV-[CN:Ecp:17]-MN210343	244	418	AYVCV-[CN:Aeg:18]-AJ495813	SiLCV-[VI:Abu:07]-DQ641706	NS	NS	8.03E-05	1.06E-04	5.79E-08	3.94E-03
AYVV-[CN:Aeg:12]-MN233600	2549	2692	ClGMV-[VE:Cle:06]-DQ641692	AYVHV-[CN:Aeg:00]-DQ866132	NS	NS	1.39E-04	NS	1.33E-12	NS
ToYSV-[BR:Leo:10]-KX348166	2013	2119	ErYMV-[VI:Ere:07]-DQ641698	SiMAV-[BZ:Sid:10]-KX896420	NS	NS	9.89E-05	NS	7.29E-21	NS
SiChVV-[BZ:Sid:08]-NC_031454	370	649	DMV-[TZ:Dei:15]-NC029803	SiBYMV-[BZ:Sid:10]-NC040191	NS	NS	NS	NS	NS	NS
AYVHV-[TA:Aeg:00]-NC038913	1284	1345	ToLCJoV-[IN:Sol:11]-KJ028211	AYVV-[CN:Aeg:12]-MN233600	NS	NS	NS	NS	NS	NS
ToLCJoV-[IN:Sol:11]-KJ028211	2555	2727	ClGMV-[VE:Cle:06]-DQ641692	AEV-[IN:Aeg:09]-FN794198	NS	NS	2.35E-04	1.18E-02	NS	9.33E-04
ToLCGuV-[PK:Xan:11]-FR819708	79	425	SyLCV-[CN:Sig:17]-MN210353	EcYVV-[PK:Ecp:06]-NC016660	NS	NS	2.53E-04	NS	6.84E-10	4.72E-02
MYSV-[BR:Mac:10]-JN419015	152	1635	MYSV-[BR:Mac:10]-JN419019	MYSV-[BR:Mac:11]-KJ939897	NS	NS	NS	NS	NS	NS
WChSV-[SU:Dat:16]-KY488568	100	486	ClGMV-[VE:Cle:06]-DQ641692	SChSV-[BR:Mac:12]-KJ939918	NS	NS	NS	NS	NS	NS

TABLE 4.7 *(Continued)*

Recombinant	Recombination		Minor Parents	Major Parents	Detection Methods					
	Begin	End			RDP	GENECONV	Maxchi	Chimaera	SiSscan	3Seq
SiMV-[BZ:Sid:10]-JX871378	490	1017	SiAMV-[BZ:Sid:14]-KX691408	SiYMV-[BZ:Sid:10]-JX871369	NS	NS	NS	NS	NS	NS
AMBV-[BO:Abu:07]-NC015045	1393	1526	SiYBV-[BZ:Sid:10]-JX871382	SiMAV-[BZ:Sid:10]-KX896420	NS	NS	NS	NS	NS	NS
SiYMAV-[BZ:Sid:10]-JX871383	2633	188	LAYVV-[CN:Lin:12]-AY795900	MYSV-[BR:Mac:10]-JN419018	NS	NS	NS	NS	NS	NS
SiYMV-[BZ:Sid:10]-JX871369	1324	1493	SiYMAV-[BZ:Sid:10]-JX871383	SiMAV-[BZ:Sid:10]-KX896415	NS	NS	NS	NS	NS	NS
SiChYVV-[BZ:Sid:10]-NC031455	15	76	LEMV-[PR:Leo:07]-KC683374	SiYMV-[BZ:Sid:15]-MF163258	NS	NS	NS	NS	NS	NS
CLCuRV-[Dig:09]-FM202328	222	411	SiLCV-[VI:Abu:07]-DQ641706	SyLCV-[CN:Ecp:16]-KU933258	NS	NS	1.55E-04	NS	3.45E-07	NS
DiYMV-[FL:Dic:05]-NC003856	1936	2170	SiYMAV-[BZ:Sid:10]-JX871383	SiChLV-[MX:Sid:18]-MN013784	NS	NS	NS	NS	NS	NS
AlYVCNV-[CN:Ecp:12]-MN218667	2583	2668	MYSV-[BR:Mac:09]-NC016999	SiChLV-[MX:Sid:18]-MN013784	3.98E-03	NS	4.79E-04	NS	2.16E-31	NS
CYSV-[BR:Mac:09]-JN419002	15	119	AlYVV-[CN:Ecp:06]-DQ375456	AMV-[FR:Abu:09]-LN611622	NS	NS	NS	NS	NS	NS
SiYMV-[IN:Sid:16]-KX513859	35	180	SyLCV-[CN:Age:17]-MN210356	SiLCV-[VI:Abu:07]-DQ641707	NS	NS	NS	NS	NS	NS
MYSV-[BR:Mac:10]-JN419018	2593	89	SChSV-[BR:Mac:12]-KJ939918	MYSV-[BR:Mac:10]-JN419007	NS	NS	NS	NS	NS	NS
ToLCPaV-[IN:Xan:116]-KY612435	411	573	MBMV-[BR:Mac:14]-KX691399	SBYVGV-[CN:Sig:05]-NC_008317	NS	NS	NS	NS	NS	NS
SiLCV-[VI:Abu:07]-DQ641706	150	410	EcYVV-[IN:And:10]-KC476655	LAYVV-[CN:Lin:12]-AY795900	NS	NS	NS	NS	NS	NS
MCMV-[BR:Mac:09]-NC031453	2511	2547	MYVV-[BR:Mac:10]-JN419021	MBMV-[BR:Mac:14]-KX691399	NS	NS	NS	NS	NS	NS
PLCV-[IN:Cro:14]-KM525657	2054	2123	LAYVV-[CN:Lin:12]-KC172827	PLCV-[IN:Cro:08]-JN817517	NS	NS	NS	NS	NS	NS
AMBzV-[BZ:Sid:12]-JF694482	477	1066	SiBYMV-[BZ:Sid:10]-KX348184	AMBzV-[BZ:Sid:12]-JF694481	NS	NS	7.76E-03	NS	1.01E-09	NS
SiMAV-[BZ:Sid:10]-KX896415	1990	2087	ErYMV-[VI:Ere:07]-DQ641698	SiAMV-[BZ:Sid:14]-KX691407	NS	NS	2.92E-03	NS	1.64E-10	NS
SiChVV-[BZ:Sid:08]-NC_031454	1466	1905	ToMMV-[BZ:Sid:10]-KC706608	SiBYMV-[BZ:Sid:10]-NC040191	NS	NS	NS	NS	NS	NS
DaLDV-[VZ:Dat:07]-NC018717	384	1057	SiYMV-[BZ:Sid:15]-MF163258	AMYV-[MX:Abu:07]-KC430935	NS	NS	1.29E-02	4.61E-03	NS	NS
MBMV-[BR:Mac:14]-KX691400	1164	1880	SiCMOV-[BZ:Sid:10]-NC038990	SChSV-[BR:Mac:12]-KJ939918	NS	NS	NS	NS	NS	NS
SYVV-[VE:Spi:06]-DQ641694	1452	1555	LAYVV-[CN:Lin:12]-AY795900	AYVV-[CN:Aeg:12]-MN233600	NS	NS	NS	NS	NS	NS
ToLCNDV-[PK:Par:13]-KF002409	17	37	ToLCV-[CN:Abu:12]-KC852149	ToLCNDV-[PK:Sol:06]-AJ620187	NS	NS	NS	NS	NS	NS
SiMCV-[BZ:Sid:10]-JX871373	337	853	SiYMV-[BZ:Sid:10]-JX871369	SiMCV-[BZ:Sid:07]-NC038457	NS	NS	2.27E-06	NS	6.86E-12	NS

TABLE 4.7 *(Continued)*

Recombinant	Recombination		Minor Parents	Major Parents	Detection Methods					
	Begin	End			RDP	GENECONV	Maxchi	Chimaera	SiSscan	3Seq
PLCrV-[IN:Sol:11]-KJ028210	1419	1629	SiYMAV-[BZ:Sid:10]-JX871383	ToMMV-[BZ:Sid:10]-KC706608	NS	NS	3.60E-04	NS	4.90E-22	NS
AlYVV-[CN:Ecp:06]-DQ375456	823	942	SiYBV-[BZ:Sid:10]-JX871382	SiYMCV-[Sid:14]-AJ810096	NS	NS	NS	NS	NS	NS
AlYVCNV-[CN:Ecp:12]-MN218667	1950	2065	SyLCV-[CN:Ecp:16]-MN210350	AYVCV-[CN:Aeg:18]-NC004090	NS	NS	NS	NS	NS	NS
LuYVV-[VE:Lud:06]-DQ641708	792	1357	AYVCV-[CN:Aeg:14]-MN218666	LAYVV-[VI:Lin:12]-NC009550	NS	NS	NS	NS	NS	NS
ToLCPaV-[IN:Xan:116]-KY612435	1099	1260	ToYLCCV-[CN:Mal:09]-KC189893	ToLCJoV-[IN:Sol:11]-KJ028211	NS	NS	NS	NS	NS	NS
EmYVV-[CN:Cra:17]-MN201999	955	1053	MCMV-[BR:Mac:09]-NC031448	TYLCCNV-[CN:Mal:12]-JX679252	NS	NS	NS	NS	NS	NS

TABLE 4.8 Details of Recombination Analysis of DNA-B Sequences of Bipartite Begomoviruses Infecting Weed Hosts Using RDP5

Recombinant	Recombination		Minor Parents	Major Parents	Detection Methods					
	Begin	End			RDP	GENECONV	Maxchi	Chimaera	SiSscan	3Seq
ToYSV-[BR:Leo:10]-KX348205	55	1504	ToYSV-[BR:Leo:10]-KX348211	ToYSV-[BR:Leo:10]-KX348212	4.36E-04	NS	1.47E-09	NS	1.55E-04	6.86E-18
AMV-[Abu:15]-X15984	186	2435	AMV-[GR:Abu:09]-LN611624	AMV-[FR:Abu:07]-LN611625	NS	NS	4.91E-08	4.33E-02	5.78E-18	8.67E-16
RhRuGMV-[CU:Rhy:09]-NC038804	2109	2555	ToYSV-[BR:Leo:10]-KX348210	CLCV-[VZ:Rhy:17]-OK044493	NS	NS	2.48E-15	6.21E-08	NS	1.62E-11
SiChVV-[BZ:Sid:08]-KX691413	83	157	MBMV-[BR:Mac:10]-JN419017	SiChVV-[BZ:Sid:14]-KX691414	1.76E-13	4.29E-12	9.81E-05	1.60E-03	NS	8.05E-09
ToLCNDV-[PK:Par:13]-KF002410	2177	2611	TYCV-[JM:Rhy:15]-KP641346	RhGMSiV-[MX:Rhy:08]-MK618663	NS	NS	3.67E-13	9.90E-06	NS	NS
ToLCNDV-[EX:Dat:15]-MH577654	2157	2591	TYCV-[JM:Rhy:15]-KP641346	RhGMSiV-[MX:Rhy:08]-MK618663	1.86E-02	NS	5.29E-13	2.16E-06	NS	6.39E-03
AMBV-[BO:Abu:07]-HM585446	105	2222	EuYMV-[BR:Eup:08]-FJ619508	DMV-[KY:Dei:14]-KT878830	NS	NS	3.63E-08	4.50E-04	2.72E-36	8.12E-12
TYCV-[JM:Rhy:15]-KP641346	2345	2543	SiChLV-[MX:Sid:18]-NC_060102	CLCV-[VZ:Rhy:17]-OK044493	1.19E-09	1.02E-06	1.10E-05	1.73E-06	4.73E-03	8.12E-12
MBMV-[BR:Mac:10]-JN419017	2131	2433	DMV-[KY:Dei:14]-KT878827	MCMV-[BR:Mac:09]-NC031453	NS	NS	4.17E-11	2.63E-04	NS	NS
EuMV-[CU:Eup:07]-HQ896202	521	2230	MBMV-[BR:Mac:10]-JN419017	DMV-[KY:Dei:14]-KT878830	NS	NS	2.88E-02	NS	8.70E-19	8.12E-12
EuYMV-[BR:Eup:08]-NC012554	466	2198	MBMV-[BR:Mac:10]-JN419017	DMV-[KY:Dei:14]-KT878830	NS	NS	3.71E-04	NS	1.05E-18	8.12E-12
RhGMV-[MX:Rhy:19]-MK634539	203	2092	DiYMV-[Dic:05]-NC003857	SiMSiV-[MX:Sid:15]-MK643154	NS	NS	5.64E-03	NS	7.51E-04	8.12E-12
ToYSV-[BR:Leo:10]-KX348211	430	1299	ToYSV-[BR:Leo:10]-KX348208	ToYSV-[BR:Leo:10]-KX348222	1.94E-06	1.02E-09	1.56E-08	6.46E-03	2.13E-11	1.18E-02
TYCV-[JM:Rhy:15]-KP641346	2100	2333	WChSV-[SU:Dat:16]-KY488569	RhMMV-[PR:Rhy:1997]-NC015489	NS	NS	5.08E-09	2.62E-02	NS	NS
WChSV-[SU:Dat:16]-KY488569	2311	2757	AMBzV-[BZ:Bla:12]-JF694479	RhGMSiV-[MX:Rhy:08]-MK618663	4.52E-02	NS	3.78E-08	4.77E-04	NS	2.87E-03
ToMMV-[BZ:Sid:10]-KC706609	2232	2532	DMV-[TZ:Dei:14]-KT878825	DaLDV-[VZ:Dat:07]-NC018715	NS	NS	1.54E-07	4.15E-03	NS	NS
SiMSiV-[MX:Sid:15]-MK643154	2466	2523	DaLDV-[VZ:Dat:07]-NC018715	SiChLV-[MX:Sid:18]-NC_060102	1.71E-07	1.41E-04	NS	NS	NS	3.62E-03
DMV-[TZ:Dei:14]-KT878825	2298	2538	SiMSiV-[MX:Sid:15]-MK643154	AMV-[SA:Abu:03]-AM886130	NS	NS	1.44E-07	4.54E-02	NS	NS
ToYSV-[BR:Leo:10]-KX348214	1392	2555	ToYSV-[BR:Leo:10]-KX348207	ToYSV-[BR:Leo:10]-KX348213	2.47E-02	NS	4.62E-07	8.79E-06	NS	2.43E-03
WChSV-[SU:Dat:16]-KY488569	218	639	ToYSV-[BR:Leo:10]-KX348210	BGMV-[BR:Mac:10]-JN419008	NS	NS	8.47E-07	6.34E-04	NS	1.77E-02
ToLCNDV-[PK:Sol:06]-AJ620188	56	234	ToMMV-[BZ:Sid:10]-KC706609	ToLCNDV-[PK:Par:13]-KF002410	4.12E-08	2.64E-02	6.18E-11	NS	NS	1.18E-03
SiChVV-[BZ:Sid:14]-KX691414	2382	2430	SiChVV-[BZ:Sid:08]-KX691415	AMBzV-[BZ:Sid:12]-JF694483	NS	1.61E-06	NS	NS	1.20E-04	NS
ToCrLYV-[BR:Mac:10]-JN419011	287	630	ToCrLYV-[BR:Mac:10]-JN419010	MYNV-[BR:Mac:09]-JN418999	8.54E-06	6.55E-04	7.24E-03	1.01E-05	1.58E-04	7.32E-08
TYCV-[JM:Rhy:15]-KP641346	147	199	RhGMCV-[CO:Rhy:14]-KX216481	RhRuGMV-[CU:Rhy:09]-NC038804	2.05E-02	1.20E-05	NS	NS	0.015969	3.27E-03
AMV-[Abu:15]-X15984	2436	2514	AMV-[FR:Abu:07]-LN611625	SiMSiV-[MX:Sid:15]-MK643154	NS	NS	NS	NS	5.94E-07	1.38E-05
DaLDV-[VZ:Dat:07]-NC018715	43	1104	ToYSV-[BR:Leo:10]-KX348213	SiChLV-[MX:Sid:18]-NC_060102	NS	NS	2.63E-04	3.02E-06	6.11E-04	2.19E-05

TABLE 4.8 *(Continued)*

Recombinant	Recombination		Minor Parents	Major Parents	Detection Methods					
	Begin	End			RDP	GENECONV	Maxchi	Chimaera	SiSscan	3Seq
SiYGMV-[BZ:Sid:15]-NC_060100	2480	2558	DaLDV-[VZ:Dat:07]-NC018715	ToCrLYV-[BR:Mac:10]-JN419010	NS	NS	3.29E-05	NS	1.52E-05	NS
DMV-[KY:Dei:14]-KT878827	448	2275	SLCV-[JO:Mal:08]-EF532621	DiYMV-[Dic:05]-NC003857	NS	NS	3.64E-06	NS	6.21E-25	3.73E-05
SiMIMV-[BZ:Sid:11]-KX348199	2288	2437	SiMIMV-[BZ:Sid:11]-KX348200	SiMIMV-[BZ:Sid:11]-KX348201	NS	NS	7.56E-05	6.24E-03	NS	7.36E-05
ToLCNDV-[EX:Dat:15]-MH577654	328	417	ToMMV-[BZ:Sid:10]-KC706609	ToLCNDV-[PK:Sol:06]-AJ620188	1.17E-04	NS	1.60E-04	NS	NS	3.54E-02
WChSV-[SU:Dat:16]-KY488569	1291	1333	SiChVV-[BZ:Sid:14]-KX691414	ToMMV-[BZ:Sid:10]-KC706609	NS	NS	1.75E-04	NS	1.65E-31	NS
SiChVV-[BZ:Sid:08]-KX691415	1989	2494	AMBzV-[BZ:Bla:12]-JF694479	ToYSV-[BR:Leo:10]-KX348207	NS	NS	1.19E-03	2.21E-04	NS	NS
AMBzV-[BZ:Sid:12]-JF694483	90	188	AMV-[GR:Abu:09]-LN611624	ToYSV-[BR:Leo:10]-KX348221	NS	NS	2.30E-04	4.44E-02	NS	NS
SiChVV-[BZ:Sid:14]-KX691414	747	2263	ToYSV-[BR:Leo:10]-KX348218	AMBzV-[BZ:Sid:12]-JF694483	NS	NS	1.23E-05	6.45E-03	1.18E-10	2.68E-05
ToYSV-[BR:Leo:10]-KX348210	2434	2617	ToYSV-[BR:Leo:10]-KX348222	ToYSV-[BR:Leo:10]-KX348218	4.54E-04	2.62E-04	4.57E-05	NS	4.55E-06	NS
AMBzV-[BZ:Sid:12]-JF694483	857	2211	ToYSV-[BR:Leo:10]-KX348218	ToMMV-[BZ:Sid:10]-KC706609	NS	NS	4.23E-04	1.52E-02	2.24E-09	8.12E-12
AMBzV-[BZ:Bla:12]-JF694479	41	157	ToLCNDV-[PK:Par:13]-KF002410	ToCrLYV-[BR:Mac:10]-JN419010	2.51E-02	6.64E-04	NS	NS	NS	NS
RhMoV-[VZ:Rhy:17]-OK044489	103	2070	CLCV-[VZ:Rhy:17]-OK044493	BGMV-[BR:Mac:10]-JN419008	NS	NS	NS	NS	1.17E-11	1.45E-03
SiMIMV-[BZ:Sid:11]-KX348201	170	351	AMBzV-[BZ:Sid:12]-JF694483	SiMiMVV-[BZ:Sid:11]-KX348204	NS	NS	NS	NS	1.05E-07	2.39E-02
ToCrLYV-[BR:Mac:10]-JN419011	2276	2350	RhMMV-[PR:Rhy:1997]-NC015489	ToCrLYV-[BR:Mac:10]-JN419010	NS	NS	2.27E-03	NS	1.12E-08	NS
ToYSV-[BR:Leo:10]-KX348218	2542	2585	MYNV-[BR:Mac:09]-NC017002	ToYSV-[BR:Leo:10]-KX348222	NS	3.10E-03	NS	NS	4.09E-06	NS
SiMIMV-[BZ:Sid:11]-KX348202	792	1073	SiMIMV-[BZ:Sid:11]-KX348201	SiMIMV-[BZ:Sid:11]-KX348199	NS	1.33E-02	5.06E-03	NS	NS	NS
ToLCNDV-[PK:Sol:06]-AJ620188	238	453	ToYSV-[BR:Leo:10]-KX348216	MCMV-[BR:Mac:09]-NC031453	NS	NS	9.53E-03	NS	1.49E-05	NS
DiYMV-[Dic:05]-NC003857	29	193	ToLCNDV-[EX:Dat:15]-MH577654	BlIChV-[MX:Ble:11]-NC019036	1.47E-02	NS	NS	NS	NS	3.35E-02
ToYSV-[BR:Leo:10]-KX348219	2468	2509	MCMV-[BR:Mac:09]-NC031453	ToYSV-[BR:Leo:10]-KX348206	NS	1.58E-02	NS	NS	1.98E-56	NS
AMBzV-[BZ:Bla:12]-JF694479	1116	1815	SiMSiV-[MX:Sid:15]-MK643154	ToYSV-[BR:Leo:10]-KX348207	NS	NS	NS	1.34E-02	NS	2.25E-03
BGMV-[BR:Mac:10]-JN419008	1289	1821	ToYSV-[BR:Leo:10]-KX348213	AMV-[SA:Abu:03]-AM886130	NS	NS	3.44E-02	8.61E-03	NS	NS
SLCV-[JO:Mal:08]-EF532621	2458	2538	RhGMSiV-[MX:Rhy:08]-MK618663	DiYMV-[Dic:05]-NC003857	1.66E-02	2.38E-02	NS	NS	NS	NS
ToCrLYV-[BR:Mac:10]-JN419010	126	252	SLCV-[JO:Mal:08]-EF532621	SiBzV-[BZ:Sid:14]-FN436002	NS	NS	1.07E-03	NS	5.14E-03	NS
RhMMV-[PR:Rhy:1997]-NC015489	59	2122	RhGMSiV-[MX:Rhy:08]-MK618663	AMV-[GR:Abu:09]-LN611624	NS	NS	NS	NS	5.20E-04	2.02E-02
SiYGMV-[BZ:Sid:15]-NC_060100	1537	1720	RhMMV-[PR:Rhy:1997]-NC015489	SiMSiV-[MX:Sid:15]-MK643154	NS	NS	8.43E-03	1.15E-03	NS	NS

4.5.1 RECOMBINATION ANALYSIS OF WEED INFECTING BEGOMOVIRUSES ASSOCIATED WITH DNA SATELLITES

Neighbor-net analysis of 21 alphasatellites associated with begomoviruses infecting weeds showed evidence of non-tree-like evolution (possibly indicative of recombination). Around 11 out of 21 alphasatellite sequences have shown significant intra- and inter-specific recombination and are detected by more than three methods in alphasatellites associated with begomoviruses infecting weeds. The analysis also showed that the majority of the recombination takes place in the Rep region of alphasatellites (Table 4.9).

Similarly, an analysis of 72 complete sequences of betasatellites associated with weed-infecting begomoviruses revealed that only 20 betasatellite sequences showed significant recombination. The analysis also showed that the majority of the recombination takes place in the C1 (βC1 region), satellite conserved region (SCR), and AT-rich region of the betasatellites (Table 4.10).

4.6 CONCLUSION

The genus *Begomovirus* belongs to the family *Geminiviridae* and comprises a diverse group of plant viruses that infect crop plants and weed hosts, causing severe losses to crops. Weed hosts serve as virus/vector reservoirs in the absence of the main host under natural conditions. Diversity analysis of weed-infecting begomoviruses and DNA satellites from all over the world has revealed their geographical distribution, similar to the begomoviruses distributed on crop plants across the world. The percentage of monopartite viruses reported on weed hosts is higher compared to bipartite viruses. Furthermore, most of the viruses reported on weed hosts across the world have shown to have intra and interspecies recombination, which might have led to the emergence of novel species or strains of viruses that later adapted to cultivated crops, resulting in crop loss. Knowing the source of inoculum is key for designing management strategies for diseases. Determining the diversity of begomoviruses associated with weed hosts, which serve as virus/vector reservoirs, will be useful in studying the epidemiology of plant diseases associated with begomoviruses and designing management strategies in crop plants.

TABLE 4.9 Details of Recombination Analysis of Alphasatellites Associated with Begomoviruses Infecting Weed Hosts Using RDP5

Recombinants	Recombinant		Minor Parents	Major Parents	Detection Methods					
	Begin	End			RDP	GENECONV	Maxchi	Chimaera	SiSscan	3Seq
AYVIA-[IN:Mal:17]-LT716983	123	968	AYVIA-[PK:Age:12]-HG515071	AYVIA-[PK:weed:15]-MN191626	1.33E-19	1.25E-22	1.65E-19	1.86E-18	1.93E-19	5.33E-40
AYVIA-[PK:Age:12]-HG515071	1103	1312	MeYVMA-[IN:Luda:OK:10]JX183090	AYVIA-[IN:Par:13]-KU852743	3.98E-17	1.48E-04	8.18E-14	7.87E-09	3.14E-03	3.80E-13
AYVIA-[PK: Par:12]-MF683835	1044	1212	MeYVMA-[IN:Luda:OK:10]JX183090	PLCA-[PK:Par:12]-MF683834	4.95E-12	6.09E-13	6.26E-09	4.55E-09	4.39E-06	2.53E-13
CrYVMA-[IN:Har:Acaly:07].FN658711	1	935	ToYSA-[BZ:Leo:11]-NC_038993	HoYVA-[PK:17-5:06].FR772086	0.022337	NS	4.68E-14	2.36E-04	2.69E-22	7.59E-12
SiLCA-[VE:Sid:06]-DQ641717	169	1010	AYVIA-[IN:Dig:17]-MH973694	AYVIA-[IN:Par:12]-NC_019546	2.58E-02	1.46E-04	2.17E-09	3.63E-02	2.56E-17	7.47E-09
AAYVSGA-[CN:Age:01].AJ416153	357	606	CA-[CU:Sid:07]-NC_021708	AYVIA-[IN:Par:12]-NC_019546	3.44E-08	7.86E-07	6.61E-07	7.28E-06	2.15E-06	1.94E-05
PLCA-[PK:Par:12]-MF683834	695	755	AYVIA-[PK:weed:15]-MN191627	PLCA-[PK:Par:16]-NC_030226	2.57E-07	7.10E-06	NS	8.79E-03	NS	6.32E-04
AYVIA-[PK:weed:15]-MN191627	34	80	ToYSA-[BZ:Leo:11]-NC_038993	AYVIA-[PK:Age:12]-HG515071	NS	5.69E-07	4.36E-05	NS	NS	NS
ToYSA-[BZ:Leo:11]-NC_038993	1076	1258	HoYVA-[PK:17-5:06].FR772086	PLCA-[PK:Par:16]-NC_030226	1.42E-06	9.46E-03	2.03E-02	3.12E-02	1.40E-05	8.98E-03
AYVIA-[IN:Mal:17]-LT716983	176	466	SiLCA-[VE:Sid:06]-DQ641717	AYVIA-[IN:Par:13]-KU852743	2.12E-03	1.55E-02	NS	7.76E-04	NS	3.52E-02
PLCA-[PK:Par:12]-MF683834	1250	1335	OEnLCuA-[IN:GJ:OK:11].HF546575	SiYVVA-[VE:Sid06]-DQ641718	2.62E-02	4.84E-05	1.04E-05	2.38E-02	4.24E-30	1.68E-03
AAYVSGA-[CN:Age:01].AJ416153	91	167	ToYSA-[BZ:Leo:11]-NC_038993	HoYVA-[PK:17-5:06].FR772086	6.03E-03	NS	NS	3.41E-02	8.38E-12	NS
HoYVA-[PK:17-5:06].FR772086	578	614	PLCA-[PK:Par:12]-MF683834	MeYVMA-[IN:Luda:OK:10] JX183090	3.83E-02	1.44E-02	NS	NS	NS	NS

TABLE 4.10 Details of Recombination Analysis of Betasatellites Associated with Begomoviruses Infecting Weed Hosts Using RDP5

Recombinant Sequences	Recombination		Minor Parents	Major Parents	Detection Methods					
	Begin	End			RDP	GENECONV	Maxchi	Chimaera	SiSscan	3Seq
ChYMBB-[NG:Sid:14]-KT454830	198	560	AgYLCB-[CM:Age:06]-FM164738	CYVMB-[PK:Cro:14]-LT674472	2.76E-29	1.36E-22	1.54E-17	2.61E-14	6.69E-22	1.93E-12
SiYVB-[CN:Sig;07]-AM238695	27	1128	SiYVB-[CN:Sig;07]-KF499590	ErYMB-[VE:Ere:06]-DQ641713	2.47E-05	1.45E-03	7.52E-03	1.48E-06	7.44E-43	5.50E-23
PaLCuB-[IN:Par:12]-JX987089	1255	1367	CoLCuGB-[SU:Dat:09]-DQ644564	PaLCuB-[PK:Par:16]-LN906595	1.73E-16	3.67E-15	2.30E-04	2.56E-04	NS	5.69E-09
AgYLCB-[IN:Sol:11]-KJ028213	1258	1336	CoLCuGB-[SU:Dat:09]-DQ644564	AgYLCB-[PK:Age:03]-AJ316026	1.21E-12	3.76E-16	3.63E-04	1.07E-03	1.01E-04	6.08E-10
AgYLCB-[CM:Age:06]-FM164738	1129	1258	PaLCuB-[PK:Par:12]-MF683837	CoLCuGB-[SU:Dat:09]-DQ644564	5.37E-07	3.28E-15	1.73E-07	2.17E-05	1.50E-13	6.54E-04
AgYLCB-[PK:Age:03]-AJ316026	1245	1338	SiYVVB-[CN:Lin:22]-KF990602	AgYLCB-[PK:Urt:13]-LN880825	4.96E-13	1.58E-13	7.15E-04	3.55E-04	NS	2.48E-09
CYVMB-[PK:Cro:14]-LT674481	211	335	CYVMB-[PK:Cro:14]-LT674472	CYVMB-[PK:Cro:14]-LT674484	1.37E-07	5.49E-06	4.53E-04	1.67E-04	3.98E-06	6.84E-11
SiYVB-[CN:Sig;07]-AM230643	69	755	SiYVB-[CN:Sig;07]-AM230644	SiYVB-[CN:Sig;07]-KF499590	1.08E-04	9.49E-11	1.32E-09	3.88E-08	4.73E-26	4.49E-06
AgYLCB-[PK:Age:03]-AJ316026	819	994	AgYLCB-[PK:Urt:13]-LN880825	AgYLCB-[IN:Sol:11]-KJ028213	2.17E-08	1.02E-09	1.23E-05	1.47E-04	1.39E-12	NS
CYVMB-[PK:Cro:14]-JN831448	604	1050	CYVMB-[PK:Cro:14]-LT674468	CYVMB-[PK:Cro:14]-JN831447	NS	NS	7.76E-03	NS	NS	1.09E-09
CYVMB-[PK:Cro:14]-JN831447	1064	1270	CYVMB-[PK:Cro:14]-LT674468	CYVMB-[PK:Cro:14]-JN831448	6.38E-06	5.28E-05	2.97E-06	1.35E-07	9.79E-09	4.65E-09
MaYVKHB-[CA:Mal:14]-KP188832	260	627	PaLCuB-[PK:Par:12]-MF683837	LuYVMB-[CN:Lud:18]-NC007212	NS	NS	9.10E-04	2.70E-06	1.61E-11	7.32E-21
CYVMB-[PK:Cro:14]-LT674472	172	1160	CYVMB-[PK:Cro:14]-LT674479	CYVMB-[PK:Cro:05]-AM410551	6.83E-03	2.30E-02	7.14E-06	NS	1.95E-05	3.14E-07
AgYLCB-[IN:Age:10]-JF728868	677	927	AgLCB-[IN:Cal:15]-KU376492	AgYLCB-[PK:Age:06]-AM698010	1.76E-03	2.35E-06	6.19E-03	9.16E-03	2.12E-04	3.44E-03
LuLDB-[IN:cal:16]-KX513857	48	146	SiYVB-[CN:Sig;07]-AM230645	MaYVYuB-[CN:Lin:08]-FN806781	NS	NS	3.56E-05	3.32E-03	NS	1.77E-02
PaLCuB-[IN:Par:12]-JX987089	61	540	AgYLCB-[PK:Age:06]-AM698010	SiYVB-[CN:Sig;07]-AM230644	NS	NS	3.87E-02	1.27E-04	4.02E-10	NS
SiYVB-[CN:Sig;07]-AM238695	1246	1355	CoLCuGB-[SU:Dat:09]-DQ644564	CYVMB-[PK:Cro:14]-LT674468	NS	1.59E-04	2.26E-02	NS	NS	NS
CYVMB-[PK:Cro:14]-LT674469	359	1007	CYVMB-[PK:Cro:14]-LT674484	CYVMB-[PK:Cro:14]-LT674482	NS	NS	NS	NS	7.85E-10	2.14E-04
AgYLCB-[CM:Age:06]-FM164738	930	1122	CoYVMB-[IN:Dat:16]-KX513863	MaYVYuB-[CN:Lin:08]-FN806781	2.30E-04	1.39E-02	NS	1.15E-03	7.73E-12	NS
SiYVB-[CN:Sig;07]-AM230644	929	1020	SiYVB-[CN:Sig;07]-AM230643	SiYVB-[CN:Sig;07]-NC_008237	NS	4.96E-04	NS	NS	NS	6.45E-04
ErYMB-[VE:Ere:06]-DQ641713	1153	1341	SiYVB-[CN:Sig;07]-KF499590	LiAYVB-[VE:Lin:06]-DQ641715	7.72E-04	2.04E-02	NS	NS	3.95E-06	NS
ToCSB-[CN:Age:06]-AM260736	107	202	SiYVB-[CN:Sig;07]-KF499590	TYLCCNB-[CN:Mal:09]-AM980512	1.25E-03	1.80E-02	NS	NS	3.65E-06	NS
LuLDB-[IN:Sid:16]-KX513860	17	81	CYVMB-[PK:Cro:14]-JN831447	MaYVYuB-[CN:Lin:08]-FN806781	2.75E-03	NS	NS	NS	NS	1.13E-02

TABLE 4.10 *(Continued)*

Recombinant Sequences	Recombination		Minor Parents	Major Parents	Detection Methods					
	Begin	End			RDP	GENECONV	Maxchi	Chimaera	SiSscan	3Seq
CYVMB-[PK:Cro:14]-LT674468	17	25	MiYLCCB-[VE:Mim:06]-DQ641710	CYVMB-[PK:Cro:14]-LT674467	NS	3.07E-03	NS	NS	9.14E-70	NS
LuYVMB-[CN:Lud:06]-AJ965541	929	1204	LuLDB-[IN:Sid:16]-KX513860	TYLCCNB-[CN:Mal:09]-AM980512	NS	NS	5.10E-03	4.16E-02	NS	3.80E-03
SiYVB-[VE:Sid:06]-DQ641711	1234	1304	ChYMBB-[NG:Sid:14]-KT454830	SiYVB-[CN:Sig;07]-AM230643	4.53E-03	NS	NS	NS	1.48E-04	NS
SiYVB-[CN:Sig;07]-KF499590	1090	1147	CoLCuGB-[SU:Dat:09]-DQ644564	SiYVB-[CN:Sig;07]-AM230643	NS	3.55E-03	NS	NS	5.17E-07	NS
CoLCuGB-[SU:Dat:09]-DQ644564	914	1175	CoYVMB-[IN:Dat:16]-KX513863	MaYVYuB-[CN:Lin:08]-FN806781	1.19E-02	2.54E-02	NS	NS	NS	NS
AgYLCB-[CM:Age:06]-FM164738	188	451	ToLCB-[CN:Age:05]-AM691551	CYVMB-[IN:Cro:07]-JQ354987	NS	NS	NS	NS	2.64E-05	1.35E-02
AgLCB-[IN:Cal:15]-KU376492	10	404	AgYLCB-[PK:Age:06]-AM698010	AgYLCB-[PK:Urt:13]-KT716083	3.89E-02	0.010474	1.62E-03	NS	2.52E-06	0.036281
ErYMB-[VE:Ere:06]-DQ641713	84	413	TYLCCNB-[CN:Mal:09]-AM980512	AgYVInB-[IN:Age:18]-AJ557441	NS	NS	5.57E-04	NS	3.70E-09	NS
SiYVB-[VE:Sid:06]-DQ641711	541	621	SiYVB-[CN:Sig;07]-AM230645	ToCSB-[CN:Age:06]-AM260736	NS	NS	1.70E-02	NS	1.67E-02	4.24E-02
CYVMB-[PK:Cro:14]-LT674484	337	1007	CYVMB-[PK:Cro:14]-LT674477	CYVMB-[PK:Cro:14]-LT674482	NS	NS	NS	NS	2.12E-09	4.96E-02

KEYWORDS

- **begomoviruses**
- **bipartite begomoviruses**
- **DNA satellites**
- **genomes**
- **monopartite begomoviruses**
- **phylogenetic analysis**
- **recombination analysis**
- **weed**
- **weed hosts**

REFERENCES

Adams, C. D., & Baker, H. G., (1962). Weeds of cultivation and grazing lands. In: Wills, J. B., (ed.), *Agriculture and Land Use in Ghana* (Vol. 4, p. 504). London.

Akbar, M., Javaid, A., Ahmed, E., Javed, T., & Clary, J., (2014). Holadysenterine, a natural herbicidal constituent from *Drechslera australiensis* for management of *Rumex dentatus*. *J. Agric. Food Chem.*, *62*(2), 368–372.

Al-Aqeel, H. A., Iqbal, Z., & Polston, J. E., (2018). Characterization of *Sida golden mottle virus* isolated from *Sida santaremensis* Monteiro in Florida. *Arch. Virol.*, *163*(10), 2907–2911.

Albu, S., Schneider, R. W., Price, P. P., & Doyle, V. P., (2016). *Cercospora* cf. *flagellaris* and *Cercospora* cf. *sigesbeckiae* are associated with *Cercospora* leaf blight and purple seed stain on soybean in North America. *Phytopathology*, *106*(11), 1376–1385.

Ambrozevicius, L. P., Calegario, R. F., Fontes, E. P., Carvalho, M. G., & Zerbini, F. M., (2002). Genetic diversity of *Begomovirus* infecting tomato and associated weeds in Southeastern Brazil. *Fitopatol. Bras.*, *27*(4), 372–377.

Annapoorani, S. G., (2013). An eco-friendly antimicrobial finish using *Datura inoxia* and *Leucas aspera* on cotton fabric. *Int. J. Sci. Res.*, *2*(4), 199–201.

Ansar, M., Agnihotri, A. K., Ranjan, T., Karn, M., Srinivasaraghavan, A., Kumar, R. R., & Bhagat, A. P., (2021). Nightshade (*Solanum nigrum*), an intermediate host between tomato and cucurbits of *Tomato leaf curl New Delhi virus*. *Phytopathol. Mediterr.*, *60*(3), 409–419.

Baker, H. G., (1965). *The Genetics of Colonizing Species* (pp. 147–168). Academic Press: New York and London.

Barbosa, J. C., Eckstein, B., Filho, A. B., Rezende, J. A., & Dallagnol, L. J., (2013). First report of *Tomato yellow spot virus* infecting *Leonurus sibiricus* in Brazil. *Plant Dis.*, *97*(2), 289–293.

Barros, E. M., Torres, J. B., Ruberson, J. R., & Oliveira, M. D., (2010). Development of *Spodoptera frugiperda* on different hosts and damage to reproductive structures in cotton. *Entomol. Exp. Appl.*, *137*(3), 237–245.

Briddon, R. W., Bull, S. E., Amin, I., Idris, A. M., Mansoor, S., Bedford, I. D., Dhawan, P., et al., (2003). Diversity of DNA β, a satellite molecule associated with some monopartite *Begomoviruses*. *Virol.*, *312*(1), 106–121.

Briddon, R. W., Bull, S. E., Amin, I., Mansoor, S., Bedford, I. D., Rishi, N., Siwatch, S. S., et al., (2004). Diversity of DNA 1: A satellite-like molecule associated with monopartite *Begomovirus*–DNA β complexes. *Virol.*, *324*(2), 462–474.

Brown, J. K., Fauquet, C. M., Briddon, R. W., & Zerbini, F. M., (2012). Family *Geminiviridae: Virus Taxonomy 9th Report of the International Committee on Taxonomy of Viruses* (pp. 351–373). Elsevier, Academic Press, London.

Chahwala, F. D., Singh, D., Kumar, R. V., Rathore, S., Yadav, B. K., Singh, B., & Singh, A. K., (2021). A novel monopartite *Begomovirus* and satellites associated with yellow mosaic disease of *Sida* spp. in India. *Arch. Virol.*, *166*(1), 299–302.

Colariccio, A., Eiras, M., Chaves, A. L. R., Bergmann, J. C., Zerbini, F. M., Harakava, R., & Chagas, C. M., (2007). *Tomato yellow vein streak virus*, a new *Begomovirus* on tomato from Brazil: Complete DNA-A sequence and some molecular and biological features. *J. Plant Pathol.*, *89*, 385–390.

Costa, A. S., & Betti, J. A., (1978). Summa *Phytopathol.*, *4*, 3.

Echemendía, A. L., Ramos, P. L., Peral, R., Fuentes, A., Pujol, M., & Gonzalez, G., (2003). First report of *Dicliptera yellow mottle virus* (DiYMoV) infecting *Dicliptera vahliana* in Cuba. *Plant Pathol.*, *52*(6), 787–787.

Faria, J. C., & Maxwell, D. P., (1999). Variability in geminivirus isolates associated with *Phaseolus* spp. in Brazil. *Phytopathology*, *89*(3), 262–268.

Fernandes, F. R., Albuquerque, L. C., De Oliveira, C. L., Cruz, A. R., Da Rocha, W. B., Pereira, T. G., & Inoue-Nagata, A. K., (2011). Molecular and biological characterization of a new Brazilian *begomovirus*, *Euphorbia yellow mosaic virus* (EuYMV), infecting *Euphorbia heterophylla* plants. *Arch. Virol.*, *156*(11), 2063–2069.

Fernandes-Acioli, N. A., Pereira-Carvalho, R. C., Fontenele, R. S., Lacorte, C., Ribeiro, S. G., Fonseca, M. E., & Boiteux, L. S., (2011). First report of *Sida micrantha mosaic virus* in *Phaseolus vulgaris* in Brazil. *Plant Dis.*, *95*(9), 1196.

Fiallo-Olive, E., Martínez-Zubiaur, Y., Moriones, E., & Navas-Castillo, J., (2021). A novel class of DNA satellites associated with new world *Begomoviruses*. *Virology*, *26*(1), 1–6.

Fontenele, R. S., Ribeiro, G. C., Lamas, N. S., Ribeiro, S. G., Costa, A. F., Boiteux, L. S., & Fonseca, M. E., (2018). First report of *Sida micrantha mosaic virus* infecting *Oxalis* species in Brazil. *Plant Dis.*, *102*(9), 1862.

Garcia-Andres, S., Accotto, G. P., Navas-Castillo, J., & Moriones, E., (2007). Founder effect, plant host, and recombination shape the emergent population of *Begomoviruses* that cause the tomato yellow leaf curl disease in the Mediterranean basin. *Virol.*, *15*, 302–312.

Gnanasekaran, P., & Chakraborty, S., (2018). Biology of viral satellites and their role in pathogenesis. *Curr. Opinion Virol.*, *33*, 96–105.

Ha, C., Coombs, S., Revill, P., Harding, R., Vu, M., & Dale, J., (2008). Molecular characterization of *Begomoviruses* and DNA satellites from Vietnam: Additional evidence that the new world geminiviruses were present in the old world prior to continental separation. *J. Gen. Virol.*, *89*(1), 312–326.

Haider, M. S., Liu, S., Ahmed, W., Evans, A. A., & Markham, P. G., (2005). Detection and relationship among *Begomoviruses* from five different host plants, based on ELISA and Western Blot analysis. *Pakistan J. Zool.*, *37*(4), 69.

Haider, M. S., Tahir, M., Latif, S., & Briddon, R. W., (2006). First report of *Tomato leaf curl New Delhi virus* infecting *Eclipta prostrata* in Pakistan. *Plant Pathol., 55*(2), 232–239.

Heinrich, M. P., Jenkinson, M., Bhushan, M., Matin, T., Gleeson, F. V., Brady, M., & Schnabel, J. A., (2012). Modality independent neighborhood descriptor for multi-modal deformable registration. *Med. Image Anal., 16*(7), 1423–1435.

Hiremath, S., Kumar, H. D., Nandan, M., Mantesh, M., Shankarappa, K. S., Venkataravanappa, V., Basha, C. R., & Reddy, C. N., (2021). *In silico* docking analysis revealed the potential of phytochemicals present in *Phyllanthus amarus* and *Andrographis paniculata*, used in Ayurveda medicine in inhibiting SARS-CoV-2. *3 Biotech., 11*(2), 1–18.

Holm, L. G., Plucknett, D. L., Pancho, J. V., & Herberger, J. P., (1997). *The World's Worst Weeds: Distribution and Biology* (Vol. 20, pp. 424–436). University Press of Hawaii.

Holzner, W., (1982). Concepts, categories, and characteristics of weeds: In: *Biology and Ecology of Weeds* (Vol. 2, pp. 3–20). Springer: Dordrecht.

Huang, J. F., & Zhou, X. P., (2006). First report of *Papaya leaf curl China virus* infecting *Corchoropsis timentosa* in China. *Plant Pathol., 55*(2), 1–8.

Hussain, K., Mazhar, H., Shahid, M., & Rob, W., B., (2011). Complete nucleotide sequence of a *begomovirus* and associated betasatellite infecting croton (*Croton bonplandianus)* in Pakistan. *Arch. Virol., 156*(6), 1101–1105.

Jauzein, P., (1991). *Eclipta prostrata* (L.) a weed of rice fields in the Camargue. *Monde Des Plantes., 86*(4), 15, 16.

Jiang, T., & Zhou, X., (2005). Molecular characterization of a distinct *Begomovirus* species and its associated satellite DNA isolated from *Malvastrum coromandelianum* in China. *Virus Genes., 31*(1), 43–48.

Jovel, J., Reski, G., Rothenstein, D., Ringel, M., Frischmuth, T., & Jeske, H., (2004). *Sida micrantha* mosaic is associated with a complex infection of *Begomoviruses* different from *Abutilon mosaic virus*. *Arch, Virol., 149*(4), 829–841.

Juarez, M., Rabadan, M. P., Martinez, L. D., Tayahi, M., Grande-Perez, A., & Gomez, P., (2019). Natural hosts and genetic diversity of the emerging *Tomato leaf curl New Delhi virus* in Spain. *Front. Microbiol., 10*, 140.

Khan, A., & Samad, A., (2014). A new isolate of *Eclipta yellow vein virus* along with a betasatellite associated with yellow vein leaf curl disease of *Andrographis paniculata* in India. *Plant Dis., 98*(5), 698–698.

Khatun, M. F., Hwang, H. S., Shim, J. K., Kil, E. J., Lee, S., & Lee, K. Y., (2020). Identification of *Begomoviruses* from different cryptic species of *Bemisia tabaci* in Bangladesh. *Microb. Pathog., 42*, 104069.

Kumar, S., Srivastava, A., Jaidi, M., Chauhan, P. S., & Raj, S. K., (2016). Molecular characterization of a *Begomovirus*, α-satellite, and β-satellite associated with leaf curl disease of *Parthenium hysterophorus* in India. *Plant Dis., 100*(11), 2299–2305.

Kumar, S., Stecher, G., Li, M., Knyaz, C., & Tamura, K., (2018). MEGA X: Molecular evolutionary genetics analysis across computing platforms. *Mol. Biol. Evol., 35*(2), 1547–1549.

Kwon, S. J., Cho, I. S., Yoon, J. Y., & Chung, B. N., (2018). Incidence and occurrence pattern of viruses on peppers growing in fields in Korea. *Res. Plant Dis., 24*(1), 66–74.

Kyallo, M., Sseruwagi, P., Skilton, R. A., Ochwo-Ssemakula, M., Wasswa, P., & Ndunguru, J., (2017). *Deinbollia mosaic virus*: A novel *Begomovirus* infecting the sapindaceous weed *Deinbollia borbonica* in Kenya and Tanzania. *Arch. Virol., 162*(5), 1393–1396.

Lefeuvre, P., Lett, J. M., Reynaud, B., & Martin, D. P., (2007). Avoidance of protein fold disruption in natural virus recombinants. *PLoS Pathog.*, *3*(11), 181.

Li, P., Jing, C., Ren, H., Jia, Z., Ghanem, H., Wu, G., Li, M., & Qing, L., (2020). Analysis of pathogenicity and virulence factors of *Ageratum leaf curl Sichuan virus*. *Front. Plant Sci.*, *11*, 527–530.

Liu, P., Xie, Y., & Zhou, X. P., (2009). *Malvastrum coromandelianum* is an alternative host of *Tomato yellow leaf curl China virus*. *Plant Pathol.*, *58*(2), 403.

Lotrakul, P., Valverde, R. A., & Landry, A. D., (2000). Biological and molecular properties of a *begomovirus* from *Dicliptera sexangularis*. *Phytopathology*, *90*(7), 723–729.

Lozano, G., Trenado, H. P., Fiallo-Olivé, E., Chirinos, D., Geraud-Pouey, F., Briddon, R. W., & Navas-Castillo, J., (2016). Characterization of non-coding DNA satellites associated with sweepoviruses (genus *Begomovirus*, *Geminiviridae*) definition of a distinct class of *Begomovirus*-associated satellites. *Front. Microbiol.*, *17*(7), 162.

Martin, D. P., Varsani, A., Roumagnac, P., Botha, G., Maslamoney, S., Schwab, T., Kelz, Z., Kumar, V., & Murrell, B., (2021). RDP5: A computer program for analyzing recombination in, and removing signals of recombination from, nucleotide sequence datasets. *Virus Evol.*, *7*(1), 087.

Marwal, A., Sahu, A., Prajapat, R., Choudhary, D. K., & Gaur, R. K., (2012). First report of association of a *Begomovirus* with the leaf curl disease of a common weed, *Datura inoxia*. *Indian J. Virol.*, *23*(1), 83, 84.

Melgarejo, T. A., Kon, T., Rojas, M. R., Paz-Carrasco, L., Zerbini, F. M., & Gilbertson, R. L., (2013). Characterization of a new world monopartite *Begomovirus* causing leaf curl disease of tomato in Ecuador and Peru reveals a new direction in geminivirus evolution. *J. Virol.*, *87*(10), 5397–5413.

Mohammed, H. S., El Siddig, M. A., El Hussein, A. A., Navas-Castillo, J., & Fiallo-Olive, E., (2017). First report of *Datura innoxia* as a natural host of *Watermelon chlorotic stunt virus* in Sudan. *Plant Dis.*, *101*(7), 1334–1334.

Mondal, D., Mandal, S., Shil, S., Sahana, N., Pandit, G. K., & Choudhury, A., (2019). Genome wide molecular evolution analysis of *Begomoviruses* reveals unique diversification pattern in coat protein gene of old world and new world viruses. *Virus Dis.*, *30*(1), 74–83.

Mubin, M., Akhtar, S., Amin, I., Briddon, R. W., & Mansoor, S., (2012). *Xanthium strumarium*: A weed host of components of *Begomovirus*–betasatellite complexes affecting crops. *Virus Genes.*, *44*(1), 112–119.

Mubin, M., Briddon, R. W., & Mansoor, S., (2009). Diverse and recombinant DNA betasatellites are associated with a *Begomovirus* disease complex of *Digera arvensis*, a weed host. *Virus Res.*, *142*(2), 208–212.

Muhire, B. M., Varsani, A., & Martin, D. P., (2014). SDT: A virus classification tool based on pairwise sequence alignment and identity calculation. *PLoS One*, *9*(9), e108277.

Nesom, G. L., (2006). *Grindelia: En Flora of North America Editorial Committee* (Vol. 20, pp. 424–436). Flora of North America North of Mexico.

Ogawa, T., Sharma, P., & Ikegami, M., (2008). The *Begomoviruses Honeysuckle yellow vein mosaic virus* and *Tobacco leaf curl Japan virus* with DNAβ satellites cause yellow dwarf disease of tomato. *Virus Res.*, *137*(2), 235–244.

Paprotka, T., Boiteux, L. S., Fonseca, M. E., Resende, R. O., Jeske, H., Faria, J. C., & Ribeiro, S. G., (2010). Genomic diversity of sweet potato geminiviruses in a Brazilian germplasm bank. *Virus Res.*, *149*(2), 224–233.

Passos, L. S., Teixeira, J. W., Teixeira, K. J., Xavier, C. A., Zerbini, F. M., Araújo, A. S., & Beserra, J. E., (2017). Two new *Begomoviruses* that infect non-cultivated malvaceae in Brazil. *Arch. Virol.*, *162*(6), 1795–1797.

Prajapat, R., Marwal, A., & Gaur, R. K., (2013). Molecular and *in silico* characterization of two coat protein of weed associated *Begomovirus* isolates. *Sci. Int.*, *1*(12), 408–414.

Rahman, S. U., Zubair, M., Ahmed, N., Khan, M. Z., Raza, G., Amin, I., & Mansoor, S., (2019). Identification of *Malvastrum yellow vein Lahore virus* a proposed new species of *Begomovirus* in association with cotton leaf curl Multan betasatellite infecting green bean (*Phaseolus vulgaris*) in Pakistan. *Australas. Plant Dis. Notes.*, *14*(1), 1–3.

Rao, A. N., & Moody, K., (1990). Weed seed contamination in rice seed. *Seed Sci. Technol.*, *18*(1), 139–146.

Rodríguez-Negrete, E. A., Morales-Aguilar, J. J., Domínguez-Duran, G., Torres-Devora, G., Camacho-Beltrán, E., Leyva-López, N. E., Voloudakis, A. E., et al., (2019). High-throughput sequencing reveals differential *Begomovirus* species diversity in non-cultivated plants in Northern-Pacific Mexico. *Viruses*, *11*(7), 594.

Sattar, M. N., Qurashi, F., Iqbal, Z., & Saleem, H. M., (2017). Molecular characterization of *Hollyhock leaf curl virus* and associated DNA-satellites infecting *Malva parviflora* in Pakistan. *Canadian J. Plant Pathol.*, *39*(7), 229–234.

Saunders, K., Bedford, I. D., Yahara, T., & Stanley, J., (2003). The earliest recorded plant virus disease. *Nature*, *422*(6934), 831–831.

Seal, S. E., VandenBosch, F., & Jeger, M. J., (2006). Factors influencing *Begomovirus* evolution and their increasing global significance: Implications for sustainable control. *Crit. Rev. Plant Sci.*, *25*(1), 23–46.

Silva, R. L., & Krumholz, M. R., (2011). Stochastic star formation and a (nearly) uniform stellar initial mass function. *Astrophys. J. Lett.*, *741*(2), 26.

Sobrinho, R. R., Xavier, C. A., De Barros, Pereira, H. M., De Andrade, Assunçao, I. P., Mizubuti, E. S., et al., (2014). Contrasting genetic structure between two *Begomoviruses* infecting the same leguminous hosts. *J. Gen. Virol.*, *95*(11), 2540–2452.

Sravya, G., Prabhudas, S. K., Sharma, S. K., Mandal, B., Roy, A., & Swapnageetanjali, A., (2022). *Croton Yellow Vein Leaf Curl Virus: A Novel Begomovirus Infecting Croton Bonplandianum in India.* Preprint. doi: 10.21203/rs.3.rs-1645288/v1.

Srivastava, A., Kumar, S., & Raj, S. K., (2015). Molecular characterization of a *Begomovirus* and betasatellite causing yellow vein net disease of *Ageratum houstonianum*. *Plant Dis.*, *99*(5), 627–631.

Stewart, C. S., Kon, T., Gilbertson, R. L., & Roye, M. E., (2011). First report of the complete sequence of *Sida golden yellow vein virus* from Jamaica. *Arch. Virol.*, *156*(8), 1481–1485.

Stewart, C., Kon, T., Rojas, M., Graham, A., Martin, D., Gilbertson, R., & Roye, M., (2014). The molecular characterization of a sida-infecting *Begomovirus* from Jamaica. *Arch. Virol.*, *159*(2), 375–378.

Tahir, M., Amin, I., Haider, M. S., Mansoor, S., & Briddon, R. W., (2015). *Ageratum enation virus*-A *Begomovirus* of weeds with the potential to infect crops. *Viruses*, *7*(2), 647–665.

Tanaka, Y., & Nguyen, V. K., (2007). *Edible Wild Plants of Vietnam*. Orchid Press.

Tang, Y., Li, Z., She, X., Yu, L., Lan, G., Lv, L., Brown, J. K., & He, Z., (2021). Complete genome sequence of a previously undescribed monopartite *Begomovirus* and betasatellite infecting *Malvastrum coromandelianum* in Cambodia. *Arch. Virol.*, *166*(6), 1789–1793.

Torres-Herrera, S. I., Romero-Osorio, A., Moreno-Valenzuela, O., Pastor-Palacios, G., Cardenas-Conejo, Y., Ramírez-Prado, J. H., Riego-Ruiz, L., et al., (2019). A lineage of

Begomoviruses encode Rep and AC4 proteins of enigmatic ancestry: Hints on the evolution of geminiviruses in the New World. *Viruses*, *11*(7), 644.

Tsai, W. S., Shih, S. L., Kenyon, L., Green, S. K., & Jan, F. J., (2011). Temporal distribution and pathogenicity of the predominant tomato-infecting *Begomoviruses* in Taiwan. *Plant Pathol.*, *60*(4), 787–799.

Walker, P. J., Siddell, S. G., Lefkowitz, E. J., Mushegian, A. R., Adriaenssens, E. M., Alfenas-Zerbini, P., & Zerbini, F. M., (2021). Changes to virus taxonomy and to the international code of virus classification and nomenclature ratified by the international committee on taxonomy of viruses. *Arch. Virol.*, *166*(9), 2633–2648.

Wilson, A. K., (1981). *Euphorbia heterophylla*: A review of distribution, importance, and control. *Int. J. Pest Manag.*, *27*(1), 32–38.

Wisler, G. C., & Norris, R. F., (2005). Interactions between weeds and cultivated plants as related to management of plant pathogens. *Weed Sci.*, *53*(6), 914–917.

Wu, J. B., & Zhou, X. P., (2007). *Siegesbeckia yellow vein virus* is a distinct *Begomovirus* associated with a satellite DNA molecule. *Arch. Virol.*, *152*(4), 791–796.

Wu, J. B., Cai, J. H., & Zhou, X. P., (2007). A new *Begomovirus* associated with yellow vein disease of *Siegesbeckia glabrescens*. *Plant Pathol.*, *56*(2), 223–226.

Wyant, P. S., Gotthardt, D., Schäfer, B., Krenz, B., & Jeske, H., (2011). The genomes of four novel *Begomoviruses* and a new *Sida micrantha mosaic virus* strain from Bolivian weeds. *Arch. Virol.*, *156*(2), 347–352.

Xiong, Q., Fan, S., Wu, J., & Zhou, X., (2005). *Ageratum yellow vein China virus* is a distinct *Begomovirus* species associated with a DNAβ molecule. *Phytopathology*, *97*(4), 405–411.

Yang, C. X., Luo, J. S., Zheng, L. M., Wu, Z. J., & Xie, L. H., (2012). First report of the occurrence of *Emilia yellow vein virus* in *Crassocephalum crepidioides* in China. *J. Plant Pathol.*, *94*(4), 105.

Zaidi, S. S. E. A., Martin, D. P., Amin, I., Farooq, M., & Mansoor, S., (2016). *Tomato leaf curl New Delhi virus*: A widespread bipartite *Begomovirus* in the territory of monopartite *Begomoviruses*. *Mol Plant Pathol.*, *18*, 901–911.

Zhang, J., Jia, S. P., Yang, C. X., Liu, Z., & Wu, Z. J., (2015). Detection and molecular characterization of three *Begomoviruses* associated with yellow vein disease of *Eclipta Prostrata* in Fujian, China. *J. Plant Pathol.*, *1*(3), 161–165.

CHAPTER 5

Diversity of Arthropod Vectors of Plant Viruses

LALIT MAHATMA[1] and ABHISHEK SHUKLA[2]

[1]*Department of Plant Pathology, N.M. College of Agriculture, Navsari Agricultural University, Navsari, Gujarat, India*

[2]*Department of Entomology, N.M. College of Agriculture, Navsari Agricultural University, Navsari, Gujarat, India*

ABSTRACT

The adaptation of living organisms to various environments, owing to their biological and ecological competence, enhances their diversity and stability. Highly diverse organisms are capable of efficiently surviving in the ecosystem. Viruses are highly diverse and have also chosen diverse vectors, particularly arthropods such as members of the class Insecta and Arachnida, for their own horizontal movement. The high diversity of arthropod vectors, along with other factors such as their specificity, association of symbionts, and changes in vector behavior, distinguishes them as the most potential vectors of plant viruses for better existence in the ecological niche. This chapter discusses the diversity of arthropod vectors and these factors.

5.1 INTRODUCTION

Our understanding of viruses is only 140 years old, which originated from the rejection of the belief that bacteria are the smallest disease-causing organisms. In 1882, Adolf Mayer from the Netherlands observed mosaic symptoms on tobacco leaves, which he named tobacco mosaic disease. Through experiments, he demonstrated that the disease could be transmitted

New and Emerging Plant Viruses: The Threat to Food Security.
Alireza Golnaraghi and Rajarshi Kumar Gaur (Eds.)

to healthy tobacco plants even when the sap was filtered through capillary glass tubes. This proposal was further supported by Dmitri Ivanovsky in 1892 and Beijerinck in 1898, who used the term "*contagium vivum fluidum*" (contagious living liquid) to describe the virus, in contrast to "*contagium vivum fixum*" (contagious solid particle). Loeffler and Frosch also made a similar observation in 1898 regarding foot and mouth disease in cattle, marking the first description of an animal viral disease.

In the study of microorganisms, animal science has taken the lead in unraveling complexities, while plant science has been at the forefront in understanding viruses. The first instance of vector transmission was observed in *Rice stunt virus* by Takata (1885, 1886). He discovered that the leafhopper *Deltocephalus* (Inazuma) *dorsalis* Motsch could transmit the *Rice stunt virus*. Subsequently, Reed (1902) observed the transmission of the yellow-fever virus by mosquitoes. Seed transmission allows the virus to persist from one season to another and over long distances. Vector transmission, on the other hand, facilitates both primary and secondary spread of the disease.

In vector-borne viruses, if a vector feeds on a virus-infected plant and subsequently feeds on a susceptible host plant, it transmits the disease, which is evident through symptoms. Species richness and vector stability in the ecosystem provide additional benefits to the virus for its existence. The occurrence and severity of plant viral diseases depend on four factors. Three of them, namely the availability of the susceptible host, virulent pathogen, and congenial environmental conditions, form the disease triangle. The fourth additional factor is the availability of an efficient vector, which transforms it into a disease quadrate instead of a disease triangle. If the viruliferous vector feeds on a non-host plant, it may release the virus. However, due to the non-establishment of host-pathogen interaction, the disease is not exhibited.

Therefore, the biological and ecological competence of the vector determines the pace of virus disease spread and disease epidemics. Vectors that have high fecundity, a wide host range, fast and long-distance movement, persistent transmission, and easy adaptability in diverse environmental conditions are categorized as the most efficient. Depending on the time of acquisition and release of the virus by the vector (non-persistent, persistent, semi-persistent), the movement of the virus in the vector (circulatory), the multiplication of the virus in the vector (propagative), and the movement of the virus from generation to generation (transovarially), different viruses have been categorized. Non-persistent virus transmission is also marked by very short acquisition and inoculation times, requiring just seconds to minutes, and is primarily mediated by aphids. Crucially, retention for these viruses is limited to the external surfaces of the aphid, occurring on the chitin lining of the aphid's stylets within the food and salivary canal, which are completely lost upon molting (Watson & Roberts, 1939). Vectors of plant viruses are taxonomically highly diverse and can be found among arthropods, nematodes, fungi, and plasmodiophorids (Hull, 2002). Out of the nearly 550 vector-transmitted viruses recorded so far, 88% of viruses are transmitted by arthropods (classes Insecta and Acari), 7% of viruses by nematodes. Whereas merely 5% of viruses have been vectored by fungi and plasmodiophorids. The adaptability of the virus to different vectors and the adaption of any organism to particular environments, owing to their biological and ecological competence, enhance diversity and maintain a high ecological niche. The diversity of both the vector and the virus has led the virus to evolve as one of the most intriguing pathogens of plants. Highly diverse organisms are capable of efficiently surviving in the ecosystem. Arthropods, being the most diverse and important vector group, are thoroughly discussed in succeeding sections.

5.2 ARTHROPODS VECTORS OF PLANT VIRUSES

Most of the viruses are transmitted by arthropod vectors. Arthropods have jointed appendages and a chitinous exoskeleton. Two subphyla, Hexapoda (class Insecta) and Chelicerata (class Arachnida), contain members that transmit plant viruses. Arthropoda is the largest phylum, with the class Insect alone having more than 5.5 million species, of which only 1 million insect species have been named (Stork, 2018). Among the arthropod virus-transmitting members, aphids transmit 55% of the viruses, while leafhoppers and beetles transmit 11% each, and whiteflies transmit 7%. The remaining 2% of viruses are transmitted by thrips, mites, mirids, or mealybugs. Among these, thrips and mites are the main insects compared to mirids and mealybugs (Astier et al., 2001; Link & Fuchs, 2005). Arthropod vectors of plant viruses belong to the class Insecta (six orders) and Arachnida (one order).

5.2.1 CLASS INSECTA AND ARACHNIDA AS VECTOR OF PLANT VIRUSES

Class Insecta has been further classified into two subclasses, namely Apterygota and Pterygota, based on the presence and absence of wings. Subclass Apterygota (=Ametabola) consists of primitive wingless insects that are slow in movement, do not undergo metamorphosis, and are not known to transmit plant viruses. On the other hand, subclass Pterygota (=Metabola) includes winged or secondarily wingless insects that are fast in movement, have prognathous or hypognathous heads, and possess plant virus vectors. This subclass Pterygota is divided into two superorders or divisions, namely *Exopterygota* and *Endopterygota.* Members of Superorder *Exopterygota* have wings that develop externally on the young nymphs and do not have a pupal stage. In addition to the wings, nymphs also possess typical compound eyes. The immature insects have habits and habitats similar to the adults. Superorder *Exopterygota* is of great importance in terms of viral disease transmission, including four orders that transmit viruses: Hemiptera (bugs, hoppers, cicadas, whiteflies, scale insects), Thysanoptera (thrips), Orthoptera (crickets, mole crickets, locusts, grasshoppers), and Dermaptera (earwigs). In contrast to the members of superorder *Exopterygota*, members of superorder *Endopterygota* have wings that develop inside the pupal case. The immature stage of these insects is the larva, which is entirely different in appearance and habits. They have simple eyes. Two orders of superorder *Endopterygota,* namely Coleoptera (beetles and weevils) and Diptera (true flies), possess insects that transmit plant viruses.

Two orders (number of species in parentheses): Hemiptera (300) and Thysanoptera (6) with piercing-sucking mouthparts pose the majority of vectors and transmit the maximum number of viruses. Among the four orders of chewing insects, Coleoptera (30), Orthoptera (10), Diptera (2), and Dermaptera (1) possess virus-transmitting insects. Mites belonging to the subphylum Chelicerata and Class Arachnida, order Acari (or Acarina), and three families Eriophyidae (worm mites), Tenuipalpidae (false spider mites), and Tetranychidae (spider mites) are known as vectors of viral diseases.

5.2.1.1 ORDER HEMIPTERA

Insects belonging to the order Hemiptera, commonly known as true bugs, transmit plant viruses. As the name indicates, members of the order Hemiptera (*hemi* 'half' + *pteron* 'wing') have the first half pair of wings extremely hard, while the remaining one and a half pair, i.e., the rest of the first half pair and the second pair of wings, are membranous, which helps them in protection as well as flying. There are three major suborders of Hemiptera, namely Auchenorrhyncha (Hoppers), Sternorrhyncha (Aphids, Whiteflies, Scales), and Heteroptera (True Bugs). The suborders Auchenorrhyncha and Sternorrhyncha include virus-transmitting insect vectors. Apart from these, there are two superfamilies, namely Lygaeoidea (family *Piesmatidae*) and

Miroidea (family Miridae), which possess virus-transmitting species of insects. These insects are equipped with piercing and sucking type of mouthparts. The concentric stylet consists of modified mandibles and maxillary laciniae. The food and salivary canal are made up of modified mandibles and maxillary laciniae, which is further covered by a multi-segmented sheet-like labium. The maxillary and labial palpi are always absent. This modification makes them efficient suckers of plant sap (Weber, 1930; Hennig, 1969, 1981; Cobben, 1978; Kristensen, 1991; Dimitri, 2008).

5.2.1.1.1 Suborder Auchenorrhyncha (Hoppers)

Members of the suborder Auchenorrhyncha (*auchen* 'neck' + *rhyncos* 'nose or snout') have mouthparts (rostrum) that arise from the back of the head near the neck region. Members of this suborder possess a composite sound-producing tymbal apparatus, hair-like antennal flagellum, and a wing-coupling apparatus consisting of a long, downturned fold on the forewing and a short, upturned lobe on the hind wing (Dietrich, 2009). Auchenorrhyncha includes insects that feed on phloem, xylem, and mesophyll. The vascular feeder insects are most strongly associated with pathogen transmission (Mitchell, 2004). The suborder Auchenorrhyncha consists of three families: *Membracidae* (treehoppers), *Cicadellidae* (leafhoppers), and *Fulgoroidea* (planthoppers), which contain different species of insects that transmit plant viruses. Treehoppers typically have a vertical face and a notably enlarged pronotum (thorax), which expands from the head to the entire thorax, forming a hood-like covering. The treehopper *Micrutalis malleifera* transmits the *tomato pseudo-curly top virus* (TPCTV), a single-stranded DNA (ssDNA) virus belonging to the virus family *Geminiviridae.* TPCTV is the only *Geminivirus* known to be transmitted by a treehopper in a persistent circulative manner (Simons & Coe, 1958). Both nymphs, which retain virus transmissibility even after molting, and adults are capable of transmitting the virus. The virus only infects dicot plants, namely tomato (*Lycopersicon esculentum*) and bean (*Phaseolus vulgaris*) (Andrew et al., 2008). It has only been identified in the southeastern part of the USA (Briddon et al., 1996). The treehopper *Spissistilus festinus* has been found to artificially acquire and transmit *Grapevine red blotch associated virus* (Bahder et al., 2016).

Leafhoppers transmit plant viruses in various ways, including semi-persistent, persistent, circulative, propagative, and transovarial transmission. Leafhopper species can also transmit bacteria, phytoplasma (Class Mollicutes), and other fastidious organisms. Viruses such as *rice tungro virus* (RTV) and *maize chlorotic dwarf virus* (MCDV) are transmitted by their leafhopper vectors in a semi-persistent manner (Ling, 1966; Nault et al., 1973) without a detectable latent period. Species in the genera *Becurtovirus, Curtovirus, Mastrevirus*, and *Turncurtovirus* are transmitted by leafhoppers. Rice tungro disease is caused by the interaction of two different viruses, namely *rice tungro spherical virus* (RTSV) with a single-stranded RNA genome, and *rice tungro bacilliform virus* (RTBV) with a double-stranded DNA genome. RTBV develops the primary disease symptoms and is transmitted by the leafhopper species *Nephotettix virescens* and *N. nigropictus* (Hibino & Kabauatan, 1987). However, for the transmission of RTBV, the presence of RTSV is required in the vector. However, RTSV can be transmitted independently (Osmat & Tim, 2002). *Wound tumor virus* (WTV) and *potato yellow dwarf virus* (PYDF) are transovarially transmitted (Sinha & Shelley, 1965; Black, 1953). With a few exceptions, such as sugar beet curly top virus (SBCTV) and PYDF, none of these leafhopper-transmitted viruses are mechanically transmitted.

Whitefly-transmitted *Begomovirus, African cassava mosaic virus*, became transmissible by a leafhopper when the CP gene was replaced by that of the leafhopper-borne beet curly top virus (BCTV) (Briddon et al., 1990). Leafhoppers can acquire the virus within 30 minutes of acquisition feeding. They have a defined latent period during which the vector cannot transmit the pathogen after acquisition. The latent period varies in the vector-virus combination, ranging from a few hours to a few weeks. Leafhoppers that transmit circulative/propagative viruses remain viruliferous throughout their life (Harris, 1979). Circulatory viruses enter the hemocoel and are subsequently transmitted via the salivary glands. An inoculation access period of more than 30 minutes is required for efficient transmission (Klein, 2001; Rabinowitch & Levy, 2001). Members of the *Marafivirus* genus (family *Tymoviridae*) are transmitted in a persistent-propagative manner by leafhopper vectors. *Marafivirus, Maize rayado fino virus* (MRFV), infects corn and teosintes exclusively. The maize leafhopper *Dalbulus maidis* efficiently transmits MRFV in a persistent-propagative manner (Madriz-Ordeneña et al., 2000; Nault et al., 1980). The minimum acquisition access period is 6 hours. After this, the vector requires a latent period of 8–37 days for transmission. Once acquired, the virus is retained in the vector for 20 days (Gámez, 2008). Both nymphs and adults are capable of acquiring and transmitting the virus, although nymphs are more efficient (Nault et al., 1980).

The genus *Phytoreovirus* comprises three accepted species (Lefkowitz et al., 2017). A true bug of the family Cicadellidae, the green rice leafhopper,

Nephotettix cincticeps, transmits *rice dwarf virus* (RDV) in a persistent-propagative manner. Virus accumulation has been observed in the epithelial cells of the filter chamber of the vector after 2 days. Subsequently, it reaches the adjacent epithelial cells of the anterior midgut and esophagus in 4 days, and finally the posterior organs by 12 days (Chen et al., 2011).

The genus *Tenuivirus,* family *Phenuiviridae*, order *Bunyavirale*s, infects cereals such as wheat, maize, rice, sorghum, oat, and barley. The virus is transmitted by planthoppers in a persistent-propagative mode. Vectors can acquire *Tenuivirus* within 15 minutes to four hours of feeding. A latent period of 4 to 31 days has been observed. However, subsequently, the planthopper vector can transmit a *Tenuivirus* for the rest of its life. As short as 30 seconds of inoculation feeding is sufficient to transmit the virus (Hull, 2014a, b). The eggs and sperm of the *Tenuivirus*-infected leafhopper do have the virus and transmit it from parents to their offspring in successive generations (Ramirez & Haenni, 1994). *Rice grassy stunt virus* (RGSV) is an exception where the virus accumulates in the midgut for one day after access to diseased plants. Subsequently, on the fourth day, the virus spreads from the epithelium of the midgut and by the sixth day, it spreads into the salivary glands.

A nucleorhabdovirus, *maize mosaic virus* (MMV), is transmitted by the planthopper vector *Peregrinus maidis*. After allowing the planthopper to acquire the virus from infected plants at different time intervals over a period of a week, researchers detected the virus disseminating along a path from the midgut and anterior diverticulum to the esophagus, nerves, nerve ganglia, and visceral muscles. After 3 weeks, MMV was also found in the hemocytes, tracheae, compound eyes, and salivary glands of its insect vector (Ammar & Hogenhout, 2008).

The *Fijivirus, Southern rice black-streaked dwarf virus* (SRBSDV), and the *Oryzavirus, rice ragged stunt virus* (RRSV), are transmitted by the rice white-backed planthopper, *Sogatella furcifera*, and the brown planthopper, *N. lugens*, respectively, in a persistent-propagative manner. Both SRBSDV and RRSV are restricted to the phloem of infected rice plants (Hibino et al., 2018; Li et al., 2017; Xu et al., 2014).

5.2.1.1.2 Suborder Sternorrhyncha (Aphids, Whiteflies, Mealybug, Scales)

Sternorrhyncha have the mouthparts positioned towards the rear of the head. Most viruses are transmitted by insects belonging to the Sternorrhyncha suborder. The Sternorrhyncha suborder includes the *Aphididae* family (aphids), *Aleyroididae* family (whitefly), *Pseudococcidae* family (mealybug), and *Coccidae* family (soft scale).

1. ***Family Aphididae (Aphids):*** Among all insects, aphids belonging to the Infra-order *Aphidomorpha,* super-family *Aphidoidea*, and family *Aphididae* are known to transmit the maximum number of viruses (Blackman & Eastop, 2000; James & Perry, 2004). Aphid vectors also belong to nine other subfamilies, although they are relatively few in number. The choice of host plant and feeding habits determine their transmission specificity (Harris & Maramorosch, 1977). The stylet of the aphid can penetrate a single sieve tube or epidermal cell without causing damage, allowing it to ingest food from the host plant. During this process, virus particles are released into the host cell, which then heals (Nault, 1997).

   Aphids have personalized mouthparts with thin and flexible stylets, enabling them to feed on different plant tissues without causing damage. Aphids feed on plants in two steps. In the first step, they sample the cell content during brief intracellular probes of the epidermal and/or mesophyll cells. The second step involves the introduction of their stylets deep into the phloem, under the epidermis. During feeding, aphids secrete two types of saliva (Harmel et al., 2008). Gel-forming saliva, secreted by the basilar gland, forms a continuous layer around the needle as it penetrates deeper into the phloem. Watery saliva, secreted by accessory salivary glands, is produced during superficial probing and the feeding phase of the phloem. The movement of sap during the feeding on plant sap is bidirectional (Brault et al., 2010).

   The characteristic features responsible for the evolution of aphids as potent vectors of plant viruses are: (i) Their wide host range and polyphagous nature (e.g., *Myzus persicae*), which is important for the dissemination of viruses that infect a wide range of plant species; (ii) Their ability to undergo parthenogenetic reproduction, facilitating rapid multiplication; and (iii) Their feeding through small stylets, which penetrate plant cells without causing lethal injury.

   Aphids (*Hemiptera: Aphididae*) transmit several genera of plant viruses. The family *Potyviridae* includes the genera *Potyvirus* and *Macluravirus*, while the family *Bromoviridae* includes the genera *Alfamovirus* and *Cucumovirus*. The genus *Carlavirus* belongs to the family *Betaflexiviridae*, and the genus *Fabavirus* belongs to the family *Secoviridae.* These viruses are transmitted non-persistently by aphids.

The *Alfamovirus* genus is represented by the species alfalfa mosaic virus (AMV), which has a wide host range, infecting more than 400 hosts from 51 plant families (Jaspars & Bos, 1980). This virus is vectored by several species of aphids. The genus *Cucumovirus*, with its type species *cucumber mosaic virus* (CMV), is another aphid-transmitted virus that has a wide host range, infecting around a 1,000 species. It is transmitted by approximately 80 species of aphids (Bujarski et al., 2012; Gallitelli, 2000; Palukaitis & Garc´ıa-Arenal, 2003).

Genus *Closterovirus*, *citrus tristeza virus* (CTV), is another important virus transmitted by several species of aphids, including the brown citrus aphid *Toxoptera citricida*. The virus is widely distributed in many citrus growing areas (Yokomi et al., 1994; Lefkowitz et al., 2017).

Lettuce necrotic leaf curl virus (LNLCV), a non-tomato infecting *Torradovirus,* is vectored by the lettuce currant aphid, *Nasonovia ribisnigri*, which retains it in the stylet (Verbeek et al., 2007). The tomato-infecting *Torradoviruses, tomato torrado virus* (ToTV), *tomato marchitez virus* (ToMarV), and *tomato chocolàte virus* (ToChV), are transmitted by whitefly species. A *Sequivirus, Parsnip yellow fleck virus*, is transmitted in a semi-persistent manner by the aphid; however, it requires the presence of a *Waikavirus, Anthriscus yellows virus*, as a helper component (HC) for transmission (Murant et al., 1976).

Members of the genus *Badnavirus* in the family *Caulimoviridae,* order *Ortervirales,* are mainly vectored by mealybugs in a semi-persistent manner. However, three members, namely *Rubus yellow net virus* (RYNV) on *Rubus* species, *gooseberry vein banding associated virus* (GVBaV) on Ribes species, and *Spiraea yellow leafspot virus* (SYLSV) on Japanese spiraea, are transmitted by aphids (Bhat et al., 2016).

Primarily, the midgut is the site of acquisition for many circulative viruses (Hunter et al., 1998; Pan et al., 2017; Reinbold et al., 2001, 2003). However, several luteoviruses are transmitted by aphids using the hindgut as the site of ingress (Gildow, 1993; Gildow et al., 2000; Reinbold et al., 2003).

Members of the family *Luteoviridae* are transmitted exclusively by aphids, generally with a high degree of vector specificity. Many *Luteoviruses* have a host range limited to one plant family (Domier, 2012).

The genus *Nanovirus* comprises the plant viruses that have a relatively narrow host range for individual species and primarily infect legumes, leading to devastating crop losses. *Nanovirus* is transmitted by various aphid species in a circulative non-propagative manner. The pea aphid *A. pisum,* the cowpea aphid *Aphis craccivora*, or the black bean aphid *A. fabae* appear to be the most important vector species (Vetten et al., 2012, 2016).

Lettuce necrotic yellows virus (LNYV) and *alfalfa dwarf virus* (ADV) of the genus *Cytorhabdovirus* are transmitted by two aphid species, *Hyperomyzus lactucae* and *Hyperomyzus carduellinus* (Randles & Carver, 1971; Stubbs & Grogan, 1963). LNYV is acquired after a day of *H. lactucae* feeding on infected plants. The latent period of LNYV is up to 18 days; however, it is highly temperature-dependent, with the latency being inversely correlated with temperature. PYDF of the genus *Nucleorhabdovirus* is transmitted by aphids. *Blueberry shoestring virus* (BBSSV) is transmitted by the aphid *Illinoia pepperi* (Morimoto et al., 1985).

Aphids can simultaneously acquire and transmit more than one virus. For example, the CMV and *papaya ringspot virus* Type W (PRSV-W) can be acquired by aphids at the same time. However, the transmission of both CMV and PRSV is reduced when they are acquired together, compared to their individual transmission. Different species of aphids also have different patterns of virus transmission. The efficiency of *A. gossypii* in transmitting the cucumovirus (CMV) and the *potyvirus zucchini yellow mosaic virus* (ZYMV) is the same when acquired alone. On the other hand, *M. persicae* is more efficient in transmitting *potyviruses* than CMV. However, when the same *M. persicae* aphid acquires both CMV and ZYMV, the transmission of ZYMV is reduced (Ng & Perry, 1999). CMV and potyviruses use different strategies for aphid transmission, but the competition for binding sites on the aphid's stylets affects the transmission rate. This competition reduces the virus load of each virus (Pinto et al., 2008). *Bean common mosaic virus* (BCMV), a potyvirus, can be effectively transmitted from infected plants to healthy plants by many aphid species, including *Macrosiphum solanifolii*, *M. pisi*, *M. ambrosiae*, *Myzus persicae*, *Aphis rumicis*, *A. gossypii*, *A. medicaginis*, *Hyalopterus atriplicis*, and *Rhopalosiphum pseudobrassicae* (Zaumeyer & Thomas, 1957).

Among aphid-vectored *Potyviruses,* the majority are transmitted in a non-persistent manner, characterized by very short acquisition and inoculation times of seconds to minutes (Pirone & Perry, 2002). Furthermore, during non-persistent transmission, the vector carries the viruses for a very short period of time. The specific site of retention for the majority of Potyviruses in the aphid stylets is near the tip (Berger & Pirone, 1986; Wang et al., 1996).

In semi-persistent transmission, acquisition of the virus can occur within minutes; however, the efficiency increases with increased feeding and is retained for hours to days (Palacios et al., 2002). The best examples of semi-persistent transmission are *Caulimovirus* and *Closterovirus*. A complex molecular mechanism governs the virus vector specificity. *Cauliflower mosaic virus* (CaMV), the type of member of the genus *Caulimovirus,* employs a helper-dependent transmission strategy requiring two viral-encoded, nonstructural proteins, P2 and P3 (Blanc et al., 2001). During transmission, a P2-P3-virion complex is formed, where P2 binds to the aphid and P3 binds to virions. *Potyvirus* transmission by aphids depends on the appropriateness of the HC and the N terminus of the coat protein (CP) (Dombrovsky et al., 2005). Unlike *Potyviruses,* AMV and *Cucumovirus* do not depend on an additional HC and exclusively rely on the CP, which interacts directly with ligands of the vector (Chen & Francki, 1990; Gera et al., 1979).

Helper transmission strategies have evolved in both non-persistently and semi-persistently transmitted viruses, as well as in viruses with both RNA and DNA genetic material. HC and the virus particles are acquired in succession. The virus transmission efficiency of circulatory transmitted viruses depends on the acquisition feeding period, which can range from hours to days. Phloem-limited viruses require a longer acquisition period. These viruses also have a longer retention period. Based on the replication and movement of the virus in the host, it has been classified as 'circulative non-propagative' or 'circulative propagative.'

After acquisition of Luteovirus, *barley yellow dwarf virus* (BYDV) from infected phloem cells, it traverses the food canal and foregut into the mid and hindgut, where they are transported across gut epithelial cells into the hemocoel. Viruses subsequently circulate in the hemolymph and enter the accessory salivary gland cells and are again inoculated into the plants along with the saliva (Gildow & Gray, 1993). Vector specificity, i.e., transmission of the virus by specific aphid species, is a characteristic feature of *Luteovirus.* Compatible interactions between the proteins of the virus and aphid result in the transmission. The aphid-transmitted circulative propagative plant viruses are restricted to the family *Rhabdoviridae.* This family includes viruses that replicate in vertebrates, invertebrates, and plants. Individual viruses can replicate in either mammals and insects or plants and insects; none has been shown to replicate in all three systems. Virus-encoded determinants for transmission have not been identified in plant *Rhabdovirus,* but by comparison with animal-infecting species, it is likely that the viral glycoprotein (G protein) governs receptor binding and infection of the insect vector (Hogenhout et al., 2003).

2. ***Family Aleyrodidae (Whiteflies):*** The family *Aleyrodidae* contains whiteflies, which are the most evolved and studied vectors of plant viruses. Whiteflies transmit various viruses, including *Begomovirus* (*Geminiviridae)*, *Crinivirus* (*Closteroviridae*), *Torradovirus (Secoviridae),* and *Ipomovirus (Potyviridae).* In the genus *Carlavirus (Betaflexiviridae)*, only *cowpea mild mottle virus* (CpMMV) is transmitted by whiteflies (Zanardo & Carvalho, 2017). Another exceptional case recently reported is *pepper whitefly-borne vein yellows virus* (PeWBVYV), a *Polerovirus* (*Luteoviridae*), which has been found to be transmitted by whiteflies. However, other members of the family *Luteoviridae* are typically vectored by aphids (Ghosh et al., 2019). Among all these viruses, *Begomovirus* (*Geminiviridae*) is the most common and significant virus transmitted by whiteflies. Over the past three decades, whiteflies have emerged as one of the most devastating pests and vectors worldwide (Navas-Castillo et al., 2011). Bemisia tabaci was first described in 1889 as a tobacco pest in Greece (Gennadius, 1889). The center of origin of whiteflies

is considered to be East Africa-Uganda (Mugerwa et al., 2018; Mugerwa et al., 2021). The potential of whiteflies to transmit *Begomovirus* was first observed in the Democratic Republic of Congo (Kufferath & Ghesquie're, 1932). The rapid adaptability of whiteflies to tropical, subtropical, temperate, and Mediterranean (MED) Basin regions is one of the factors that have made them one of the most successful vectors of plant viruses. The suitable temperature range of 28°C–32°C and relative humidity between 62% and 92% is generally found in most regions and helps whiteflies survive. Female whiteflies lay 150–300 eggs during their lifetime, which hatch within 3 to 5 days and complete their lifecycle in 14–122 days depending on environmental conditions (Mahatma et al., 2016). Due to their light weight, they can fly up to 7 km over land and water surfaces with the help of wind (Cohen, 1982; Byrne & Bellows, 1991). They acquire the virus within approximately 30 minutes of feeding and remain capable of transmitting the virus throughout their life. These features make whiteflies the most successful vector of plant viruses transmitted by insects. Generally, these viruses do not cross the lining of the ovary and are carried forward in subsequent generations. However, multiplication of *tomato yellow leaf curl virus* (TYLCV) has been observed in their whitefly vector. This virus can be transmitted transovarially for at least two generations, with up to 20% of the insects in each generation found to inoculate tomato plants (Ghanim et al., 1998).

Around 21 species of whitefly were identified based on their partial *mtCO1* sequences. Among them, seven were newly discovered species that diverged by more than 4% from any *mtCO1* sequence in GenBank (Mugerwa et al., 2021). The most destructive and widespread whitefly species is *Bemisia tabaci (*Gennadius), also known as the tobacco, cotton, or sweet potato whitefly, along with *Trialeurodes vaporariorum*, commonly known as the greenhouse whitefly.

Tomato chlorosis virus (ToCV) and *tomato infectious chlorosis virus* (TICV), both belonging to the family *Closteroviridae,* genus *Crinivirus*, exhibit similar symptoms but are genetically distinct viruses. TICV is unique among whitefly-transmitted viruses as it can be transmitted by four whitefly vectors from two genera. *Trialeurodes abutilonea* and *Bemisia tabaci* biotype B are the most efficient vectors of ToCV. *B. tabaci biotype* A and *T. vaporariorum* are comparatively less efficient vectors but still capable of transmission. ToCV can persist for up to five days in *T. abutilonea*, two days in *B. tabaci* biotype B, and only one day in *B. tabaci biotype A* and *T. vaporariorum* (Wintermantel & Wisler, 2006).

3. ***Family Pseudococcidae (Mealybug) and Coccidae (Soft Scale):*** Insects belonging to the order Hemiptera, suborder Sternorrhyncha, superfamily Coccoidea, family *Pseudococcidae*, and family *Coccidae* are milky, cottony, soft insects with piercing/sucking mouthparts. The majority of viruses transmitted by members of the family *Pseudococcidae* are also transmitted by members of the family *Coccidae.* Similar to plant scale insects and aphids, mealybugs also feed on the fluids from leaves and stems for nourishment. Mealybugs are slow-moving insects that can transmit *Ampelovirus* (*Closteroviridae*) and *Badnavirus* (*Caulimoviridae*) genera of plant viruses in a semi-persistent manner.

   Genus *Ampelovirus* causes diseases in woody plants such as blackberry, pineapple, grapes, plum, cherry, pistachio, etc. The symptoms produced by the virus depend on the host. In grapes, it produces mottling, rolling, yellowing, and reddening of the leaves. Stem pitting is a characteristic symptom of plum, whereas wilting or symptomless infections have been observed in pineapple. *Ampelovirus* is transmitted by mealybugs and soft scale insects in a semi-persistent modality. *Grapevine leafroll-associated virus* 1 (GLRaV-1) is transmitted by pseudococcid mealybug species (*Heliococcus, Phenacoccus, Pseudococcus*) and coccid soft scale insect species (*Pulvinaria, Neopulvinaria, Parthenolecanium*). *Grapevine leafroll-associated virus* 3 (GLRaV-3) is transmitted by pseudococcid mealybug species (*Planococcus, Pseudococcus, Heliococcus, Phenacoccus*) and soft scale insect species (*Pulvinaria, Neopulvinaria, Parthenolecnium, Coccus, Saissetia, Parasaissetia, Ceroplastes*). Whereas *Grapevine leafroll-associated virus* 4 (GLRaV-4)is transmitted by *Pseudoccocus* sp., *Planococcus* sp., and *Ceroplastes* sp. Two species of the genus Dysmicoccus transmit pineapple *Ampelovirus. Little cherry virus* 2 (LChV-2) is transmitted by apple and grape mealybugs (*Phenacoccus aceris* and *Pseudococcus maritimus*, respectively).

Badnavirus, BANANA streak virus (*BSV*), *Dioscorea bacilliform virus* (*DBV*), *Kalanchoë top-spotting virus* (*KTSV*), *Schefflera ringspot virus* (*SRV*) are transmitted by the mealybug *Planococcus citri*. *Cocoa swollen shoot virus* (*CSSV*) has a limited host range and is semi-persistently transmitted by *Planococcoides njalensis, Planococus citri,* and at least 12 other species of mealybugs (Roivainen et al., 1976). The virus is not mechanically transmitted. *Pineapple bacilliform comosus virus* (PBCOV) and *Pineapple bacilliform erectifolius virus* (PBERV) are transmitted by the gray pineapple mealybug, *Dysmicoccus neobrevipes*. *Sugarcane bacilliform virus* (SCBV) is transmitted by its mealybug vector *Sacharicoccus sachhari*. SCBV can be experimentally transmitted by the mealybug vector *S. sacchari* from infected sugarcane to bananas. SCBV-infected bananas also produce symptoms similar to BSV. However, natural infection of SCBV in bananas has not been reported. *Taro bacilliform virus* (TaBV) is distributed throughout the taro (*Colocasia esculenta*) growing area of Pacific Islands. The virus causes typical vein clearing, stunting, and downward curling of leaf blades. TaBV is transmitted by mealybugs (*Pseudococcus solomonensis*) as well as seeds (Devitt et al., 2005). *Aucuba bacilliform virus* (AuBV) on *Aucuba japonica* produces yellow ringspot or yellow mosaic symptoms. This virus is also transmitted by seeds as well as mealybugs.

5.2.1.1.3 Superfamilies Lygaeoidea (Family Piesmatidae)

Piesmatidae is a small family of true bugs, commonly called ash-gray leaf bugs. *Beet leaf curl virus* (BLCV), a plant infecting *Rhabdovirus*, is transmitted by *Piesma quadratum* (Proeseler, 1966, 1983). The virus is found in the phloem parenchyma of the leaf and storage parenchyma of the beet root (Eisbein, 1976), and is transmitted by the *Piesma quadratum* bug. Both nymphs and adults of *P. quadratum* can transmit the virus; however, the long latent period generally exceeds the nymphal development time. The virus can be acquired in 30 minutes and inoculated in 40 minutes of feeding by both nymphs and adults. The virus multiplies in the insect's body and moves into the salivary glands, hemolymph, and intestinal wall. European tarnished plant bugs, *Lygus rugulipennis* and *L. pratensis*, have been found to transmit potato viruses, for which aphids are the usual vector (Turka, 1978).

5.2.1.1.4 Spuperfamiles Miroidea (Family Miridae)

Members of the family *Miridae* possess a small, well-developed, triangular-shaped area commonly referred to as the cuneus on the forewing, typically pointing downwards (Edde, 2022). Only one species of mirid, *Cyrtopelts nicotianae*, transmits the satellite-bearing subgroup of the *Sobemovirus, Velvet tobacco mottle virus* (VTMoV). The acquisition threshold was less than one minute, and the rate of transmission increased with increasing acquisition time up to 30 minutes. The minimum inoculation period was between one to two hours. VTMoV is transmitted both in non-circulative (non-persistent and semi-persistent) and circulative association. The vector remains viruliferous even after molting. The virus does not penetrate the lining of the salivary gland; however, it has been observed in the gut and in small quantities in the hemolymph as well (Gibb & Randles, 1991). The mode of transmission is similar to beetles, where the virus enters the hemocoel, passes into the gut lumen, and then into the exterior through regurgitation or defecation. Fecal contamination of plant areas damaged by feeding could lead to transmission.

5.2.1.2 ORDER THYSANOPTERA (FAMILY THRIPIDAE)

Thysanoptera (*thysanos* 'fringe' + *ptera* 'wings'), commonly referred to as thrips, typically have two pairs of slender wings with a dense fringe of long hairs (Tipping, 2004). Thysanoptera is a unique order of insects that comprises fungal mycelium as well as spore-feeding insects. However, a large number of members also feed on flowers and green plant parts. A few also feed on other small arthropods. Thrips are small to minute insects, with adult body sizes ranging in most species from about 0.5 to 5 mm. Reproduction in thrips is haplodiploid, where males develop from unfertilized eggs. Several thrips' species can produce females from unfertilized eggs, a process known as thelytoky (Mound, 2009). Thrips have many unique characteristics, including asymmetric mouthparts involving only one mandible, intermediate metabolism, and a haplodiploid sex control system. Many thrips' species exhibit complex behavioral patterns, including lekking (aggregation of males in competitive courtship), fighting, and eusociality (group decision-making). In the tropics, many species induce galls on leaves. Plant-feeding thrips are believed to have evolved from fungus-feeding, leaf litter-living ancestors. Nine families are currently distinguished, with

95% of the known species being in the Thripidae and Phlaeothripidae. The relatively few thrips species that are important plant pests are mainly in the Thripidae (Lewis, 1973).

Members of the family Thripidae, specifically the subfamily Thripinae, consist of 1,400 species in 230 genera (Mound, 1996, 1997). It has been observed that species within the Thrips, *Frankliniella, Scirtothrips, Microcephalothrips*, and *Ceratothripoides* genera are capable of transmitting plant viruses. Thrips are known to transmit viruses from the *Tospovirus, Ilarvirus, Carmovirus, Sobemovirus,* and *Machlomovirus* genera to plants. Thrips is the largest genus within the family *Thripidae*, with 275 species currently documented worldwide (Mound & Kibby, 1998; Mound, 2001), while Frankliniella is the second largest genus with approximately 180 species (Nakahara, 1997). Only six species of Frankliniella, three species of *Thrips*, and one species each of *Scirtothrips* and *Ceratothripoides* have been identified as vectors. Thrips that act as vectors are among those that lay eggs, which then hatch into two larval instars that feed. The larval instars are followed by two relatively inactive pupal instars that likely do not feed.

When a larva of thrips is ready to feed, it grasps the surface of the leaf with its pretarsi, sinks down between its front legs, lifts its head, and places the tip of its mouth cone against the surface. It then shortens its mouth cone and punches a hole in the epidermis by rapidly and repeatedly protracting and retracting its left mandibular stylet. The thrips then inserts its two maxillary stylets as a unit into the wound with a series of rapid thrusts and withdrawals, salivating continuously while doing so. When a food source in the epidermis or mesophyll is found, probing and salivation stop and cibarial pumping begins. Cytoplasm is sucked into the opening at the tip of the protracted stylets, up the food canal between them and into the cibarium.

Probing and feeding can occur without mandibular intervention, but uptake of liquid seems to require use of the mutually co-adapted maxillary stylets, even when these are fully retracted (Heming, 1978).

The transmission of *Ilarvirus* by thrips is a different mechanism. It involves the physical movement of virus-carrying pollen from one plant to another and its introduction into the plant through feeding wounds. Thrips in the genera *Frankliniella, Microcephalothrips*, and *Thrips* have been implicated in the spread of Ilarvirus though more may be discovered. There has been one report of transmission of a *Carmovirus* (*maize chlorotic mottle virus* (MCMV)) and a *Sobemovirus (Sowbane mosaic virus)* utilizing the same mechanism as for *Ilarvirus.*

5.2.1.3 ORDER COLEOPTERA (BEETLES)

Coleoptera, commonly known as beetles, is the largest insect order, comprising over 3.6 lakh species, which accounts for 40% of the known insect species worldwide. While beetles have not traditionally been considered common virus vectors, a few species can cause significant damage through the transmission of plant viruses. According to Smith et al. (2017), more than 70 species of beetles have been reported to transmit viruses affecting vegetables and cereals. It is estimated that beetles transmit approximately 11% of insect-borne viruses (Smith et al., 2017; Bhat & Rao, 2020). Beetles from the families Chrysomelidae (leaf beetles), Coccinellidae (ladybirds or lady beetles), Curculionidae (snout beetles and true weevils), and Meloidae (blister beetles) have been found to transmit plant viruses (Gergerich, 2001; Fereres & Raccah, 2015). The viruses transmitted by beetles include *Machlomovirus, Bromovirus, Carmovirus, Comovirus, Sobemovirus,* and *Tymovirus* (Scott & Fulton, 1978; Gedling et al., 2018). Beetles have a unique mode of action, as they feed on plant cells between leaf veins and regurgitate during feeding, allowing virus particles to mix with sap. The viruses are then deposited into chewing wounds. Subsequently, the virus particles are transported through the xylem elements and other plant parts away from the site of release (Gergerich & Scott, 1988, 1991; Gedling et al., 2018).

One bite is enough for the acquisition of virus by the beetles; however, the efficiency of acquisition increases with the prolonged feeding period (Fulton & Scott, 1977). The duration of retention of the virus by the vector varies among different beetles. The Mexican bean beetle, *E. varivestis*, retains cowpea severe mosaic virus (CPSMV, *Secoviridae*) for one day, the banded cucumber beetle, *Diabrotica balteata* (*Chrysomelidae*), retains bean rugose mosaic virus (BRMV, *Secoviridae)* for three days, whereas the bean leaf beetle, *Cerotoma trifurcate* (*Chrysomelidae*), retains the same virus for several days (Fulton & Scott, 1977; Fereres & Raccah, 2015). Noncirculative viruses are transmitted in a semi-persistent manner by the beetle vector (Raccah & Fereres, 2009). Noncirculative viruses are retained by the chitinous anterior region of the foregut (Mauck et al., 2018). A few viruses can move into the hemolymph immediately after ingestion by the beetles and be retained for longer periods (Fulton & Scott, 1977; Scott & Fulton, 1978; Fereres & Raccah, 2015).

In some cases, larvae are more efficient in virus transmission than the adult beetle. Larvae of the cereal leaf beetle, *Oulema melanopus* (*Chrysomelidae*), transmit MCMV (*Tombusviridae*) more efficiently than adults (Nault, 1978). Moreover, larvae and adults of *O. melanopus*, as well as *Oulema gallaeciana* (Chrysomelidae), transmit a *Sobemovirus, cocksfoot mottle virus* (CfMV), even after 15 days of acquisition. Larvae are more efficient than adults in virus transmission (Catherall, 1987).

Bean pod mottle virus (BPMV, *Comovirus*) is the most destructive pathogen of soybean, causing yield losses of 36%–52% depending on the time of infection (Gergerich & Scott, 1996; Hopkins & Mueller, 1984; Gergerich, 1999). The virus can be effectively transmitted by several species of beetles from the *Chrysomelidae* family (Giesler et al., 2002).

The Mexican bean beetle, *Epilachna varivestis*, is a ladybird beetle that is recognized as a severe pest of soybean (Fan et al., 1992; Nakamura & Chavez, 2007). In addition to causing direct damage to the crop, the beetle can transmit several plant viruses, including CPMV from the family *Secoviridae, Southern bean mosaic virus* (SBMV) from the family *Solemoviridae,* and *Blackgram mottle virus* from the family *Tombusviridae* (Fulton & Scott, 1974; Scott & Fulton, 1978; Scott & Phatak, 1979; Musser et al., 2003; Gedling et al., 2018).

5.2.1.4 ORDER DIPTERA

Members of the Order Diptera, known as dipteran flies or two-winged flies or "true flies," are insects that have a pair of functional wings. One pair of wings is used for flight, while the other pair is vestigial hind wings, also known as halters, which help the insect maintain balance (Sarwar, 2020).

Order Diptera is not a common order of plant virus vectors. However, the leaf miner *Liriomyza langei* (family *Agromyzidae*) is capable of transmitting tobamovirus, *tobacco mosaic virus* (TMV), and *Sobemovirus Sowbane mosaic virus.* The relationship between the transmitted viruses and the vector is of a mechanical nature. It is suggested that the virus is probably carried as a contaminant on the insect's ovipositor (Costa et al., 1958). The vegetable leaf miner *Liriomyza sativae* has been capable of transmitting three Potyvirus in tests conducted, and leaf miner flies successfully transmitted *Celery mosaic virus* in celery (*Apium graveolens*) and *Watermelon mosaic virus* strains in squash (Zitter & Tsai, 1977).

Rice yellow mottle virus (RYMV), belonging to the genus Sobemovirus, is another important virus that is also transmitted by the stalked-eyed borer *Diopsis thoracica* (Togola et al., 2011).

5.2.1.5 ORDER ORTHOPTERA

The order Orthoptera is not a common order of plant virus vectors. It includes grasshoppers, locusts, crickets, and katydids. This order is composed of two suborders: *Caelifera* (grasshoppers, locusts, and mole crickets), which have short antennae and abdominal tympanal organs (ears); and *Ensifera* (true crickets, katydids, and bush crickets), which have antennae that reach at least to their abdomen and their ears are located in their fore tibia (front legs).

Grasshoppers such as *Conocephalus merumontanus*, *C. longipennis*, *Stenohippus aequus*, *C. merumontanus*, ground-hoppers or pygmy grass-hopper, *Paratettix sp*., painted grasshopper, *Zonocerus variegates*, and cricket species like *Euscyrtus* spp., band-winged grasshopper, *Paracinema tricolor,* have been tested and confirmed as vectors of RYMV, a member of *Solemoviridae*, in the African continent (Nwilene et al., 2009; Koudamiloro et al., 2014).

5.2.1.6 ORDER DERMAPTERA

The order Dermaptera is not a common order of plant virus vectors. Earwigs (*Dermaptera: Forficulidae*) have uniform brown and long slim bodies with light brown or yellow legs. They have pincers on the back of their abdomen, which are long and curved like forceps in males and almost straight in females. Earwigs can be beneficial as well as harmful to crops. They feed on aphids, other small insects, dead bugs, and plant debris, which helps farmers. However, they also feed on young, soft plant shoots, corn silk, flowers, and sometimes ripened fruits, causing small holes to appear on the leaves. The European earwig, *Forficula auricularia*, is a polyphagous insect pest in gardens and greenhouses where vegetables and flowers are grown. It has been found to transmit *Turnip yellow mosaic virus* (Markham & Smith, 1949). CaMV has also been found in the gut content of the major predatory earwig *Doru luteipes*, which preys on maize pests and thrips (Godinho et al., 2017).

5.2.2 CLASS ARACHNIDA

The class Arachnida includes around 1,00,000 species of mites. They have chelated mouthparts and are devoid of antennae. Another distinguishing feature of the group is the presence of eight jointed legs and simple single-lens eyes. Tagmatization varies in different groups, but most have a cephalothorax and an abdomen (=opisthosoma). The cephalothorax contains the mouthparts, sensory organs, and walking legs, while the opisthosoma possesses digestive and reproductive organs and may or may not show segmentation (Proctor et al., 2015).

Among the different Arachnids, gall or worm mites (Family *Eriophyidae*), false spider mites (Family *Tenuipalpidae*), and spider mites (Family *Tetranychidae*) are known to transmit viruses in various cereals, fruit trees, pulse crops, and coffee plants. Their typical elongated cheliceral stylet allows them to feed by piercing the host's cells. Members of the family *Tetranychidae* and family *Tenuipalpidae* have a single, long, retractable stylet. However, eriophyoid mites have many non-retractable stylets. Eriophyoid mites penetrate their stylets into the plant tissue by gradually telescoping the palpal tissue surrounding the stylet bundle (Krantz & Lindquist, 1979; Alberti & Kitajima, 2014).

The Tetranychoidea superfamily in the suborder Prostigmata contains flat mites (Tenuipalpidae) and spider mites (Tetranychidae), which are most important as virus vectors. Members of the superfamily can be easily recognizable by their chelicerae, which are modified into long curved J-shaped stylets and enclosed within a hemispherical capsule called the stylophore. Members of the family Tenuipalpidae are also known as false spider mites because they do not produce silk webs on plants. *Brevipalpus* is one of the important mite genera in the family Tenuipalpidae that transmits viruses. *Brevipalpus* is a shade-loving species often found in the interior of plant canopies. *B. phoenicis* transmits Coffee ring spot and *passion fruit green spot virus* (PFGSV), while *B. californicus* transmits Orchid fleck virus. *B. yothersi* transmits Citrus leprosis virus (Tassi et al., 2017). Citrus leprosis virus is also transmitted by *B. californicus and B. obovatus* (Beard et al., 2015; Ramos-Gonzalez et al., 2016). False spider mites pierce the mesophyll cells of plants with their interlocked stylets to feed from the host, possibly injecting saliva in the process. After withdrawing the stylet, the cell content of the punctured cells is sucked out by the pre-oral cavity, causing the affected cells to collapse. The mite's saliva is toxic to the host, resulting in conspicuous damage characterized by browning of the affected area.

Tetranychus telarius has a wide, nonspecific range of hosts and also appears to be a nonspecific or casual vector of *potato virus Y* (PVY), having a non-persistent relationship with the mite vector (Gotoh et al., 2009).

Gall mite family *Eriophyidae* feed on plants by puncturing cells with their stylets, which are held inside the groove of the rostrum. They have two pads that act as ducts for the saliva. Eriophyid mites are comparatively very small arthropods, measuring about 0.2 mm in length, and they are sluggish in nature. They typically infest buds and young leaves without causing significant damage. The mites go through two nymphal instars and a resting pseudopupal stage (Oldfield, 1996).

Aceria is a genus of mites that belongs to the family *Eriophyidae*. There are over 900 species in this genus, but only a few are known vectors of plant-infecting viruses. The wheat curl mite, *Aceria tosichella*, is the most well-known mite vector within this genus. It is capable of transmitting several plant viruses, including *wheat streak mosaic virus* (WSMV), *Wheat mosaic virus, Triticum mosaic virus* (TriMV), and others. *A. ficus* transmits *Fig mosaic virus*, while *A. tulipae* transmits onion mite-borne latent virus and Shallot mite-borne latent virus. Three important genera of the family *Potyviridae*, namely *Poacevirus* (TriMV), *Rymovirus*, and *Tritimovirus* (WSMV), are vectored by eriophyid mites *A. tosichella* in a semi-persistent manner (Wylie et al., 2017). Adult eriophyid mites can transmit viruses only if they acquire them during an immature stage (Orlob, 1966).

The genus *Emaravirus*, belonging to the family *Fimoviridae* of the order *Bunyavirales,* consists of nine species of plant viruses. The virus has four to eight segments of negative-sense ssRNA genomes.

Members of the Genus *Allexivirus*, family *Alphaflexiviridae*, have been suspected to be transmitted by the dry bulb mite (*A. tulipae*) (Adams et al., 2012a, b; Martelli et al., 2007; van Djik et al., 1991). However, the specific virus vector relationship is still unexplored. The family Secoviridae, genus Nepovirus, contains nematode-transmitted viruses. However, *black currant reversion virus* (BRV) is transmitted by an eriophyid gall mite, *Cecidophyopsis ribis. Grapevine berry inner necrosis virus* (GINV), *cherry mottle leaf virus* (CMLV), *peach mosaic virus* (PcMV), and *grapevine pinot gris virus* (GPGV), all belonging to the genus *Trichovirus,* family *Betaflexiviridae,* are transmitted by eriophyd mites *Eriophyes insidiosus* in a semi-persistent manner. *E. insidiosus* requires an AAP of 3 days, while inoculation requires between 3 and 6 hours.

5.2.3 HOST VIRUS AND VECTOR RELATIONSHIP: WHO IS PULLING THE STRINGS?

High specificity is found in vector transmission. Both the virus and vector have complementary protein sequences that recognize, attach, detach, and release the virus during different probing stages. In most cases, the virus specificity is determined by either structural proteins, such as coat protein (CP) and its derivatives, or non-structural proteins, such as HC or transmission factor. Numerous virion-binding proteins have been characterized, playing a crucial role in the virus-vector relationship. However, the most significant question is who commands the relationship between the virus, vector, and host. Most virologists agree that the virus is the main driver of this relationship. The features responsible for the dominance of insects and their evolution as successful pests help the virus survive and spread more efficiently. Many insects would not be efficient vectors of the virus without symbiotic bacteria in their bodies. Therefore, we need to consider the doctrine from a symbiotic perspective as well.

The evolution of viruses, especially RNA-containing viruses, is rapid due to the lack of proofreading activity in RNA polymerases (RdRp) (Domingo et al., 1996; Drake & Holland, 1999; Gago et al., 2009). This weakness is exploited by the virus, leading to the rapid evolution of new strains in rapidly changing ecological situations. Sometimes, the virus utilizes multiple vectors to achieve rapid and wide transmission, each with different vectoring capacities.

ToCV and TICV are two distinct viruses belonging to the genus *Crinivirus*, family *Closteroviridae.* Despite their differences, they produce nearly identical symptoms, often causing confusion. The genomes of both TICV and ToCV consist of two molecules, RNA1 and RNA2, which are encapsidated independently. These two viruses are semi-persistently transmitted by different cryptic species of whiteflies. TICV is exclusively transmitted by the greenhouse whitefly (*Trialeurodes vaporariorum*), while ToCV can be transmitted by the greenhouse whitefly (*T. vaporariorum*), banded wing whitefly (*T. abutilonea),* and all the *B. tabaci* species, including MED (formerly Q biotype), Middle East Asia Minor-1 (MEAM1, formerly B biotype), and New World (NW, formerly A biotype) (Brown et al., 1991; Costa & Brown, 1994; Olivé & Castillo, 2019). Among these different vectors, *B. tabaci* MED is the most efficient vector, followed by *B. tabaci* MEAM1 ≈ *T. abutilonea*, and then *B. tabaci* NW. *T. vaporariorum* is a less efficient vector (Wintermantel & Wisler, 2006; Olivé & Castillo, 2019). *T. vaporariorum* is a worldwide distributed whitefly of temperate regions. On the other hand, *T. abutilonea* has only been described in Cuba and the United States. *B. tabaci* (both NY and MEAM1) were originally described in tropical and subtropical regions, but they have also spread to temperate regions (Hanssen & Lapidot, 2012). In the MED basin, TICV was first observed in 1991 causing severe disease, but it was soon replaced by ToCV, possibly due to the replacement of *T. vaporariorum* with the invasive and dominating whitefly *B. tabaci* MEAM1 (Navas-Castillo et al., 2011). The difference in transmission between MED and MEAM1 has been associated with differences in virus acquisition and accumulation rate (Shi et al., 2018). The restriction of TICV to one vector and the adaptation of ToCV to four different vectors of two different species with different capabilities provided ecological advantages to ToCV.

Shifting of hosts, where a pathogen jumps from its original host into a novel species, has also been observed. An excellent example in this category is a ssDNA-containing *Circovirus* infecting mammals, which evolved from a plant-infecting single-stranded multipartite *Nanovirus.* The Reps gene of the *Circovirus* is a hybrid of the Reps gene of the plant-infecting *Nanovirus* (from the N-terminal region) and the vertebrate-infecting *Calicivirus. Calicivirus* only infects vertebrates, suggesting that the host-switch happened first, and that the recombination took place in a vertebrate (Gibbs & Georg, 1999).

5.2.4 HITCHHIKING SYMBIONTS

The role of symbionts in the survival of vectors and virus transmission is indispensable in many vector-virus relationships. Many insects, such as aphids, psyllids, planthoppers, leafhoppers, whiteflies, weevils, and certain ants, feed on poor or unbalanced food sources (Hansen & Moran, 2014; Rio et al., 2016; Wu et al., 2022). Obligate symbionts, such as mycetocytes or bacteriocytes, provide essential nutrients to these poorly fed insects (Douglas, 2011). These symbionts are integrated into large organs known as mycetomes or bacteriomes and are exclusively dependent on hosts for survival, unable to be artificially cultured.

In many cases, the proteins of the virus and the proteins lining the insect's mouthparts do not have complementary regions to recognize each other, preventing the establishment of specific vector-virus relationships. Insect symbionts provide a complementary site that is essential for the establishment of these relationships. For example, Rickettsia associated with whiteflies enhances the acquisition, retention, and transmission of TYLCV (Kliot et al., 2019). Interactions between the protein of Rickettsia (rickettsial secretary protein, BtR242) and the coat protein of *cotton leaf curl Multan virus* (CLCuMuV) help in the whitefly transmission of CLCuMuV (Lei et al., 2021). Vitellogenin (Vg), a critical insect reproductive protein, is synthesized by the fat bodies of the insect. TYLCV is internalized through endocytosis into midgut epithelial cells by hitchhiking Vg and forming a virus-Vg complex. The synthesis of Vg in *B. tabaci* with the Rick population was found to be two times higher compared to insects lacking it (He et al., 2021). *Potato leafroll virus* (PLRV) in the green peach aphid *Myzus persicae* hitchhikes the symbionin expressed by the obligate endosymbiont of *M. persicae, Buchnera aphidicola*, for binding and transmission. The application of antibiotics inhibits the growth of the endosymbiont, resulting in viral degradation, loss of infection, and reduction in transmission efficiencies (Van den Heuvel et al., 1994; Hogenhout et al., 1996). The RDV, belonging to the genus *Phytoreovirus* in the family *Reoviridae*, is transovarially transmitted by the green rice leafhopper, *Nephotettix cincticeps*. The vector possesses two types of obligate symbiotic bacteria, *Sulcia muelleri* and *Nasuia deltocephalinicola* (Honda et al., 2007; Noda et al., 2012). The virus hitchhikes these two symbionts, and a tripartite complex is formed by the RDV, *Sulcia,* and *Nasuia*. RDV enters the eggs by hitchhiking on the envelopes of *Sulcia* and *Nasuia,* allowing for vertical transmission from one generation to another.

5.2.5 HOST AND VECTOR MANIPULATION BY THE VIRUS

What the small genome of a virus can do is unfolding layer by layer with the development of science and our understanding. Until recently, it was understood that a virus can neutralize the host's resistance, can use the host's machineries at the host's own cost, can hook themselves with the vector's mouthparts protein, can invade the vector's internal tissue, and can multiply in the vector's body for its survival and spread. However, increasing evidence suggests that viruses can "brainwash" the vector's behavior in ways that enhance their own transmission and survival.

The phenomenon has been well studied in many intermediate hosts of parasites of birds and mammals. Brainwashing of ants by the fluke causes them to climb on the tip of grass blades and patiently wait for grazing sheep (Carney, 1969; Moore, 2002; Wickler, 1976; Paulin, 2010). A similar effect has been observed in wheat and potato crops infected by BYDV and *Potato leafroll virus* (PLRV), respectively. The non-infective aphid vectors, *Rhopalosiphum padi* and *Myzus persicae*, show greater settling or arrestment due to the elicitation of altered volatile organic compound profiles from the infected plants (Eigenbrode et al., 2002; Jime´nez-Martı´nez et al., 2004; Medina Ortega et al., 2009; Werner et al., 2009; Ingwell et al., 2012). The "vector manipulation hypothesis" (VMH) was proposed by Ingwell et al. (2012) to explain this phenomenon. Insect vectors feeding on virus-infected host plants exhibit manipulated behavior, expressing all the characteristics that make them efficient vectors, such as rapid development, high fecundity, greater longevity, and more winged forms. The secondary chemistry of the host plant affects vector behavior, causing non-viruliferous insects to move towards infected plants. Once the virus is acquired, the vector's preferences shift to non-infected hosts. This manipulation of vector behavior enhances virus transmission in the field. In an experiment on non-viruliferous and viruliferous beet leafhoppers (*Circulifer tenellus* (Baker)) carrying BCTV, it was observed that BCTV manipulates the probing preference and behavior of the insect (Lee et al., 2022). Non-viruliferous beet leafhoppers preferred to probe on barley and ribwort plantain, which are preferred hosts of the insect. However, once they acquire BCTV, their host preference changes and they probe more on the host of BCTV, which is not preferred by the leafhopper. Even the probing duration is altered by the virus. Leafhoppers carrying BCTV feed for a shorter time on barley compared to their non-viruliferous counterparts. All these alterations in probing preference and behavior in beet leafhoppers carrying BCTV contribute to better transmission of the virus in tomato fields.

The phenomenon indicates that olfactory signaling might have been hijacked by plant viruses. *Rice stripe virus* and SRBSDV have been found to directly regulate the gene transcripts of odorant receptor coreceptor (ORco) and odorant-binding protein (OBP) in their insect vectors, the small brown planthopper (*Laodelphax striatellus*) and the white backed plant hopper (*Sogatella furcifera*), respectively (Hu et al., 2019; Li et al., 2019). This helps in differentially preferring their host plant before and after acquiring the virus.

TYLCV suppresses the emission of repellent volatiles and enhances the olfactory attractiveness of the plant to non-viruliferous MED whiteflies (Fang et al., 2013). After acquisition, TYLCV causes caspase-dependent apoptotic neurodegeneration with severe vacuolar neuro-pathological lesions in the brain of viruliferous whiteflies. This leads to putative inflammatory signaling. The sensory defects do not affect the preference of whiteflies for virus-infected plants. This manipulation of host and vector ultimately benefits TYLCV transmission (Wang et al., 2020).

5.2.6 *CROP DIVERSIFICATION AND SURVIVAL OF PEST*

Crop diversification, use of resistant varieties, application of chemicals for pest management, and abiotic stress pose threats to the survival of pests. However, these circumstances exert selection pressure for the evolution of improved biotypes of pests with a high host range and fecundity. The feeding habits and host preferences of insects, the localization of viruses in hosts, and the modulation of viral proteins for the retention of viruses at specific sites on vectors in any ecological niche indicate that the evolution of hosts, vectors, and viruses occurred simultaneously. A high degree of specificity is found in the transmission of viruses by vectors. Nearly 60% of vector-transmitted plant viruses have a single known vector species (Richard et al., 2010). Viruses that have multiple vectors also have a wider host range than viruses with a single known vector species. Vector diversity helps vectors survive better in ecological niches through enhanced host range, high fecundity, and better tolerance to adverse environmental conditions. All these factors help virus to survive better in the ecological niche by additional host colonization opportunities, higher transmission rate and better survival of the virus in the vector. Vector diversity plays a crucial role in determining functional variations in virus survival, evolution, host range expansion, and transmission, which significantly impact viral disease epidemics. In the context of plant vector transmission, diversity extends beyond the different species within a genus, encompassing biotype variations within vector species that enhance their ecological niche adaptation.

5.3 CONCLUSION

The diversity of arthropod vectors of plant viruses, due to their biological and ecological competence, has contributed to the evolution of viruses as some of the most devastating pathogens for various crops. Arthropod vectors that transmit viruses are found in the classes Insecta and Arachnida. Among these, the most important orders in Insecta are Hemiptera, Thysanoptera, Coleoptera, and Orthoptera, while Arachnida includes Acari. These insects have either piercing and sucking or biting and chewing mouthparts. In insects with piercing and sucking mouthparts, the mandibles and maxillary laciniae are modified to pierce plant tissues and reach the sites where viruses typically multiply, allowing them to efficiently extract plant sap. In insects with biting and chewing mouthparts, they have mechanisms to feed on cells that harbor viruses, and then regurgitate and bathe their mouthparts with sap and virus particles. During this process, the virus particles adhere to their mouthparts from infected tissues, and subsequently, when they feed on healthy plants in the same manner, they release the virus. The explicit molecular mechanism involving the viral protein and the vector, either directly or through the association of symbionts, provides high specificity and efficiency to the vector transmission. There are also many viruses that can be transmitted through more than one vector. These viruses have simple mechanisms and most of them can be easily inoculated through mechanical sap as well. Viruses can "brainwash" vector behavior in ways that enhance their own transmission and survival. The "VMH" explains the phenomenon where vectors that feed on virus-infected host plants exhibit faster growth rates, higher fecundity, greater longevity, and/or enhanced production of alate forms of the vector. These factors can lead to increased virus spread and are typically attributed to indirect effects of virus infection on host quality. Viruses also alter the secondary chemistry of host plants to attract non-viruliferous insects and deter viruliferous insects from healthy plants. Crop diversification has always been a widely discussed feature of the evolution of viruses as well as vectors, and it is responsible for viral disease epidemics. In plant vector transmission, diversity does not only refer to different species within the vector genus, but also encompasses biotype variation within vector species, which helps them better exist in their ecological niche.

KEYWORDS

- **arthropods**
- **crop diversification**
- **host virus**
- **orthoptera**
- **plant viruses**
- **strings**
- **vector manipulation**
- **vectors**

REFERENCES

Adams, M. J., Candresse, T., Hammond, J., Kreuze, J. F., Martelli, G. P., & Namba, S., (2012a). Betaflexiviridae. In: King, A. M. Q., Adams, M. J., Carstens, E. B., & Lefkowitz, E. J., (eds.), *Virus Taxonomy: Classification and Nomenclature of Viruses* (pp. 920–941). Ninth Report of the International Committee on Taxonomy of Viruses. Elsevier, Waltham.

Adams, M. J., Candresse, T., Hammond, J., Kreuze, J. F., Martelli, G. P., & Namba, S., (2012a). Alphaflexiviridae. In: King, A. M. Q., Adams, M. J., Carstens, E. B., & Lefkowitz, E. J., (eds.), *Virus Taxonomy: Classification and Nomenclature of Viruses* (pp. 904–919). Ninth Report of the International Committee on Taxonomy of Viruses. Elsevier, Waltham.

Alberti, G., & Kitajima, E. W., (2014). Anatomy and fine structure of *Brevipalpus* mites (*Tenuipalpidae*) - economically important plant virus vectors - part 1: An update on the biology and economic importance of brevipalpus mites. In: *Zoologica* (Vol. 160, p. 110). Schweizerbart Science Publishers, Stuttgart.

Ammar, E. D., & Hogenhout, S. A., (2008). A neuro tropic route for *Maize mosaic virus* (*Rhabdoviridae*) in its planthopper vector *Peregrinus maidis*. *Virus Res., 131*(1), 77–85.

Andrew, M. Q., King, M. J., Adams, E. B., Carstens, E., & Lefkowitz, J., (2012). *Virus Taxonomy: Family-Geminiviridae*. Ninth Report of the International Committee on Taxonomy of Viruses.

Astier, S., Albouy, J., Maury, Y., & Lecoq, H., (2001). *Principes de Virologie Végétale*. INRA, Paris, France. Available from: https://doi.org/10.1016/j.virusres.2011.04.028.

Bahder, B. W., Zalom, F. G., Jayanth, M., & Sudarshana, M. R., (2016). Phylogeny of geminivirus coat protein sequences and digital PCR aid in identifying *Spissistilus festinus* as a vector of grapevine red blotch-associated virus. *Phytopathology, 106*(10), 1223–1230.

Beard, J. J., Ochoa, R., Braswell, W. E., & Bauchan, G. R., (2015). *Brevipalpus* phoenicis (Geijskes) species complex (Acari: *Tenuipalpidae*)-a closer look. *Zootaxa, 3944*, 1–67.

Beijerinck, M. W., (1898). *Ueber ein contagium vivum fluidum als Ursache der Fleckenkrankheit der Tabaksblatter. Verhandelingen der Koninklyke akademie van Wettenschappen te Amsterdam, 65*(2), 3–21. (With 2 plates).

Berger, P. H., & Pirone, T. P., (1986). The effect of helper component on the uptake and localization of potyviruses in *Myzus persicae*. *Virology, 153*, 256–261.

Bhat, A. I., & Rao, G. P., (2020). Virus transmission through pollen. In: *Characterization of Plant Viruses* (pp. 61–64). New York: Humana Press.

Bhat, A. I., Hohn, T., & Selvarajan, R., (2016). Badnaviruses: The current global scenario. *Viruses, 8*(6), 177.

Black, L. M., (1940). Strains of potato yellow dwarf virus. *Am. J. Bot., 27*, 386–392.

Black, L. M., (1953). Viruses that reproduce in plants and insects. *Ann. N.Y. Acad. Sci., 56*, 398–413.

Blackman, R. L., & Eastop, V. F., (2000). *Aphids of the World's Crops – An Identification and Information Guide* (p. 466). England: Wiley & Sons.

Blanc, S., Hébrard, E., Drucker, M., & Froissart, R., (2001). Molecular basis of vector transmission: *Caulimoviruses*. In: Harris, K., Smith, O. P., & Duffus, J. E., (eds.), *Virus–Insect–Plant Interactions* (pp. 143–166). San Diego: Academic Press.

Brault, V., Uzest, M., Monsion, B., & Blanc, E. J. S., (2010). Aphids as transport devices for plant viruses. *Comptes Rendus Biologies, 333*(6, 7), 524–538.

Briddon, R. W., Bedford, I. D., Tsai, J. H., & Markham, P. G., (1996). Analysis of the nucleotide sequence of the treehopper-transmitted geminivirus, tomato pseudo-curly top virus, suggests a recombinant origin. *Virology*, *15*(2), 387–394.

Briddon, R. W., Pinner, M. S., Stanley, J., & Markham, P. G., (1990). Geminivirus coat protein gene replacement alters insect specificity. *Virology, 177*, 85–94.

Brown, J. K., Frohlich, D., & Rosell, R., (1995). The sweetpotato/silverleaf whiteflies: Biotypes of *Bemisia tabaci* (Genn.), or a species complex? *Annu. Rev. Entomol.*, *40*, 511–534.

Bujarski, J., Figlerowicz, M., Gallitelli, D., Roossinck, M. J., & Scott, S. W., (2012). Family *bromoviridae*. In: King, A. M. Q., Adams, M. J., Carstens, E. B., & Lefkowitz, E. J., (eds.), *Virus Taxonomy: Classification and Nomenclature of Viruses*; Ninth *Report of the International Committee on Taxonomy of Viruses* (pp. 965−976). Elsevier Academic Press, Waltham, MA USA.

Byrne, D. N., & Bellows, Jr. T. S., (1991). Whitefly biology. *Annual Review of Entomology, 36*, 431–457.

Carney, W. P., (1969). Behavioral and morphological changes in carpenter ants harboring dicrocoelid metacercariae. *Am. Midl. Nat.*, *82*, 605–611.

Chen, B., & Francki, R. I. B., (1990). Cucumovirus transmission by the aphid *Myzus persicae* is determined solely by the viral coat protein. *J. Gen. Virol.*, *71*, 939–944.

Chen, H., Chen, Q., Omura, T., Uehara-Ichiki, T., & Wei, T., (2011). Sequential infection of *Rice dwarf virus* in the internal organs of its insect vector after ingestion of virus. *Virus Res.*, *160*(1), 389–394.

Cobben, R. H., ((1978). *Evolutionary Trends in Heteroptera*; *Part 2. Mouthpart-Structures and Feeding Strategies* (p. 407). Mededelingen Landbouwhogeschool 78-5. H. Veeman, Wageningen, the Netherlands.

Cobben, R. H., (1978). Notes on *Auchenorrhyncha* (homoptera) from pitfall traps in the gerendal reserve (Southern part of the Limburg Province). *Entomologische Berichten*, *38*, 9–16.

Cohen, S., (1982). Control of whitefly vectors of viruses by mulches. In: Harris, K. F., & Maramorosch, K., (eds.), *Pathogens, Vectors and Plant Diseases, Approaches to Control* (pp. 46–56). Academic, New York.

Costa, A. S., De Silva, D. M., & Duffus, J. E., (1958). Plant virus transmission by a leaf-miner fly. *Virology*, *5*(1), 145–149.

Costa, H. S., & Brown, J. K., (1991). Variation in biological characteristics and in esterase patterns among populations of *Bemisia tabaci* (Genn.) and the association of one population with silver leaf symptom development. *Entomol. Exp. Appl.*, *61*, 211–219.

David, R. J., (2005). *Plant Viruses Transmitted by Thrips.*, *113*(2), 119–157.

Devitt, L., Ebenebe, A., Gregory, H., Harding, R., Hunter, D., & Macanawai, A., (2005). Investigations into the seed and mealybug transmission of *Taro bacilliform virus*. *Aust. Plant Pathol.*, *34*, 73–76.

Dietrich, C. H., (2009). *Auchenorrhyncha. Encyclopedia of Insects*, 56–64. doi: 10.1016/b978-0-12-374144-8.00015-1.

Dimitri, F., (2008). The systematics of the Hemiptera. *Revista Colombiana de Entomologia.* 34. 1–21. WEBER, H. 1930. Biologie der hemipteren. Julius Springer, Berlin, Germany. 543 p HENNIG, W. 1969. Die Stammesgeschichte der Insekten. Waldemar Kramer, Frankfurt am Main, Germany. 436 p.

Dombrovsky, A., Huet, H., Chejanovsky, N., & Raccah, B., (2005). Aphid transmission of a potyvirus depends on suitability of the helper component and the N terminus of the coat protein. *Arch. Virol., 150*(2), 287–298.

Domier, L. L., (2012). Family *Luteoviridae*. In: King, A. M. Q., Adams, M. J., Carstens, E. B., & Lefkowitz, E. J., (eds.), *Virus Taxonomy: Ninth Report of the International Committee on the Taxonomy of Viruses* (pp. 1045–1053). Amsterdam: Elsevier Academic Press.

Domingo, E., Escarmis, C., Sevilla, N., Moya, A., Elena, S., Quer, J., et al., (1996). Basic concepts in RNA virus evolution. *FASEB J., 10*, 859–864.

Douglas, A. E., (2011). Lessons from studying insect symbioses. *Cell Host Microbe., 10*, 359–367.

Drake, J. W., & Holland, J. J., (1999). Mutation rates among RNA viruses. *Proc. Natl. Acad. Sci. U. S. A., 96*, 13910–13913.

Edde, P. A., (2022). Arthropod pests of cotton (*Gossypium hirsutum* L.). In: Peter, A. E., (ed.), Field *Crop Arthropod Pests of Economic Importance* (pp. 208–274). Academic Press.

Eigenbrode, S. D., Ding, H., Shiel, P., & Berger, P. H., (2002). Volatiles from potato plants infected with *Potato leafroll virus* attract and arrest the virus vector, *Myzus persicae* (*Homoptera: Aphididae*). *Proc. R. Soc. Lond. Ser. B. Biol. Sci., 269*, 455–460.

Eisbein, K., (1976). Untersuchungen zum elektronenmikroskopischen nachwies des rübenkräusel-virus (*Beta virus* 3) in *Beta vulgaris* L. und *Piesma quadratum* Fieb. *Arch. Phytopathol. Pflanzenschutz, Berlin., 12*, 299–313.

Fan, Y., Groden, E., & Drummond, F. A., (1992). Temperature-dependent development of Mexican bean beetle (*Coleoptera: Coccinellidae*) under constant and variable temperatures. *J. Econ. Entomol., 85*, 1762–1770.

Fang, Y., Jiao, X., Xie, W., Wang, S., Wu, Q., Shi, X., Chen, G., et al., (2013). Tomato yellow leaf curl virus alters the host preferences of its vector *Bemisia tabaci. Scientific Reports, 3*, 2876.

Fereres, A., & Raccah, B., (2015). Plant virus transmission by insects. In: *The Major Transmission Modes. eLS* (pp. 1–12).

Forbes, A. R., (1977). *Aphids As Virus Vectors || The Mouthparts and Feeding Mechanism of Aphids, 83*, 103.

Fulton, J. P., & Scott, H. A., (1974). Virus vectoring efficiencies of two species of leaf-feeding beetles. *Proc. Amer. Phytopath. Soc. 1*, 159.

Fulton, J. P., & Scott, H. A., (1977). Bean rugose mosaic and related viruses and their transmission by beetles. *Fitopatol. Bras., 2*, 9–16.

Ga´mez, R., (2008). Transmission of *Rayado fino virus of maize* (*Zeamays*) by *Dalbulus tnaidis*. *Ann. Appl. Biol., 73*(3), 285–292.

Gago, S., Elena, S. F., Flores, R., & Sanjuan, R., (2009). Extremely high mutation rate of a hammer head viroid. *Science, 323*, 1308.

Gallitelli, D., (2000). The ecology of *Cucumber mosaic virus* and sustainable agriculture. *Virus Res., 71*, 9–21.

Gedling, C. R., Smith, C. M., LeMoine, C. M. R., & Cassone, B. J., (2018). The Mexican bean beetle (*Epilachna varivestis*) regurgitome and insights into beetle-borne virus specificity. *PLoS One, 13*, e0192003.

Gennadius, P., (1889). Disease of tobacco plantations in the Trikonia. The aleurodid of tobacco. *Ellenike Georgia, 5*, 1–3.

Gera, A., Loebenstein, G., & Raccah, B., (1979). Protein coats of two strains of *Cucumber mosaic virus* affect transmission of *Aphisgossypii*. *Phytopathology, 69*, 396–399.

Gergerich, R. C., & Scott, H. A., (1988). Evidence that virus translocation and virus infection of non-wounded cells are associated with transmissibility by leaf-feeding beetles. *J. Gen. Virol., 69*, 2935–2938.

Gergerich, R. C., & Scott, H. A., (1991). Determinants in the specificity of virus transmission by leaf-feeding beetles. In: *Advances in Disease Vector Research* (pp. 1–13). Springer, New York.

Gergerich, R. C., & Scott, H. A., (1996). Comoviruses: Transmission, epidemiology, and control. In: *The Plant Viruses* (pp. 77–98). Springer, Boston, MA.

Gergerich, R. C., (1999). Comoviruses: Bean pod mottle comovirus. In: Hartman G. L., Sinclair J. B., & Rupe J. C., (eds.), *Compendium of Soybean Diseases* (4th edn., pp. 61, 62). St. Paul, Minnesota: APS Press.

Gergerich, R. C., (2001). Mechanism of virus transmission by leaf-feeding beetles. In: *Virus-Insect-Plant Interactions* (pp 133–142). Elsevier.

Ghanim, M., Morin, S., Zeidan, M., & Czosnek, H., (1998). Evidence for transovarial transmission of tomato yellow leaf curl virus by its vector, the whitefly *Bemisia tabaci*. *Virology, 240*, 295–303.

Ghosh, S., Kanakala, S., Lebedev, G., Kontsedalov, S., Silverman, D., Alon, T., Mor, N., et al., (2019). Transmission of a new polerovirus infesting pepper by the whitefly *Bemisia tabaci*. *J. Virol., 93*, e00488–19.

Gibb, K., & Randles, J., (1991). *Transmission of Velvet Tobacco Mottle Virus and Related Viruses by the Mirid Cyrtopeltis Nicotianae.* 10.1007/978-1-4613-9044-2_1.

Gibbs, M. J., & Weiller, G. F., (1999). Evidence that a plant virus switched hosts to infect a vertebrate and then recombined with a vertebrate-infecting virus. *Proceedings of the National Academy of Sciences of the United States of America, 96*(14), 8022–8027.

Giesler, L. J., Ghabrial, S. A., Hunt, T. E., & Hill, J. H., (2002). Bean pod mottle virus: A threat to U.S. soybean production. *Plant Dis., 86*, 1280–1289.

Gildow, F. E., & Gray, S. M., (1993). The aphid salivary gland basal lamina as a selective barrier associated with vector-specific transmission of barley yellow dwarf luteoviruses. *Phytopathology, 83*, 1293–1302.

Gildow, F. E., (1993). Evidence for receptor mediated endocytosis regulating luteovirus acquisition by aphids. *Phytopathology, 83*, 270–277.

Gildow, F. E., Damsteegt, V. D., Stone, A. L., Smith, O. P., & Gray, S. M., (2000). Virus-vector cell interactions regulating transmission specificity of soybean dwarf luteovirus. *J. Phytopathol., 148*, 333–342.

Godinho, M. T., Paula, D. P., Varsani, A., & Ribeiro, S. G., (2017). Genome sequence of cauliflower mosaic virus identified in earwigs (*Doru luteipes*) through a metagenomic approach. *Genome Announc.,* 5, e00043–17.

Gotoh, T., Araki, R., Boubou, A., Migeon, A., Ferragut, F., & Navajas, M., (2009). Evidence of co-specificity between *Tetranychus evansi* and *Tetranychus takafujii* (*Acari*: *Prostigmata, Tetranychidae*): Comments on taxonomic and agricultural aspects. *Int. J. Acarol., 35*, 485–501.

Hansen, A. K., & Moran, N. A., (2014). The impact of microbial symbionts on host plant utilization by herbivorous insects. *Mol. Ecol., 23*, 1473–1496.

Hanssen, I. M., & Lapidot, M., (2012). Chapter 2 – major tomato viruses in the mediterranean basin. In: Gad, L., & Hervé, L., (eds.), *Advances in Virus Research* (Vol. 84, pp. 31–66). Academic Press.

Harmel, N., Létocart, E., Cherqui, A., Giordanengo, P., Mazzucchelli, G., et al., (2008). Identification of aphid salivary proteins: A proteomic investigation of *Myzus persicae*. *Insect Mol. Biol., 17*, 165–174.

Harris, K. F., & Maramorosch, K., (1977). *Aphids as Virus Vectors*. Academic Press, New York.

Harris, K. F., (1979). Leafhoppers and aphids as biological vectors: Vector-virus relationships. In: Maramorosch, K., & Harris, K. F., (eds.), *Leafhopper Vectors and Plant Disease Agents* (pp. 217–308). Academic Press, New York.

He, Y. Z., Wang, Y. M., Yin, T. Y., Cuellar, W. J., Liu, S. S., & Wang, X. W., (2021). Gut-expressed vitellogenin facilitates the movement of a plant virus across the midgut wall in its insect vector. *mSystems, 6*, e0058121. doi: 10.1128/mSystems. 00581-21.

Heming, B. S., (1978). Structure and function of the mouthparts in larvae of *Haplothrips verbasci* (Osborn) (*Thysanoptera, Tubulifera, Phlaeothripidae*). *J. Morphol., 156*, 1–37.

Hennig, W., (1981). *Insect Phylogeny* (p. 514). John Wiley & Sons, New York, USA.

Hibino, H., Roechan, M., Sudarisman, S., & Tantera, D. M., (2018). *A Virus Disease of Rice ("Kerdil Hampa") Transmitted by Brown Plant hopper Nilaparvata lugens Stalin Indonesia.* Contributions Central Research Institute for Agriculture (Indonesia).

Hogenhout, S. A., Redinbaugh, M. G., & Ammar, E. D., (2003). Plant and animal rhabdovirus host range: A bug's view. *Trends Microbiol., 11*, 264–271.

Hogenhout, S. A., Verbeek, M., Hans, F., Houterman, P. M., Fortass, M., & Van, D. W. F., (1996). Molecular bases of the interactions between *Luteoviruses* and aphids. *Agronomie, 16*, 167–173.

Honda, K., Wei, T. Y., Hagiwara, K., Higashi, T., Kimura, I., & Akutsu, K., (2007). Retention of rice dwarf virus by descendants of pairs of *viruliferous* vector insects after rearing for 6 years. *Phytopathology, 97*, 712–716.

Hopkins, J. D., & Mueller, A. J., (1984). Effect of bean pod mottle virus on soybean yield. *J. Econ. Entomol., 77*, 943–947.

Hu, K., Yang, H., Liu, S., He, H., Ding, W., Qiu, L., & Li, Y., (2019). Odorant-binding protein 2 is involved in the preference of *Sogatella furcifera* (*Hemiptera*: *Delphacidae*) for rice plants infected with the southern rice black-streaked dwarf virus. *Florida Entomologist, 102*, 353.

Hull, R., (2002). *Matthew's Plant Virology*. Academic Press, New York, NY, USA.

Hull, R., (2014). *Plant Virology || Plant to Plant Movement,* 669–751.

Hull, R., (2014a). *Plant Virology* (5th edn.). Academic press, San Diego, CA.

Hull, R., (2014b). *Chapter 12 – Plant to Plant Movement Plant Virology* (5th edn.). *Academic Press*, Boston, MA.

Hunter, W. B., Hiebert, E., Webb, S. E., Tsai, J. H., & Polstin, J. E., (1998). Location of geminiviruses in the whitefly *Bemisia tabaci* (*Homoptera: Aleyrododae*). *Plant Dis., 82*(10), 1143–1151.

Hunter, W. B., Hiebert, E., Webb, S. E., Tsai, J. H., & Polston, J. E., (1998). Location of geminiviruses in the whitefly *Bemisia tabaci* (*Homoptera: Aleyrodidae*). *Plant Dis., 82*(10), 1147—1151.

Ingwell, L. L., Eigenbrode, S. D., & Bosque-Pe´rez, N. A., (2012). Plant viruses alter insect behavior to enhance their spread. *Sci. Rep., 2*, 578.

Islam, W., Noman, A., Naveed, H., Alamri, S. A., Hashem, M., Huang, Z., & Chen, H. Y. H., (2019). Plant-insect vector-virus interactions under environmental change. *Science of the Total Environment.*

Iwanowski, D., (1892). Über die Mosaikkrankheit der Tabakspflanze. Bulletin Scientifique Publié Par l'Académie Impériale des Sciences de Saint-Pétersbourg / Nouvelle Serie III (in German and Russian). *35*, 67–70. Translated into English in Johnson, J., Ed., (1942) Phytopathological classics (St. Paul, Minnesota: American Phytopathological Society) No. 7, pp. 27–30.

James, C. K. N., & Perry, K. L., (2004). Transmission of plant viruses by aphid vectors. *Molecular Plant Pathology, 5*, 505–511.

Jaspars, E. M., & Bos, L., (1980). Descriptions of plant viruses. *Commonw. Mycol. Inst. Assoc. Appl. Biol. Kew, UK* (p. 229). Alfalfa mosaic virus.

Jime´nez-Martı´nez, E. S., (2004). Volatile cues influence the response of *Rhopalosiphum padi* (*Homoptera*: *Aphididae*) to Barley yellow dwarf virus-infected transgenic and untransformed wheat. *Environ. Entomol., 33*, 1207–1216.

Klein, M., (2001). Transmission of viruses by leafhoppers and thrips. In: Loebenstein, G., Berger, P. H., Brunt, A. A., & Lawson, R. H., (eds.), *Virus and Virus-Like Diseases of Potatoes and Production of Seed-Potatoes.* Springer, Dordrecht. https://doi.org/10.1007/978-94-007-0842-6_21.

Kliot, A., Kontsedalov, S., Lebedev, G., Czosnek, H., & Ghanim, M., (2019). Combined infection with *Tomato yellow leaf curl virus* and rickettsia influences fecundity, attraction to infected plants and expression of immunity-related genes in the whitefly *Bemisia tabaci. J. Gen. Virol., 100*, 721–731.

Koudamiloro, A., Nwilene, F. E., & Silue, S., (2014). Identification of insect vectors of *Rice yellow mottle virus* (RYMV) in Benin. *Journal of Entomology, 11*(3), 153–162.

Krantz, G. W., & Lindquist, E. E., (1979). Evolution of phytophagous mites (Acari). *Annu. Rev. Entomol., 24*, 121–158.

Kristensen, N. P., (1991). Chapter 5 – Phylogeny of extant hexapods. In: *CSIRO, Division of Entomology. Insects of Australia* (2nd edn., Vol. 2, pp. 125–140, 1137). Cornell University Press, Ithaca, USA.

Kufferath, H., & Ghesquière, J., (1932). La mosaïque du manioc. *Compte Rendu Soc. Biol. Belge, 109*, 1146–1148.

Lee, H., Stephanus, A. P., Fowles, T. M., et al., (2022). Insect vector manipulation by a plant virus and simulation modeling of its potential impact on crop infection. *Sci. Rep., 12*, 8429.

Lefkowitz, E. J., Davison, A. J., Sabanadzovic, S., Siddell, S. G., & Simmonds, P., (2017). *Virus Taxonomy: Classification and Nomenclature of Viruses.* Online (10th) Report of the International Committee on Taxonomy of Viruses.

Lei, T., Zhao, J., Wang, H. L., Liu, Y. Q., & Liu, S. S., (2021). Impact of a novel rickettsia symbiont on the life history and virus transmission capacity of its host whitefly (*Bemisia tabaci*). *Insect Sci., 28*, 377–391.

Lewis, T., (1973). *Thrips, Their Biology, Ecology and Economic Importance.* Academic Press London, New York, pp. 349.

Li, S., Zhang, T., Zhu, Y., & Zhou, G., (2017). Co-infection of two reoviruses increases both viruses accumulation in rice by up-regulating of viroplasm components and movement proteins bilaterally and RNA silencing suppressor unilaterally. *Virol. J., 14*(1), 150.

Li, S., Zhou, C., & Zhou, Y., (2019). *Olfactory co-receptor orco* stimulated by *Rice stripe virus* is essential for host seeking behavior in small brown planthopper. *Pest Management Science. 75*, 187–194.

Ling, K. C., (1966). Non persistence of the *Tungrovirus* of rice in its leafhopper vector, *Nephoteuix impicticeps. Phytopathol., 56*, 1252–1256.

Link, A., & Fuchs, M., (2005). Transmission specificity of plant viruses by vectors. *Journal of Plant Pathology*, *87*(3), 153–165.

Loeffler, F., & Frosch, P., (1898). Report of the commission for research on foot-and-mouth disease. *Zentrabl. Bacteriol. Parastenkunde Infektionkrankh. 23*, 371–391.

Madriz-Orden˜ana, K., Rojas-Montero, R., Lundsgaard, T., Ram´ırez, P., Thordal-Christensen, H., & Collinge, D. B., (2000). Mechanical transmission of *Maize rayado fino marafivirus* (MRFV) to maize and barley by means of the vascular puncture technique. *Plant Pathol., 49*(2), 302–307.

Mahatma, L., Mahatma, M. K., Pandya, J. R., Solanki, R. K., & Solanki, V. A., (2016). Epidemiology of *Begomoviruses*: A global perspective. *Plant Viruses: Evolution and Management* (pp 171–188). Springer, Singapore.

Mark, J. B., & Georg, F. W., (1999). Evolution evidence that a plant virus switched hosts to infect a vertebrate and then recombined with a vertebrate-infecting virus. *Proc. Natl. Acad. Sci. USA, 96*, 8022–8027.

Markham, R., & Smith, K. M., (1949). Studies on the virus of turnip yellow mosaic. *Parasitology*, *39*(3, 4), 330–342.

Martelli, G. P., Adams, M. J., Kreuze, J. F., & Dolja, V. V., (2007). Family *Flexiviridae*: A case study in virion and genome plasticity. *Annu. Rev. Phytopathol.*, *45*(1), 73–100.

Mauck, K. E., Chesnais, Q., & Shapiro, L. R., (2018). Evolutionary determinants of host and vector manipulation by plant viruses. *Adv. Virus Res., 101*, 189–250.

Medina, O. K. J., Bosque-Pe´rez, N. A., Ngumbi, E., Jime´nez-Martı´nez, E. S., & Eigenbrode, S. D., (2009). *Rhopalosiphum padi* (Hemiptera: *Aphididae*) responses to volatile cues from *Barley yellow dwarf virus*-infected wheat. *Environ. Entomol., 38*, 836–845.

Mitchell, P. L., (2004). *Heteroptera* as vectors of plant pathogens. *Neotropical Entomology, 33*, 519–545.

Moore, J., (2002). *Parasites and the Behavior of Animals.* Oxford University Press, Oxford.

Morimoto, K. M., & Ramsdell, D. C., (1985). Aphid vector population dynamics and movement relative to field transmission of blueberry shoestring virus. *Phytopathology*, *75*, 1217–1222.

Mound, L. A., & Kibby, G., (1998). *Thysanoptera: An Identification Guide* (2nd edn., p. 70). CAB Int., Wallingford, United Kingdom.

Mound, L. A., & Marullo, R., (1996). The thrips of central and south America: An introduction (*Insecta: Thysanoptera*). *Memoirs on Entomology, International, 6*, 1–487.

Mound, L. A., (1996). The thysanoptera vector species of *Tospoviruses*. *Acta Horticulturae, 431*, 298–309.

Mound, L. A., (1997). Biological diversity. In: Lewis, T., (ed.), *Thrips as Crop Pest* (pp. 197–215). CAB International, London.

Mound, L. A., (2001). So many thrips–so few *tospoviruses*? Thrips and *Tospoviruses*. In: Marullo, R., & Mound, L., (eds.), *Thrips and Tospoviruses*: *Proceedings of the 7th International Symposium on Thysanoptera* (pp. 3–6). Reggio Calabria, Italy. ANIC, Reggio, Calabria, Italy.

Mound, L. A., (2009). Chapter 254 – thysanoptera. Resh, V. H., & Cardé, R. T., (eds.), *Encyclopedia of Insects* (2nd edn., pp. 999–1003). Academic Press. https://doi.org/10.1016/B978-0-12-374144-8.00263-0.

Mugerwa, H., Colvin, J., & Alicai, T., (2021). Genetic diversity of whitefly (*Bemisia* spp.) on crop and uncultivated plants in Uganda: Implications for the control of this devastating pest species complex in Africa. *J. Pest Sci., 94*, 1307–1330.

Mugerwa, H., Seal, S., & Wang, H. L., (2018). *African Ancestry of New World. Sci. Rep., Bemisia Tabaci Whitefy Species.* https://doi. org/10.1038/s41598-018-20956-3.

Muraint, A. F., Roberts, I. M., & Elngar, S., (1997). Association of virus-like particles with the foregut of the aphid *Cavariella aegopodii* transmitting the semi-persistent viruses Anthriscus yellow and parsnip yellow fleck. *J. Gen. Virol., 31*(1), 47–57.

Murant, A. F., Roberts, I. M., & Elngar, S., (1976). Associayion of virus-like particles with the foregut of the aphid *Cavariella aegopodii* transmitting the semi persistent viruses Anthriscus yellow and parsnip yellow. *J. Gen. Virol., 31*(1), 47–57.

Musser, R. O., Hum-Musser, S. M., Felton, G. W., & Gergerich, R. C., (2003). Increased larval growth and preference for virus-infected leaves by the Mexican bean beetle, *Epilachna varivestis* Mulsant, a plant virus vector. *J. Insect Behav., 16*, 247–256.

Nakahara, S., (1997). Annotated list of the frankliniella species of the world (*Thysanoptera*: *Thripidae*). *Contributions on Entomology, International, 2*, 355–389.

Nakamura, H., & Chavez, F. G., (2007). *Report of the Investigation on Epilachna varivestis Defoliation in Guatemala High Land.* Bulletin Shinshu University Alpine Field Center.

Nault, L. R., & Ammar, E. D., (1989). Leafhopper and planthopper transmission of plant viruses. *Annual Review of Entomology, 34*(1), 503–529.

Nault, L. R., (1978). Transmission of *Maize chlorotic mottle virus* by chrysomelid beetles. *Phytopathology, 68*, 1071.

Nault, L. R., (1997). Arthropod transmission of plant viruses: A new synthesis. *Ann. Entomol. Soc. Am., 90*, 521–541.

Nault, L. R., Gingery, R. E., & Gordon, D. T., (1980). Leaf hopper transmission and host range of *Maize rayado fino virus*. *Phytopathology, 70*(8), 709–712.

Nault, L. R., Styer, T. W. E., Knoke, J. K., & Pree, H. N., (1973). Semipersistent transmission of leafhopper-borne *Maize chlorotic dwarf virus*. *J. Econ. Entomol.*, *66*(6), 1271–1273.

Navas-Castillo, J., Fiallo-Olivé, E., & Sánchez-Campos, S., (2011). Emerging virus diseases transmitted by whiteflies. *Ann. Rev. Phytopathol. 49*, 219–248.

Ng, J. C. K., & Perry, K. L., (1999). Stability of the aphid transmission phenotype in *Cucumber mosaic virus*. *Plant Pathology, 48*(3), 388–394.

Noda, H., Watanabe, K., Kawai, S., Yukuhiro, F., Miyoshi, T., & Tomizawa, M., (2012). Bacteriome-associated endosymbionts of the green rice leafhopper *Nephotettix cincticeps (Hemiptera: Cicadellidae)*. *Appl. Entomol. Zool., 47*, 217–225.

Nwilene, F. E., Traore, A. K., Asidi, A. N., Sere, Y., Onasanya, A., & Abo, M. E., (2009). New records of insect vectors of *Rice yellow mottle virus* (RYMV) in cote d'Ivoire, West Africa. *Journal of Entomology*, *6*(4), 189–197.

Oldfield, G. N., (1996). Diversity and host plant specificity. In: Lindquist, E. E., Sabelis, M. W., & Bruin, J., (eds.), *Eriophyoid Mites—Their Biology, Natural Enemies and Control* (Vol. 6, pp. 199–216). Elsevier Science Publishing, Amsterdam, The Netherlands, World Crop Pests.

Olivé, E. F., & Castillo, J. N., (2019). Pathogen profile *Tomato chlorosis virus*, an emergent plant virus still expanding its geographical and host ranges. *Molecular Plant Pathology, 20*(9), 1307–1320.

Orlob, G. B., (1966). Feeding and transmission characteristics of *Aceria tulupae* Keifer as vector of *Wheat streak mosaic virus*. *Phytopathol.*, *Z*(55), 218–238.

Osmat, A., & Tim, C. B., (2002). The biology, epidemiology, and management of Rice tungro disease in Asia. *Plant Disease, 86*(2), 14–16.

Ostfeld, R. S., Keesing, F., & Eviner, V. T., (2010). *Infectious Disease Ecology (Effects of Ecosystems on Disease and of Disease on Ecosystems) – Chapter 2*. The Role of Vector Diversity in Disease Dynamics. doi: 10.1515/9781400837885.30.

Palacios, I., Drucker, M., Blanc, S., Leite, S., Moreno, A., & Fereres, A., (2002). *Cauliflower mosaic virus* is preferentially acquired from the phloem by its aphid vectors. *J. Gen. Virol.*, *83*, 3163–3171.

Palukaitis, P., & García-Arenal, F., (2003a). *Cucumber Mosaic Virus*. Descr. Plant Viruses No. 400.

Pan, L. L., Chen, Q. F., Zhao, J. J., Guo, T., Wang, X. W., Hariton-Shalev, A., Czosnek, H., & Liu, S. S., (2017). Clathrin-mediated endocytosis is involved in *Tomato yellow leaf curl virus* transport across the midgut barrier of its whitefly vector. *Virology, 502* 152–159.

Pinto, Z. V., Rezende, J. A. M., Yuki, V. A., & Piedade, S. M. S., (2008). Ability of *Aphis gossypii* and *Myzus persicae* to transmit *Cucumber mosaic virus* in single and mixed infection with two potyviruses to zucchini squash. *Summa Phytopathologica, 34*(2), 183–185.

Pirone, T. P., & Blanc, S., (1996). Helper-dependent vector transmission of plant viruses. *Annual Review of Phytopathology, 34*, 227–247.

Pirone, T. P., & Perry, K. L., (2002). Aphids-nonpersistent transmission. *Adv. Bot. Res., 36*, 1–19.

Pollard, D. G., (1973). Plant penetration by feeding aphids (*Hemiptera: Aphidoidea*): A review. *Bull. Entomol. Res., 62*, 631–714.

Poulin, R., (2010). Parasite manipulation of host behavior: An update and frequently asked questions. In: Jane, B. H., (ed.), *Advances in the Study of Behavior* (Vol. 41, pp. 151–186). Burlington: Academic Press. doi: 10.1016/S0065-3454(10), 41005-0.

Proctor, H. C., Smith, I. M., Cook, D. R., & Smith, B. P., (2015). Chapter 25 – *Subphylum chelicerata*, class *Arachnida*. In: James, H. T., & Christopher, R. D., (eds.), *Thorp and Covich's Freshwater Invertebrates* (4th edn., pp. 599–660). Academic Press.

Proeseler, G., (1966a). Relationships between *Piesma quadratum* and *Beet leaf curl virus*. I. Experiments of virus transmissibility and rearing of the vector. *Phytopathologische Zeitschrift, 56*, 191–211.

Proeseler, G., (1983). *Beet Leaf Curl Virus*. CMI/AAB Descriptions of Plant Viruses No. 268. Association of Applied Biologists, Wellesbourne, UK.

Rabinowitch, H. D., & Levy, D., (2001). Biology and physiology of the potato. *Virus and Virus-Like Diseases of Potatoes and Production of Seed-Potatoes,* 19–37. doi: 10.1007/978-94-007-0842-6_2.

Raccah, B., & Fereres, A., (2009). Plant virus transmission by insects. In: *Encyclopedia of Life Sciences (ELS).* Chichester: John Wiley and Sons Ltd.

Ramirez, B. C., & Haenni, A. L., (2004). Molecular biology of tenuiviruses, a remarkable group of plant viruses. *J. Gen. Virol., 75*, 467–475.

Ramos-González, P. L., Chabi-Jesus, C., Guerra-Peraza, O., Breton, M. C., Arena, G. D., & Nunes, M. A., (2016). Phylogenetic and molecular variability studies reveal a new genetic clade of *Citrus leprosis virus* C. *Viruses, 8*, 153.

Randles, J. W., & Carver, M., (1971). Epidemiology of *Lettuce necrotic yellows virus* in South Australia. II. Distribution of virus, host plants, and vectors. *Aust. J. Agric. Res., 22*(2), 231–237.

Reed, W., Carroll, J., & Agramonte, A., (1901). The etiology of yellow fever an additional note. *JAMA, XXXVI*(7), 431–440.

Reinbold, C., Gildow, F. E., Herbach, E., Ziegler-Graff, V., Goncalves, M. C., & Van, D. H. J. E., (2001). Studies on the role of the minor capsid protein in transport of *Beet western yellow virus* through *Myzus persicae* and *Aphis gossypii*. *J. Gen. Virol., 84*(12), 3473–3484.

Reinbold, C., Herrback, E., & Brault, V., (2003). Posterior midgut and hindgut are both sites of acquisition of *Cucurbit aphid born yellow virus* in *Myzus persicae* and *Aphis gossypii*. *J. Gen. Virol., 84*(12), 3473–3484.

Richard, O. S., Felicia, K., & Valerie, E. T., (2010). Infectious disease ecology (effects of ecosystems on disease and of disease on ecosystems) – Chapter 2. *The Role of Vector Diversity in Disease Dynamics*. doi: 10.1515/9781400837885.30.

Rio, R. V. M., Attardo, G. M., & Weiss, B. L., (2016). Grandeur alliances: Symbiont metabolic integration and obligate arthropod hematophagy. *Trends Parasitol.*, *32*, 739–749.

Roivainen, O., (1976). Transmission of cocoa viruses by mealy bugs (*Homoptera: Pseudococcidae*). *J. Sci. Agric. Soc., 48*, 433–453.

Sarwar, M., (2020). Insects as transport devices of plant viruses. In: Awasthi, L. P., (ed.), *Applied Plant Virology* (pp. 381–402). Academic Press: Cambridge, MA, USA.

Scott, H. A., & Fulton, J. P., (1978). Comparison of the relationships of *Southern bean mosaic virus* and the cowpea strain of tobacco mosaic virus with the bean leaf beetle. *Virology, 84*, 207–209.

Scott, H. A., & Phatak, H. C., (1979). Properties of *Blackgram mottle virus*. *Phytopathology, 69*, 346–348.

Shi, X., Tang, X., Zhang, X., Zhang, D., Li, F., Yan, F., & Liu, Y., (2018). Transmission efficiency, preference, and behavior of *Bemisia tabaci* MEAM1 and MED under the influence of *Tomato chlorosis virus*. *Frontiers in Plant Science*, *8*, 1–9.

Simons, J. N., & Coe, D. M., (1958). Transmission of *Pseudo-curly top virus* in Florida by a treehopper. *Virology, 6*, 43–48.

Sinha, R. C., & Shelley, S., (1965). The transovarial transmission of wound-tumor virus. *Phytopathology, 55*, 324–327.

Smith, C. M., Gedling, C. R., Wiebe, K. F., & Cassone, B. J., (2017). A sweet story: *Bean pod mottle virus* transmission dynamics by Mexican bean beetles (*Epilachna varivestis*). *Genome Biol. Evol., 9*, 714–725.

Stork, N. E., (2018). How many species of insects and other terrestrial arthropods are there on earth? *Annu. Rev. Entomol., 63*, 31–45.

Stubbs, L. L., & Grogan, R. G., (1963). Necrotic yellows: A newly recognized virus disease of lettuce. *Aust. J. Agric. Res.*, *14*(4), 439–459.

Takata, K. J., (1895). Japan. *Agr. Soc., 171*, 1–4. Cited by Katsura, S., (1936). In: *Phytopathology, 26*, 887–895.

Takata, K. J., (1896). Japan. *Agr. Soc., 172*, 13–32. Cited by Katsura, S., (1936). In: *Phytopathology*, *26*, 887–895.

Tassi, A. D., Garita-Salazar, L. C., Amorim, L., Novelli, V. M., Freitas-Astúa, J., Childers, C. C., & Kitajima, E. W., (2017). Virus-vector relationship in the Citrus leprosis pathosystem. *Experimental & Applied Acarology*, *71*(3), 227–241.

Tipping, C., (2004). Thrips (*Thysanoptera*). In: *Encyclopedia of Entomology*. Springer, Dordrecht. https://doi.org/10.1007/0-306-48380-7_4300.

Togola, A., Nwilene, F. E., Agbaka, A., Degila, F., Tolulope, A., & Chougourou, D., (2011). Screening upland varieties of NERICA and its parents for resistance to stalk-eyed fly, *Diopsis* sp. (*Diptera, Diopsidae*) in Benin. *Journal of Applied Sciences*, *11*(1), 145–150.

Turka, I., (1978). *Lygus rugulipennis* Popp. (*Heteroptera: Miridae*) vector of potato viruses. Trudy Latv. *Lauksaimn. Akad., 164*, 65–73.

Van, D. H. J. F., Verbeek, M., & Van, D. W. F., (1994). Endosymbiotic bacteria associated with circulative transmission of *Potato leafroll virus* by *Myzus persicae. J. Gen. Virol., 75*, 2559–2565.

Van, D. P., Veerbek, M., & Bos, L., (1991). Mite-borne viruses isolates from cultivated *Allium*s species and their classification into two new rymoviruses in the family *Potyviridae. J. Plant Pathol., 97*, 381—399.

Verbeek, M., Dullemans, A. M., Van, D. H. J. F. J. M., Maris, P. C., & Van, D. V. R. A. A., (2007). Identification and characterization of *Tomato torrado virus*: A new plant picorna-like virus from tomato. *Arch. Virol., 152*, 881–890.

Vetten, H. J., Dale, J. L., Grigoras, I., Gronenborn, B., Harding, R., Randles, J. W., & Yeh, H. H., (2012). Family *Nanoviridae*. In: King, A. M. Q., Adams, M. J., Carstens, E. B., & Lefkowitz, E. J., (eds.), *Virus Taxonomy—Ninth Report of the International Committee on Taxonomy of Viruses* (pp. 395–404). Amsterdam, Elsevier, The Netherlands.

Vetten, H. J., Gronenborn, B., & Bressan, A., (2016). Chapter 34: Aphid transmission of viruses of the family *Nanoviridae*. In: Brown, J. K., (ed.), *Vector-Mediated Transmission of Plant* Pathogens (pp. 453–462). St. Paul, MN: APS Press. https://doi.org/10.1094/9780890545355.034.

Wang, R. Y., Ammar, E. D., Thornbury, D. W., Lopez-Moya, J. J., & Pirone, T. P., (1996). Loss of potyvirus transmissibility and helper-component activity correlate with non-retention of virions in aphid stylets. *J. Gen. Virol., 77*, 861–867.

Wang, S., Guo, H., Ge, F., & Sun, Y., (2020). Apoptotic neuro degeneration in whitefly promotes the spread of TYLCV. *eLife, 9*, e56168.

Watson, M. A., & Roberts, F. M., (1939). A comparative study of the transmission of hyocyamus virus 3, *Potato virus Y* and *Cucumber virus* by the vector *Myzus persicae* (Sulz), *M. circumflexus* (Buckton) and *Macrosiphum gei* (Koch). *Proc. R. Soc. London B., 127*, 543–576.

Weber, H., (1930). *Biologie Der Hemipteran* (p. 543). Julius Springer, Berlin, Germany.

Werner, B. J., Mowry, T. M., Bosque-Pe´rez, N. A., Ding, H., & Eigenbrode, S., (2009). Changes in green peach aphid responses to *Potato leafroll virus*-induced volatiles emitted during disease progression. *Environ. Entomol., 38*, 1429–1438.

Wickler, W., (1976). Evolution-oriented ethology, kin selection, and altruistic parasites. *Z. Tierpsychol., 42*, 206–214.

Wintermantel, W. M., & Wisler, G. C., (2006). Vector specificity, host range, and genetic diversity of *Tomato chlorosis virus*. *Plant Dis., 90*, 814–819.

Wu, W., Shan, H. W., Li, J. M., Zhang, C. X., Chen, J. P., & Mao, Q., (2022). Roles of bacterial symbionts in transmission of plant virus by hemipteran vectors. *Front. Microbiol., 13*, 805352.

Wylie, S. J., Adams, M., Chalam, C., Kreuzue, J., Lopez-Maya, J. J., & Ohshima, K., (2017). ICTV virus taxonomy profile: Potyviridae. *J. Gen. Virol., 98*(3), 352–354.

Xu, H., He, X., Zheng, X., Yang, Y., Tian, J., & Lu, Z., (2014). *Southern rice black-streaked dwarf virus* (SRBSDV) directly affects the feeding and reproduction behavior of its vector, *Sogatella furcifera* (Horva´th) (*Hemiptera: Delphacidae*). *Virol. J., 11*(1), 55.

Yokomi, R. K., Lastra, R., Stoetzel, M. B., Lee, R. F., Garnsey, S. M., Gottwald, T. R., Rocha-Pena, M. A., & Niblett, C. L., (1994). Establishment of the brown citrus aphid (*Homoptera*:

Aphididae) in central America and the Caribbean basin and transmission of citrus tristeza virus. *J. Econ. Entomol.*, *87*, 1078–1085.
Zaumeyer, W. J., & Thomas, H. R., (1957). *A Monographic Study of Bean Diseases and Methods for Their Control.* U.S.D.A. Agr. Tech. Bull. No. 868, 255.
Zitter, T. A., & Tsai, J. H., (1977). Transmission of three potyviruses by the leaf miner *Liriomyza sativae* (*Diptera: Agromyzidae*). *Plant Disease Reporter*, *61*(12), 1025–1029.

CHAPTER 6

The Pomegranate Viruses and Their Pathogenic Expression

KHUSHBOO JAIN,[1] SHALINI TAILOR,[1] MUKESH MEENA,[2] CHITRA NEHRA,[3] RAJARSHI KUMAR GAUR,[4] and AVINASH MARWAL[1]

[1]*Department of Biotechnology, Vigyan Bhawan, New Campus, Mohanlal Sukhadia University, Udaipur, Rajasthan, India*

[2]*Laboratory of Phytopathology and Microbial Biotechnology, Department of Botany, Mohanlal Sukhadia University, Udaipur, Rajasthan, India*

[3]*Department of Biotechnology, School of Life Sciences, Mody University, Lakshmangarh, Sikar, Rajasthan, India*

[4]*Department of Biotechnology, Deen Dayal Upadhyaya Gorakhpur University, Gorakhpur, Uttar Pradesh, India*

ABSTRACT

Pomegranate is one of the most important edible fruit crops in the Mediterranean (MED) basin, with pharmaceutical and therapeutic applications. The increase in production is simultaneously affected by various insect pests, bacterial diseases, fungal diseases, and even viral diseases, which also have a negative impact on this vital cash crop. Some important plant viruses, such as cherry leaf roll virus (CLRV), cucumber mosaic virus (CMV), tomato ringspot virus (ToRSV), and hop stunt viroid (HSVd), are majorly responsible for economic losses in pomegranate trees. Pathogenicity is the potential ability of the pathogen to cause the disease. The sequential process of pathogenesis is not only related to the etiology of the disease but also dependent

New and Emerging Plant Viruses: The Threat to Food Security.
Alireza Golnaraghi and Rajarshi Kumar Gaur (Eds.)

on the disease progression and maintenance. The successful pathogenesis of viruses in plants is influenced by the replication of viruses at the cellular level, the transport of viruses from one cell to another, the long-distance spread of infection, the type of tissue infected, the host environment, and the defense mechanism of the plants. Due to the infection-defense equilibrium, plant viruses generally do not cause plant death or may be in asymptomatic association to provide selective advantage to the host. This chapter mainly focuses on the pathogenesis of different plant viruses on pomegranate trees, the genes involved in the regulation of pathogenesis, the factors that influence this process, and the defense mechanism of the plant.

6.1 INTRODUCTION

Pomegranate (*Punica granatum* L.) is a popular fruit crop in arid and semi-arid climates. Pomegranates are native to Central Asia, mainly Iran, and are grown in a variety of temperatures throughout the Mediterranean (MED), Asia, and the United States. The root, bark, wood, sprouts, leaves, flowers, fruits, rind, and seeds are all economically valuable parts of the pomegranate (Jain & Desai, 2018). Bioactive chemicals found in the fruit, flowers, bark, and leaves act as antimicrobials, lower blood pressure, and fight major diseases including diabetes and cancer (Jain et al., 2020). It is one of the richest sources of antioxidants and has tremendous therapeutic and nutritional potential. The pomegranate is renowned as a "Superfood" because of its medical and health benefits. The expansion of pomegranate orchards has been driven by pomegranate processing companies. Due to its ability to thrive in slightly saline soils, it is characterized as a saline-tolerant plant. Pomegranate production is expanding at a faster rate due to its high market demand, long storage life, high nutritional and medicinal value, heat and drought tolerance, low-cost management, ability to regulate blooming, and wider adaptability to varied soils (Jain & Desai, 2018).

India is the world's largest producer of pomegranates, with a production of 743.1000 tons. Iran, Turkey, the United States, Afghanistan, and Spain follow as the top pomegranate producers. In India, pomegranates are extensively cultivated in Maharashtra, Gujarat, Karnataka, Andhra Pradesh, Tamil Nadu, Madhya Pradesh, and Rajasthan. In the year 2020–2021, pomegranates were cultivated on 2.22 lakh hectares in India, resulting in an annual production of 26.34 lakh tons and a productivity of 11.86 tons per hectare (NHB Database, 2021).

The commercial marketability of pomegranate fruit is threatened by pathogens that cause postharvest rots and degradation. Pomegranates are susceptible to disease both before and after harvest, which significantly affects fruit quality and storage life. The three major forms of these diseases are fruit surface rots, interior fruit infections, and stem end rots (Ammar & El-Naggar, 2014). Fruit infections can impact all stages of production and lead to significant yield losses. A severe disease epidemic can quickly devastate an entire orchard or fruit shipment (Petersen et al., 2010; Day & Wilkins, 2011).

Pathogens such as *Botrytis cinerea* (gray mold rot), *Alternaria alternate* (heart rot), *Aspergillus niger* (heart rot), blue/green mold rot (*Penicillium* spp., *P. implicatum Biourge*, and *P. expansum* Link), *Coniella granati*, *Colletotrichum gloeosporioides*, and *Pestalotiopsis* versicolor are all important fungal diseases affecting pomegranates. Bacterial blight caused by *Xanthomonas axanopodis* pv *punicae* has also been implicated in causing severe preharvest infections of pomegranates, along with *B. cinerea*, *Penicillium* spp., *A. alternata, A. niger, Aureobasidium pullulans G. arnaud, C. granati,* and *Cytosporapunicae Sacc.,* which causes wood canker and branch dieback. Pomegranate disease infections may remain dormant until they become symptomatic during postharvest storage. These pathogens contribute to fruit spoilage and accelerate the deterioration of the fruit rind and arils during storage.

There are several types of research on various pathogens that affect pomegranates, including fungi, bacteria, and phytoplasmas (Gazel et al., 2015). However, there are few studies on viral and virus-like diseases that infect this crop. Viruses pose a significant threat to the long-term viability and productivity of agriculture, costing billions of dollars annually (Muthupandi et al., 2019). The impact of emerging diseases is particularly significant due to their rapid increase in disease incidence, geographic range, and/or pathogenicity (Verma et al., 2020). Factors such as monocrop culture, viruses, host and vector transmission through germplasm and live plants, climate change, and the ability of viruses to evolve and adapt quickly all contribute to pomegranate viral infections.

To date, only cucumber mosaic virus (CMV), ToRSV, CLRV, and hop stunt viroid (HSVd) have been identified in pomegranate trees (Table 6.1) (Horvath, 1984; Gomez & Pallas, 2001; Astruc et al., 1996; Gazel et al., 2009). This chapter provides a comprehensive summary of viral pathogenesis, the molecular mechanisms involved, factors influencing this process, and the defense mechanisms adopted by the plants.

TABLE 6.1 List of Plant Viruses Associated with Pomegranate

Virus Name	Vector	Place	Detection Technique	Genome Organization	References
CLRV	Seeds and pollen	Emilia-Romagna and Italy	ELISA, RT-PCR	Bipartite single-stranded positive-sense RNA (RNA-1 and RNA-2).	Bichcheri et al. (2015); Carlo et al. (2014)
ToRSV	Nematode vector, *Xiphinema americanum* sensulato	Italy	ELISA, RT-PCR	Bipartite single-stranded, positive-sense RNA.	Matteo et al. (2016); EPPO (2015)
CMV	Aphids and seeds	Yugoslavia	RT-PCR	Composed of a CP and three single-stranded positive-sense RNAs, RNA 1, RNA 2 and RNA 3.	Horvath (1984)
HsVD	Planting material and mechanical material	Turkey	RT-PCR	Five structural domains with terminal conserved hairpin.	Gazel et al. (2009); Astruc et al. (1996)
PeSV	Arthropods	Turkey	HTS, RT-PCR, electron microscopy	Monopartite or bipartite single-stranded positive-sense RNA genome ranging from 8.2 up to 11.3 kb in size.	Caglayan et al. (2020)

6.2 GENOME ORGANIZATION AND HOST RANGE

6.2.1 *CUCUMBER MOSAIC VIRUS (CMV)*

Along with CMV, tomato aspermy virus (TAV), peanut stunt virus (PSV), and gayfeather moderate mottle virus (GMV) are all members of the Cucumovirus genus (GMMV). CMV infects 1,241 host species over 101 plant groups, including monocots, dicots, feed products, ornamentals, and a variety of wild species, in contrast to TAV and PSV, which have a very limited host range (Jacquemond, 2012). Horvath was the first to report a relation between CMV and pomegranate in Yugoslavia in 1984.

CMV virions are 29-nm-diameter icosahedral particles composed of 180 subunits of a single capsid protein (CP) and 18% RNA. The genome consists of three single-stranded positive-sense RNAs, RNA 1, RNA 2, and RNA 3, with lengths of 3,300 nt, 3,000 nt, and 2,200 nt, respectively. RNA1 is monocistronic and encodes protein 1a, which contains a putative methyltransferase domain in its N-terminal part and a helicase motif in its C-terminal part. RNA2 encodes the large 2a protein, which has an RNA-dependent RNA polymerase (RdRp) motif, and the small 2b protein that disrupts the host's RNA interfering (RNAi) pathway. RNA3 is bicistronic and codes for the viral 3a protein (or movement protein (MP)) and the CP. The overall length of each RNA may vary slightly depending on the strain or strain grouping. CMV particles can also encapsidate small RNAs, including RNA5 and satellite RNA, which have been extensively studied. Mutations or deletions in the RNA5 gene affect protein production but do not impact virus accumulation or symptom development. Some viral progenies have faulty RNA3, with a few 100 nucleotide deletion in the MP or CP genes, or both. These RNAs were discovered in a natural CMV isolate but had no effect on the viral infection cycle (Jacquemond, 2012).

6.2.2 CHERRY LEAF ROLL VIRUSES (CLRVS)

Posnette & Cropley characterized cherry leaf roll virus (CLRV) as the cause of a disease in sweet cherry (*Prunus avium* L.) in England in 1955 (Cropley, 1961). Since then, it has been discovered to infect a wide range of natural hosts, including both herbaceous and woody plants. Common birch (*Betula pendula Roth*), black elderberry (*Sambucus nigra* L.), English walnut (*Juglans regia* L.), and sweet cherry are among the most common natural hosts of CLRV. The virus is widespread, having been found in Europe, the former Soviet Union, North America, Chile, New Zealand, Australia, China, and Japan (Hetrera & Madariaga, 2001; Jones, 1986).

The bipartite single-stranded positive-sense RNA genome of CLRV is estimated to be about 15 kb in total, with RNA-1 and RNA-2 sizes of approximately 8.2 and 6.8 kb, respectively (Murant et al., 1981). Both RNAs are individually encapsulated in isometric particles (Jones Loc.cit). Genomic RNAs have a covalently attached genome-encoded protein (VPg) at their 5′-terminus and are polyadenylated at their 3′-terminus. CLRV belongs to the C nepovirus subgroup. They possess a large, individually encapsulated RNA-2 with a long (1.2 to 1.6 kb) 3′-noncoding section that is identical or nearly identical to RNA-1. This significant conservation of the 3′-NCR between the two genomic RNAs has been attributed to an RNA recombination mechanism that operates as part of the RNA-2 replication process of these viruses (Rebenstorf, 2006).

6.2.3 HOP STUNT VIROID (HSVD)

Viroids are the smallest known pathogens with self-replicating, naturally infecting plants (Flores et al., 2015). The mature form is a single-stranded small RNA molecule (246–434 nucleotides [nt]) with a highly compact shape that is covalently closed (Steger & Perreault, 2016). It was first discovered in hop plants (*Humulus lupulus*) with abnormal dwarfing of bines (Sasaki & Shikata, 1977). This hop disease was first described in Japan in the 1940s–1950s, and in the 1970s, it was estimated that between 10% and 20% of hop gardens were affected, resulting in significant economic losses (Hataya et al., 2017). HSVd was sequenced and found to have a 297-nt circular RNA with a rod-like shape that was 95% identical to CPFVd, making the latter an HSVd isolate rather than a distinct viroid species (Sano et al., 1984). Sano et al. (2001) postulated a grapevine origin for the HSVd isolate that causes hop stunt disease based on phylogenetic research, which was later verified in a 15-year evolutionary study (Kawaguchi-Ito et al., 2009).

HSVd has been found in plants belonging to the Moraceae, Rosaceae, Anacardiaceae, Lythraceae (Maddahian et al., 2019; Astruc et al., 1996), Malvaceae (Luigi et al., 2013), and Rutaceae families (Hataya et al., 2017; Vamenani et al., 2019).

The genome of Pospiviroidae members is classified into five structural domains (Keese & Symons, 1985): two terminal regions, left (TL) and right (TR), pathogenic (P), variable (V), and a central domain (C) that contains a CCR, which is the characteristic feature of all members of the family (Di Serio et al., 2014). HSVd also has a terminal conserved hairpin (TCH), which is found in members of the genus Cocadviroid as well as the type species of the genus *Coleviroid* (*Coleus blumei viroid* 1) (Di Serio et al., 2017).

In combination with its host range, the low-fidelity replication of RNA polymerase II, which is compelled to take viroid RNAs as a template, may explain the diversity of HSVd isolates. HSVd sequences have been classified into five categories based on phylogenetic analysis (Amari et al., 2001; Jo et al., 2017; Kofalvi et al., 1997). Plum-type, hop-type, and citrus-type sequence variants are found in three of these groups. On the other hand, the other two groups have signatures that point to a recombinant.

6.2.4 TOMATO RINGSPOT VIRUS (TORSV)

Tomato ringspot virus (ToRSV) is a bipartite single-stranded positive-sense RNA virus (genus Nepovirus, subgroup III, family Comoviridae). At its 5′-terminus, every RNA is covalently bonded to a small virus-encoded protein (VPg), which encodes a single large polyprotein. The genetic organization of nepoviruses is similar to that of animal picornaviruses (Mayo & Robinson, 1996; Sanfacon, 1995). The domains for proteins expected to be involved in replication are found in the RNA1-encoded polyprotein (P1), which includes a putative nucleoside triphosphate binding (NTB) protein, the VPg, the 3C-like proteinase (Pro), and the RNA-dependent RNA polymerase (Pol). *In vitro*, the proteinase cleaves P1 at five different sites, and the exact positions of the NTB-VPg, VPg-Pro, and Pro-Pol cleavage sites have been determined (Wang & Sanfacon, 2000). The NTB protein shares sequence elements with known RNA helicases (Gorbalenya & Koonin, 1989) and is homologous to the 2C and 3A proteins from picornaviruses. The C-terminus of the NTB protein, which contains a stretch of hydrophobic acids, has been identified as a transmembrane domain. This suggests that the NTB protein, or a larger polyprotein incorporating this protein, may function as a membrane anchor for the replication complex. Previous research on the cytopathology of nepovirus-infected cells has indicated the presence of diffuse inclusions, which are complex membranous structures. Some of these structures form vesicles, often found around the nuclei (Francki et al., 1985). Similarly, for a comovirus (Carette et al., 2002) and a nepovirus (grapevine fanleaf virus (GFLV), subgroup I), perinuclear membrane structures formed by significant ER proliferation have been shown to be the source of viral replication (Ritzenthaler et al., 2002).

6.3 SYMPTOMS AND TRANSMISSION

For the systemic invasion of their hosts, viruses exploit the phloem-assimilate flow pathway, as demonstrated in cucumber plants (Moreno et al., 2004). Long-distance transportation involves three distinct steps: loading into sieve elements, movement within them, and finally unloading from these elements, all of which may act as barriers to invasion. CMV is transmitted by aphids in a nonpersistent manner in the field, as well as by seeds in certain host species. CMV has been detected in various wild and weedy plants in nature, where it causes a latent infection without visible symptoms. These plants are believed to be the primary external sources of infection for cultivated crops and, in turn, can be contaminated from cultivated crops and serve as successional/seasonal survival hosts for the virus between crops (Nehra et al., 2014; Mishra et al., 2020).

Sequence variations generating mild and severe symptoms have been observed in the case of HSVd (Xia et al., 2017). In citrus, a five- or six-nucleotide motif discovered in the variable domain was linked to cachexia symptoms (Reanwarakorn & Semancik, 1998, 1999). The ability to turn a severe strain into a mild or asymptomatic strain via site-directed mutagenesis of this cachexia motif demonstrates that it efficiently controls symptom severity in citrus and may even inhibit symptom expression (Serra et al., 2008).

CLRV is spread naturally by seeds and pollen (Bandte & Büttner, 2001; Jones, 1986). CLRV belongs to the Nepovirus genus (Wellink et al., 2000), however, unlike the majority of other members of this genus, it is not thought to be transmitted by soil-borne nematodes. Typical signs include yellow vein netting, chlorotic ring spots and mottling, deformities including leaf roll, and branch or tree dieback (Bandte & Büttner, 2001). Due to environmental pressures and interactions with biotic variables, symptom development and intensity in CLRV-infected plants vary by host species, CLRV strain, and plant susceptibility.

The ToRSV is caused by a plant virus that is thought to be spread across the garden by dagger nematodes and spread from diseased plants to healthy ones via pollen. These small roundworms reside in the soil and move around freely, albeit slowly, between plants. Tomato ringspot symptoms range from clearly evident yellow ringspots, mottling, or general yellowing of leaves to less noticeable signs such as slow overall deterioration and lower fruit size in plants (Han & Sanfacon, 2003).

6.4 PATHOGENESIS

Viruses are obligatory parasites that reproduce only by infecting host cells. Effective infection occurs only when the virus identifies and exploits supportive host elements while avoiding host defense responses (Marwal et al., 2019). That is why the majority of plant-virus combinations fail to produce productive infections. It is not always harmful for a virus to be able to replicate. When a virus infects a host, it disrupts the physiology of the host, resulting in developmental defects and other phenotypic manifestations that are characterized as disease signs.

6.4.1 PATHOGENESIS OF CMV

Virus particles are uncoated upon entering the cell, and genomic RNAs are translated to produce viral proteins. Among these proteins, 1a and 2a are involved in the replication complex along with host proteins. The initial stage of replication involves the formation of minus strand RNA, which is then used to generate plus strands. These (+) strands serve three main purposes: (i) mRNA transcription; (ii) serving as a template for subsequent transcription; and (iii) viron generation. Replication is asymmetric, resulting in the accumulation of double-stranded RNA (dsRNA) in infected cells. Positive-stranded RNA viruses are known for their asymmetric replication. In the case of CMV, minus strand accumulation plateaus shortly after infection, while (+) strand accumulation continues to rise and can reach a level almost 100 times that of (–) strands (Seo et al., 2009). Although both proteins 1a and 2a are required for the synthesis of the minus strand, researchers have discovered that protein 2a can generate (+) strands from a (–) template of either genomic or subgenomic RNAs (Seo et al., Loc. Cit.). This finding could explain the presence of free protein 2a in the cytoplasm and the increased fraction of (+) strands in infected cells.

Following replication in infected cells, the virus spreads to neighboring cells via plasmodesmata. Although other viral proteins may contribute to this movement, the 3a protein encoded by RNA3 is considered the MP due to its distinct characteristics: (i) localization to plasmodesmata; (ii) ability to increase the plasmodesmal size exclusion limit (SEL); (iii) promotion of trafficking not only of RNA but also of itself through plasmodesmata; (iv) binding to single-stranded RNAs without sequence specificity and cooperative binding to CMV viral RNAs via a domain near its C-terminal region (positions 174–233) or two additional smaller domains; and (v) ability to act in trans and complement a CMV isolate deficient in virus movement when expressed in transgenic plants (Jacquemond, 2012).

6.4.2 PATHOGENESIS OF HSVD

Viroids must engage with host factors and functionally subvert their activity in order to replicate in host cells and subsequently travel to distant areas of the plant (Ding, 2009). HSVd, like the rest of the Pospiviroidae family, replicates in the nucleus via an asymmetric rolling circle pathway (Flores et al., 2004). The mature viroid is transcribed to linear oligomeric RNA intermediates, which are then employed as templates for the transcription of RNAs with the same polarity as the starting molecule (+ polarity), which are then cleaved and circularized into mature monomers. Viroid replication in highly pure nuclei was found to be particularly inhibited by α-amanitin, a toxin specific to RNA polymerase II. As a result, an RNA-dependent polymerase is not used for replication; instead, host DNA-dependent RNA polymerase II is subverted to accept viroid RNA as a template. In 2001, two separate research groups identified the *in vitro* interactions between HSVd and phloem protein 2 (PP2), a dimeric lectin that is the most abundant component of cucumber phloem exudate (Gómez & Pallás, 2001; Owens et al., 2001). *Nicotiana benthamiana* is frequently used as an experimental model in the study of plant-pathogen interactions. While HSVd infectivity in this host is conceivable, it appears to be less efficient than in other hosts. Inoculation experiments with nucleic acid preparations from viroid-infected plants or HSVd dimeric transcripts failed to establish a systemic infection that was efficient and repeatable. Transgenic plants of *N. benthamiana* producing dimeric HSVd (+) RNAs (HSVd-Nb plants) were obtained to investigate if this scenario could be related to a replication or systemic movement deficiency (Gómez & Pallás, 2006). The discovery that viroids activate RNA silencing pathways in infected cells in recent decades has strengthened an emerging theory linking viroid-induced RNA silencing with disease (Marquez-Molins et al., 2021).

6.4.3 PATHOGENESIS OF CLRV

Cell-to-cell migration via plasmodesmata and long-distance transport through the vascular system are used to infect the host plant, particularly the reproductive organs. Viruses have developed a variety of movement mechanisms to facilitate their spread within their host plants. Virus-encoded MPs are specialized in interacting with plasmodesmata and transporting viral genomes or complete virions to neighboring cells from the site of infection. They are also implicated in long-distance movement in certain situations (Fernandez et al., 2011).

Two methods can be identified in this regard: tubule-guided and non-tubule guided movement (Niehl & Heinlein, 2011). Within plasmodesmata, the viral MP causes tubular formations in CLRV. The specific interaction between the viral coat protein (CP) and MP formation is necessary for virions to move to neighboring cells in a coordinated manner in members of the Secoviridae family. Virus-like particles (VLPs) have been discovered within tubules of anther cells and pollen grains of virus-infected birch and walnut trees. However, the molecular interactions of CLRV-CP and MP involved in cell-to-cell migration and gametophyte infection have not been investigated.

The Yeast two-hybrid system was used to explore the dimerization of CLRV-MP and CP, as well as the specific interaction between both proteins, under physiological conditions in a eukaryotic setting. Interaction tests of the relevant CLRV proteins with the Arabidopsis thaliana protein At-4/1, which has been found to interact particularly with the movement protein (NSm) of tomato spotted wilt virus (TSWV) (Von Bargen et al., 2001; Dierker et al., 2015), were also performed.

There is very little information available on ToRSV infection and the life cycle of the transmission vector of the virus. Adults and three larval stages can transmit the virus, acquiring it in one hour and inoculating healthy plants in another. The life cycle of *X. americanum* takes at least a year to complete, and it can mature and survive (but not multiply) in soil in the absence of a host plant. It cannot survive in frozen soil for lengthy periods of time, and its numbers decrease at both high and low moisture levels. Temperatures of 20°C–24°C are ideal for reproduction (EPPO, 2015).

6.5 FACTORS AFFECTING INFECTION PROCESS

Plant virus replication involves multiple interactions between viral and host components, which can have favorable or negative consequences for the infection outcome. Pathogenesis can be triggered by the consequences of virus replication in the host, as well as the host's defense responses to the infectious agent (Marwal & Gaur, 2020). A single viral component could be the primary cause of the pathogenic process. The origin of the disruption effect in this scenario can be organically linked to the role of the factor in virus growth, as it is with several RNA silencing suppressors (Sahu et al., 2014). In certain circumstances, however, the actions of the virus factor that support viral infection and promote pathogenesis are independent, and only a small number of viruses isolates with specific forms of the viral pathogenicity factor induce definitive clinical signs. The elicitors of most hypersensitivity response (HR)-related symptoms are examples of this type of pathogenicity determinant. Although HR-mediated resistance is often seen in conjunction with necrosis, the two can be separated. The following factors have been previously reported (Marwal & Gaur, 2017; Nagy & Pogany, 2012):

- RNA replicase-related proteins;
- Coat proteins;

- Movement proteins;
- RNA silencing suppressor;
- Other factors: This includes the detection of viruses in aphids, increasing the sampling distance in weeds, using plants that are immune or resistant to infection, meristem tip culture, monitoring surveys, and phytosanitary measures. Good agricultural practices, including the sanitization process, removal of infected plant material, cleaning of mechanical tools, and continuous monitoring of surveys, ensure early detection in gene banks, nurseries, orchards, and public greens. Additionally, the use of certified virus-free seeds and planting material is recommended (Marquez-Molins, 2021; Jacquemond, 2012).

6.6 DEFENSE MECHANISM DEVELOPED BY PLANTS

Plant viruses, which are simple intracellular parasites with short genomes, rely solely on host machinery for their entire life cycle. This includes replication, intracellular (cell-to-cell) and systemic migration (Nelson & Citovsky, 2005). Virus infections pose a significant hazard to agriculture and result in substantial financial losses (Marwal & Gaur, 2019).

To combat virus infections, plants have developed two key strategies: resistance (R) gene-mediated defenses and RNA silencing-based defenses. Additionally, plant resistance to viruses is caused by mutations in critical genes for viral infection, known as recessive gene-mediated resistance. These methods have been used to protect crops and have demonstrated significant economic benefits (Whitham et al., 1996; Marwal & Gaur, 2021; Kang et al., 2005; Wang & Krishnaswamy, 2012).

Necrotic lesions in inoculated leaves confine the virus to a limited number of cells surrounding the infection sites and correspond to a defensive hypersensitivity reaction (HR) response in the case of CMV infection. Depending on the combination of host and strain, the HR was associated with protein 1a, 2a, or CP (Palukaitis & García-Arenal, 2003; Marwal et al., 2018). Additionally, in *A. thaliana*, the hormone spermine has recently been shown to function as a signal for the HR protective response to CMV (Mitsuya et al., 2009).

Moreover, the formation of necrotic lesions was linked to increased levels of hydrogen peroxide. Since catalase (CAT) is involved in the regulation of reactive oxygen species in cells and *A. thaliana* mutants lacking two CATs exhibit necrosis, these researchers investigated potential interactions between *A. thaliana* CAT proteins and protein 2b, which is known to be a virulence factor. The induction of multiple defense responses, including the production of pathogenesis-related (PR) proteins, along with the accumulation of hydrogen peroxide and an elevation in catalase (CAT2) levels, is associated with the development of necrosis. The impact of co-infection with viruses from other genera, such as Crinivirus, Potexvirus, Potyvirus, Tobamovirus, and Zucchini yellow mosaic virus (ZYMV), on disease severity remains uncertain.

Argonaute (AGO) is a protein complex with endonucleolytic cleavage activity that destroys target RNA in a sequence-specific manner. Small RNAs can also be duplicated by the cellular RNA-dependent RNA polymerase (RDR) to generate new dsRNAs, initiating a second gene silencing cycle. Small RNAs also function as mobile signals, enabling the spread of silencing throughout the plant (Jacquemond, 2012).

ACKNOWLEDGMENTS

The authors are thankful to the University Grants Commission (UGC) under the Start-Up Grant Scheme, New Delhi, India for the financial assistance (No. F.30-510/2020 (BSR) FD Diary No. 8839).

KEYWORDS

- **cherry leaf roll virus**
- **cucumber mosaic virus**
- **defense mechanism**
- **gene regulation**
- **hop stunt viroid**
- **plant viruses**
- **pomegranate**
- **tomato ringspot virus**

REFERENCES

Amari, K., Gomez, G., Myrta, A., Di Terlizzi, B., & Pallás, V., (2001). The molecular characterization of 16 new sequence variants of hop stunt viroid reveals the existence of invariable regions and a conserved hammerhead-like structure on the viroid molecule. *Journal of General Virology*, *82*, 953–962.

Ammar, M. I., & El-Naggar, M. A., (2014). Screening and characterization of fungi and their associated mycotoxins in some fruit crops. *Int. J. Adv. Res.*, *2*, 1216–1227.

Astruc, N., Marcos, J. F., Macquaire, G., Candresse, T., & Pallás, V., (1996). Studies on the diagnosis of hop stunt viroid in fruit trees: Identification of new hosts and application of a nucleic acid extraction procedure based on non-organic solvents. *Eur. J. Plant Pathol.*, *102*, 837–846.

Bandte, M., & Bu¨ttner, C., (2001). Occurrence, transmission, and diagnosis of cherry leaf roll nepovirus—A literature review. *Pflanzenschutzberichte*, *59*, 1–19.

Bargen, S. V., Salchert, K., Paape, M., Piechulla, B., & Kellman, J. W., (2001). Interactions between the *Tomato spotted wilt virus* movement protein and plant proteins showing homologies to myosin, kinesin, and DnaJ-like chaperones. *Plant Physiol. Biochem.*, *39*, 1083–1093.

Carette, J. E., Guhl, K., Wellink, J., & Kammen, A. V., (2002). Coalescence of the sites of *Cowpea mosaic virus* RNA replication into a cytopathic structure. *J. Virol.*, *76*, 6235–6243.

Cropley, R., (1961). *Cherry leaf-roll virus*. *Ann. Appl. Biol.*, *49*, 524–529.

Day, K. R., & Wilkins, E. D., (2011). Commercial pomegranate (*Punica granatum* L) production in California. *Acta Hortic.*, *890*, 275–286.

Di Serio, F., Flores, R., Verhoeven, J. T. J., Li, S. F., Pallás, V., Randles, J. W., et al., (2014). Current status of viroid taxonomy. *Archives of Virology*, *159*, 3467–3478.

Di Serio, F., Navarro, B., & Flores, R., (2017). Origin and evolution of viroids. In: Hadidi, et al., (eds.), *Viroids and Satellites* (pp. 125–134). Cambridge: Academic Press – Elsevier.

Dierker, L., Bargen, S. V., & Büttner, C., (2015). Interaction studies of *Cherry leaf roll virus* (CLRV)-encoded proteins involved in intercellular virus movement in host plants. *Acta Hortic.*, *1099*, 681–686.

Ding, B., (2015). The biology of viroid–host interactions. *Annual Review of Phytopathology*, *47*, 105–131.

Fernández-Calvino, L., Faulkner, C., & Maule, A., (2011). Plasmodesmata as active conduits for virus cell-to-cell movement. In: Caranta, et al., (eds.), *Recent Advances in Plant Virology* (pp. 47–74). Caister Academic Press, Norfolk.

Flores, R., Delgado, S., Gas, M. E., Carbonell, A., Molina, D., Gago, S., et al., (2004). Viroids: The minimal non-coding RNAs with autonomous replication. *FEBS Letters*, *567*, 42–48.

Flores, R., Minoia, S., Carbonell, A., Gisel, A., Delgado, S., López-Carrasco, A., et al., (2004). Viroids, the simplest RNA replicons: How they manipulate their hosts for being propagated and how their hosts react for containing the infection. *Virus Researchs*, *209*, 136–145.

Francki, R. I. B., Milne, R. G., & Hatta, T., (1985). Nepovirus group. In: *Atlas of Plant Viruses* (Vol. II, pp. 23–38). CRC Press, Boca Raton, Fla.

Gat, T., Liarzi, O., Skovorodnikova, Y., & Ezra, D., (2012). Characterization of *Alternaria alternata* causing black spot disease of pomegranate in Israel using a molecular marker. *Am. Phyotopathol. Soc.*, *96*, 1513–1518.

Gazel, M., Caglayan, K., Baspinar, H., Mejia, J. F., Paltrinieri, S., Bertaccini, A., & Contaldo, N., (2015). Detection and identification of phytoplasmas in pomegranate trees with yellows symptoms. *J. Phytopathol., 164*, 136–140.

Gazel, M., Caglayan, K., Serçe, Ç. U., & Durgac, C., (2009). Detection of hop stunt viroid in pomegranate (*Punica granatum* L.) trees in the east mediterranean region of turkey. *Acta Hortic., 818*, 273–276.

Gomez, G., & Pallás, V., (2001). Detection of viroid-like RNAs in pomegranate (*Punica granatum* L.). *Acta Hortic., 550*, 321–326.

Gómez, G., & Pallás, V., (2001). Identification of an *in vitro* ribonucleo-protein complex between a viroid RNA and a phloem protein from cucumber plants. *Molecular Plant-Microbe Interactions, 14*, 910–913.

Gómez, G., & Pallás, V., (2006). Hop stunt viroid is processed and trans-located in transgenic *Nicotiana benthamiana* plants. *Molecular Plant Pathology, 7*, 511–517.

Gorbalenya, A. E., & Koonin, E. V., (1989). Viral proteins containing the purine NTP-binding sequence pattern. *Nucleic Acids Res., 17*, 8413–8440.

Han, S., & Sanfaçon, H., (2003). *Tomato ringspot virus* proteins containing the nucleoside triphosphate binding domain are transmembrane proteins that associate with the endoplasmic reticulum and cofractionate with replication complexes. *Journal of Virology, 77*(1), 523–534.

Hand, F. P., Choudhury, R. A., & Gubler, W. D., (2014). First report of *Cytospora punicae* causing wood canker and branch dieback of pomegranate (*Punica granatum*) in the United States. *Plant Dis., 98*, 2012–2013.

Hataya, T., Tsushima, T., & Sano, T., (2017). Hop stunt viroid. In: Hadidi, et al., (eds.), *Viroids and Satellites* (pp. 199–210). Cambridge: Academic Press - Elsevier.

Herrera, G., & Madariaga, M., (2001). Presence and incidence of grapevine viruses in the central zone of Chile. *Agric. Tehc., 61,* 393–400.

Horvath, J., (1984). Isolation of *Cucumber mosaic virus* from pomegranate (*Punica granatum* L.) in Yugoslavia. *Acta Phytopathol. Acad. Sci. Hung., 4*, 309–314.

Jacquemond, M., (2012). *Cucumber mosaic virus. Adv. Virus Res., 84*, 439–504.

Jain, K., & Desai, N., (2018). Pomegranate the cash crop of India: A comprehensive review on agricultural practices and diseases. *International Journal of Health Science and Research., 8*(5), 315–336.

Jain, K., Desai, N., Sharma, K., & Marwal, A., (2021). Development and screening of by product for its secondary metabolites, antioxidant, and anti-diabetic potential from anthracnose-infected fruits of pomegranate: A sustainable approach. 3 *Biotech, 11*(2), 74.

Jo, Y., Chu, H., Kim, H., Cho, J. K., Lian, S., Choi, H., et al., (2017). Comprehensive analysis of genomic variation of hop stunt viroid. *European Journal of Plant Pathology, 148*, 119–127.

Jones, A. T., (1986). *Cherry Leaf Roll Virus. CMI*/AAB description of plant viruses no. 306. Commonwealth Mycological Institute, Kew, Surrey, England, and Association of Applied Biologists, Wellsbourne, Warwick, England.

Kang, B. C., Yeam, I., & Jahn, M. M., (2005). Genetics of plant virus resistance. *Annu. Rev. Phytopathol., 43*, 581–621.

Kawaguchi-Ito, Y., Li, S. F., Tagawa, M., Araki, H., Goshono, M., Yamamoto, S., et al., (2009). Cultivated grapevines represent a symptomless reservoir for the transmission of hop stunt viroid to hop crops: 15 years of evolutionary analysis. *PLoS One, 4*, e8386.

Keese, P., & Symons, R. H., (1985). Domains in viroids: Evidence of inter-molecular RNA rearrangements and their contribution to viroid evolution. *Proceedings of the National Academy of Sciences of the United States of America.*, *82*, 4582–4586.

Kofalvi, S. A., Marcos, J. F., Cañizares, M. C., Pallas, V., & Candresse, T., (1997). Hop stunt viroid (HSVd) sequence variants from *Prunus species*: Evidence for recombination between HSVd isolates. *Journal of General Virology*, *78*, 3177–3186.

Luigi, M., Manglli, A., Tomassoli, L., & Faggioli, F., (2013). First report of hop stunt viroid in *Hibiscus rosa sinensis* in Italy. *New Disease Reports*, *27*, 14.

Maddahian, M., Massumi, H., Heydarnejad, J., Hosseinipour, A., Khezri, A., & Sano, T., (2019). Biological and molecular characterization of hop stunt viroid variants from pistachio trees in Iran. *Journal of Phytopathology*, *9*, 1–11.

Marquez-Molins, J., Gomez, G., & Pallas, V., (2021). Hop stunt viroid: A polyphagous pathogenic RNA that has shed light on viroid–host interactions. *Mol Plant Pathol.*, *22*, 153–162.

Marwal, A., & Gaur, R. K., (2017). Understanding functional genomics of PTGS silencing mechanisms for *Tobacco streak viru*s and other ilarviruses mediated by RNAi and VIGS. In: *Plant-Microbe Interactions in Agro-ecological Perspectives* (pp. 489–499). Springer, Singapore.

Marwal, A., & Gaur, R. K., (2020). Host plant strategies to combat against viruses effector proteins. *Current Genomics*, *21*(6), 401–410.

Marwal, A., & Gaur, R. K., (2021). Plant virus identification and management. *Research Journal of Biotechnology*, *16*, 190–193.

Marwal, A., & Gaur, R., (2019). Nanophytovirology: An emerging field for disease management. In: *Plant Diseases – Current Threats and Management Trends*. IntechOpen. https://doi.org/10.5772/intechopen.86653.

Marwal, A., Gaur, R. K., & Khurana, S. M. P., (2018). Possible approaches for developing different strategies to prevent transmission of geminiviruses to important crops. In: Gaur, R. K., Khurana, S. M. P., & Dorokhov, Y., (eds.), *Plant Viruses: Diversity, Interaction and Management* (pp. 301–319). Boca Raton, Florida, US: CRC Press, Taylor & Francis Group.

Marwal, A., Kumar, R., Verma, R. K., Mishra, M., Gaur, R. K., & Khurana, S. M., (2019). Genomics and molecular mechanisms of plant's response to abiotic and biotic stresses. In: *Plant Biotechnology: Progress in Genomic Era* (pp. 131–146). Springer, Singapore.

Mayo, M. A., & Robinson, D. J., (1996). Nepoviruses: Molecular biology and replication. In: Harrison, et al., (eds.), *The Plant Viruses: Polyhedral Virions and Bipartite RNA Genomes* (Vol. 5, pp. 139–185). Plenum Press, New York, N.Y.

Mirabolfathy, M., Groenewald, J. Z., & Crous, P. W., (2012). First report of *Pilidiella granati* causing dieback and fruit rot of pomegranate (*Punica granatum*) in Iran. *Plant Dis.*, *96*, 461.

Mirzaei, S., Goltapeh, E. M., Shams-Bakhsh, M., & Safaie, N., (2008). Identification of *Botrytis* spp. on plants grown in Iran. *J. Phytopathol.*, *156*, 21–28.

Mishra, M., Verma, R. K., Marwal, A., Sharma, P., & Gaur, R. K., (2020). Biology and interaction of the natural occurrence of distinct monopartite *Begomoviruses* associated with satellites in *Capsicum annum* from India. *Frontiers in Microbiology*, *11*, 2366.

Mitsuya, Y., Takahashi, Y., Berberich, T., Miyazaki, A., Matsumura, H., Takahashi, H., Terauchi, R., & Kusano, T., (2009). Spermine signaling plays a significant role in the defense response of *Arabidopsis thaliana* to *Cucumber mosaic virus*. *J. Plant Physiol.*, *166*, 626–643.

Moreno, I. M., Thompson, J. R., & Garcı´a-Arenal, F., (2004). Analysis of the systemic colonization of cucumber plants by *Cucumber green mottle mosaic virus*. *J. Gen. Virol.*, *85*, 749–759.

Murant, A. F., Taylor, M., Duncan, G. H., & Raschke, J. H., (1981). Improved estimates of molecular weight of plant virus RNA by agarose gel electrophoresis and electron microscopy after denaturation with glyoxal. *J. Gen. Virol.*, *53*, 321–332.

Muthupandi, K., Marwal, A., & Tennyson, J., (2019). Genome of a known but distinct *begomovirus* associated with a novel satellite molecule infecting a new host bitter gourd (*Momordica charantia*). *3 Biotech*, *9*(6), 1–6.

Nagy, P. D., & Pogany, J., (2012). The dependence of viral RNA replication on co-opted host factors. *Nat Rev Microbiol.*, *10*, 137–149.

Nehra, C., Sahu, A. K., Marwal, A., & Gaur, R. K., (2014). Natural occurrence of *Clerodendron yellow mosaic virus* on *Bougainvillea* in India. *New Disease Reports*, *30*(19), 2044–0588.

Nelson, R. S., & Citovsky, V., (2005). Plant viruses. Invaders of cells and pirates of cellular pathways. *Plant Physiol.*, *138*, 1809–1814.

Owens, R. A., Blackburn, M., & Ding, B., (2001). Possible involvement of the phloem lectin in long-distance viroid movement. *Molecular Plant-Microbe Interactions*, *14*, 905–909.

Palavouzis, S. C., Tzamos, S., Paplomatas, E., & Thomidis, T., (2015a). First report of *Neofusicoccum parvum* causing shoot blight of pomegranate in Northern Greece. *N. Dis. Rep.*, *32*, 10.

Palou, L., Taberner, V., Guardado, A., Del Río, M. Á., & Montesinos-Herero, C., (2013). Incidence and etiology of postharvest fungal diseases of pomegranate (*Punicagranatum* cv: Mollar de Elche) in Spain. *Phytopathol. Mediterr.*, *52*, 478–489.

Palukaitis, P., & Garcı´a-Arenal, F., (2003). Cucumoviruses. *Adv. Virus Res.*, *62*, 242–323.

Petersen, Y., Mansvelt, E. L., Venter, E., & Langenhoven, W. E., (2010). Detection of *Xanthomonas axonopodis*pv: Punicae causing bacterial blight on pomegranate in South Africa. *Plant Pathol.*, *39*, 544–546.

Reanwarakorn, K., & Semancik, J. S., (1998). Regulation of pathogenicity in hop stunt viroid-related group II citrus viroids. *Journal of General Virology*, *79*, 3163–3171.

Reanwarakorn, K., & Semancik, J. S., (1999). Correlation of hop stunt viroid variants to cachexia and xyloporosis diseases of citrus. *Phytopathology*, *89*, 568–574.

Rebenstorf, K., Candresse, T., Dulucq, M. J., Büttner, C., & Obermeier, C., (2006). Host species-dependent population structure of a pollen-borne plant virus. *Cherry leaf roll virus*. *Journal of Virology*, *80*(5), 2453–2462.

Ritzenthaler, C., Laporte, C., Gaire, F., Dunoyer, P., Schmitt, C., Duval, S., Piéquet, A., et al., (2002). *Grapevine fanleaf virus* replication occurs on endoplasmic-derived membranes. *J. Virol.*, *76*, 8808–8819.

Sahu, A. K., Marwal, A., Nehra, C., Choudhary, D. K., Sharma, P., & Gaur, R. K., (2014). RNAi mediated gene silencing against betasatellite associated with Croton yellow vein mosaic *Begomovirus*. *Molecular Biology Reports*, *41*(11), 7631–7638.

Sanfaçon, H., (1995). Nepoviruses. In: Singh, et al., (eds.), *Pathogenesis and Host Specificity in Plant Diseases* (Vol. III, pp. 129–141). Viruses and viroids. Pergamon Press, Oxford, United Kingdom.

Sano, T., Mimura, R., & Ohshima, K., (2001). Phylogenetic analysis of hop and grapevine isolates of hop stunt viroid supports a grapevine origin for hop stunt disease. *Virus Genes*, *22*, 53–59.

Sano, T., Uyeda, I., Shikata, E., Ohno, T., & Okada, Y., (1984). Nucleotide sequence of cucumber pale fruit viroid: Homology to hop stunt viroid. *Nucleic Acids Research*, *12*, 3427–3434.

Sasaki, M., & Shikata, E., (1977). On some properties of hop stunt disease agent, a viroid. Proceedings of the Japan Academy. *Series B, Physical and Biological Sciences*, *53*, 109–112.

Seo, J. K., Kwon, S. J., Choi, H. S., & Kim, K. Y., (2009). Evidence for alternate states of *Cucumber mosaic virus* replicase assembly in positive- and negative-strand RNA synthesis. *Virology*, *383*, 248–260.

Serra, P., Gago, S., & Duran-Vila, N., (2008). A single nucleotide change in hop stunt viroid modulates citrus cachexia symptoms. *Virus Research*, *138*, 130–134.

Steger, G., & Perreault, J. P., (2016). Structure and associated biological functions of viroids. *Advances in Virus Research*, *94*, 141–172.

Thomidis, T., (2014). Fruit rots of pomegranate (cv. wonderful) in Greece. *Plant Pathol.*, *43*, 583–588.

Vamenani, R., Rahimian, H., Alavi, S. M., PakdinParizi, A., & Mirza, R. T., (2019). Genetic diversity of hop stunt viroid from symptomatic and asymptomatic citrus trees in Iran. *Journal of Phytopathology*, *167*, 484–489.

Verma, R. K., Mishra, M., Marwal, A., & Gaur, R. K., (2020). Identification, genetic diversity, and recombination analysis of *Watermelon mosaic virus* isolates. *3 Biotech*, *10*(6), 1–8.

Wang, A., & Krishnaswamy, S., (2012). Eukaryotic translation initiation factor 4E-mediated recessive resistance to plant viruses and its utility in crop improvement. *Mol. Plant Pathol.*, *13*, 795–803.

Wang, A., & Sanfaçon, H., (2000). Proteolytic processing at a novel cleavage site in the N-terminal region of the tomato ringspot nepovirus RNA-1 encoded polyprotein *in vitro*. *J. Gen. Virol.*, *82*, 1785–1790.

Wellink, J., Le Gall, O., Sanfacon, H., Ikegami, M., & Jones, A. T., (2000). Genus *Nepovirus*, In: Regenmortel, et al., (eds.) *Virus Taxonomy* (pp. 697–701). Seventh report of the International Committee on Taxonomy of Viruses. Academic Press, San Diego, California.

Whitham, S., McCormick, S., & Baker, B., (1996). The N gene of tobacco confers resistance to *Tobacco mosaic virus* in transgenic tomato. *Proc. Natl. Acad. Sci. U.S.A.*, *93*, 8776–8781.

Xia, C., Li, S., Hou, W., Fan, Z., Xiao, H., Lu, M., et al., (2017). Global transcriptomic changes induced by infection of cucumber (*Cucumis sativus* L.) with mild and severe variants of hop stunt viroid. *Frontiers in Microbiology*, *8*, 2427.

Yehia, H. M., (2013). Heart rot caused by *Aspergillus Niger* through splitting in leathery skin of pomegranate fruit. *Afr. J. Microbiol. Res.*, *7*, 834–837.

CHAPTER 7

Virome and Vectorome Analyses of Vectors: New Approaches for Evolutionary Studies of Plant Viruses

SAYYAD ALI RAZA BUKHARI[1] and MUHAMMAD SAEED[2]

[1]*Department of Biotechnology, University of Sargodha, Sargodha, Pakistan*

[2]*Technical University Kaiserslautern-Landau, Phytopathology, Department of Biology, Paul-Ehrlich-Straße 22, Kaiserslautern 67653, Germany*

ABSTRACT

Vectors are crucial in the spread of the majority of viruses from one individual to the next in sessile plant species. They may contain two different types of viruses: those that can infect them and those that are silently transferred to the plants but are non-infectious for themselves. The complete repertoire of viruses presents in an organism, such as a vector, is known as the "virome," while the subset of the virome that is transported to the non-homologous host is termed the "vectorome." The vectorome is largely non-replicative in the vector. High-throughput sequencing has made it possible to explore the virome and vectorome of an individual. Understanding the virome is decisive in unraveling the vectorome potential, host-vector interaction, and evolutionary dynamics of the virome. In this chapter, we discuss different vectors, their virome and vectorome, and strategies to study the virome and vectorome.

7.1 INTRODUCTION

The majority of plant viruses rely on biotic vectors for their transmission between plants. These vectors include whiteflies, aphids, leafhoppers, thrips,

New and Emerging Plant Viruses: The Threat to Food Security.
Alireza Golnaraghi and Rajarshi Kumar Gaur (Eds.)

planthoppers, treehoppers, nematodes, mites, and zoosporic endoparasites (Bragard et al., 2013). Among these vectors, insects, particularly whiteflies and aphids, are the most important and extensively studied (Hohn, 2007). Approximately 70% of insect-transmissible viruses, accounting for 80% of all such viruses, are transmitted by *Hemiptera*, which includes whiteflies and aphids. These two insect species are considered the primary vectors of plant viruses and are known to transmit over 500 virus species (Fereres & Raccah, 2015). Figure 7.1 provides a network graph illustrating the relationship between viruses and their vectors. Insect-transmitted viruses have been classified based on viruliferous time, virus route, and virus location. Viruses can be classified as persistent, semi-persistent, or non-persistent depending on how long they persist in the insect, ranging from several days to the insect's entire lifespan, a few days, or a few hours, respectively. The virus route can be circulative, where the virus passes through the insect's midgut, hindgut, and salivary glands before being transmitted to plants, or non-circulative, where the virus binds to the stylet or foregut and is transmitted to plants through saliva during feeding. According to virus location, they can be cuticula-borne if they are retained in the stylet or foregut, or salivary gland-borne if they are retained in the salivary glands after the circulative pathway (Fereres & Moreno, 2009; Wei & Li, 2016). All the viruses that are transmitted plant-to-plant via an insect are termed vectorome and remain largely non-replicative and non-infectious for the insects. Vectorome, along with the viruses that infect insects, make up the virome of the insect.

Viruses are the most abundant microbes on Earth that play an important role in shaping the composition and metabolism of other organisms. However, they have not been explored proportionately to their abundance and importance in the ecosystem. Most viruses are pathogenic for bacteria, archaea, eukaryotes (including humans), and important crops.

The exploration of the virus world has been hindered by technical limitations and the absence of suitable methods for analyzing large-scale data. Fortunately, recent advancements in next-generation sequencing (NGS) (Buermans & den Dunnen, 2014), metagenomics, and the development of appropriate bioinformatics methods (Antipov, Raiko, Lapidus, & Pevzner, 2020; Edgar et al., 2022; Kieft, Adams, Salamzade, Kalan, & Anantharaman, 2022; Nurk, Meleshko, Korobeynikov, & Pevzner, 2017) have led to the discovery of thousands of known and novel viruses, including both DNA and RNA viruses.

While most efforts have focused on human viruses (Nouri, Matsumura, Kuo, & Falk, 2018), there is an increasing number of reports targeting plant viruses (Adams & Fox, 2016). Recently, plant-insect vectors have been explored for the discovery of novel and known viruses, along with metagenomics of plant samples for virus discovery (Rosario et al., 2015).

7.2 DISCOVERY OF NOVEL VIRAL STRAINS

Discovery and exploration of unknown/novel species of viruses have tremendously increased with the advent of new technologies such as NGS. Hence, NGS techniques serve as a powerful tool by providing speedy and cost-effective approaches for analyzing novel species of viruses. A huge set of viral samples gathered from different resources is depicted in Figure 7.1. All different vector nodes are colored differently; however, all the virus nodes connected to a vector node are colored the same as the vector node. Similarly, edges that connect a vector node to a virus node are colored the same as the vector node, except for a virus node that is connected to multiple vector nodes. The size of the node increases as the size of its connections increases. A connection/edge means that the virus is transmitted by a connected vector(s). and locations can be sequenced at a very high rate to make discovery of novel viruses quite easy (Pecman et al., 2017). Before the advent of NGS methods, many traditional molecular approaches were used to discover and analyze pathogenic viruses of plants. These approaches covered a range of techniques, from DNA isolation to PCR-based enrichment, and from cloning of viruses to rolling-circle amplification (RCA). While these traditional methods are still used in some laboratories, they have certain drawbacks. One major drawback is that they require ample prior knowledge and significant effort to achieve successful results, which can be laborious and inconvenient. For example, designing primers for PCR requires prior sequence information from known sources of viral strains. Additionally, limited available restriction enzymes are used for restricting RCA products, which may result in the loss of new viral strains.

In contrast, NGS-based approaches do not require prior sequence information about viral pathogens. They offer a non-specific diagnostic process for novel viral strains, making the discovery of new strains easier and more effective (Jones, Baizan-Edge, MacFarlane, & Torrance, 2017).

Researchers have highlighted the importance of high-throughput sequencing (HTS) techniques in sequencing DNA and RNA from plants and insects. Recently, HTS was applied to samples of *Cyclamen persicum* from southern Italy, an important ornamental plant that is infected by several viral pathogens such as cucumoviruses, potexviruses, and tobamoviruses. The authors conducted RT-PCR-based assays but failed to detect any viral strain, which was quite surprising considering that the analyzed plants exhibited clear viral-associated symptoms such as leaf deformation, leaf yellowing, and mottling, among others. However, they successfully detected *Fig mosaic emaravirus* (FMV) when they employed HTS, which does not require prior knowledge of viral sequences (Elbeaino et al., 2018). A successful virus management plan must involve the implementation and adoption of reliable viral detection methods, which will not only contribute to improving global trade conditions but also facilitate the development of effective plant protection and management strategies. Another significant impact of HTS/NGS tools is the in-depth understanding of microbial dynamics, leading to the discovery and detection of novel plant viral strains. Therefore, it is evident from the findings of several studies that NGS tools play a crucial role in the unbiased exploration of novel viral and viral-like agents (Blawid, Silva, & Nagata, 2017; Kehoe, Coutts, Buirchell, & Jones, 2014; Massart, Olmos, Jijakli, & Candresse, 2014).

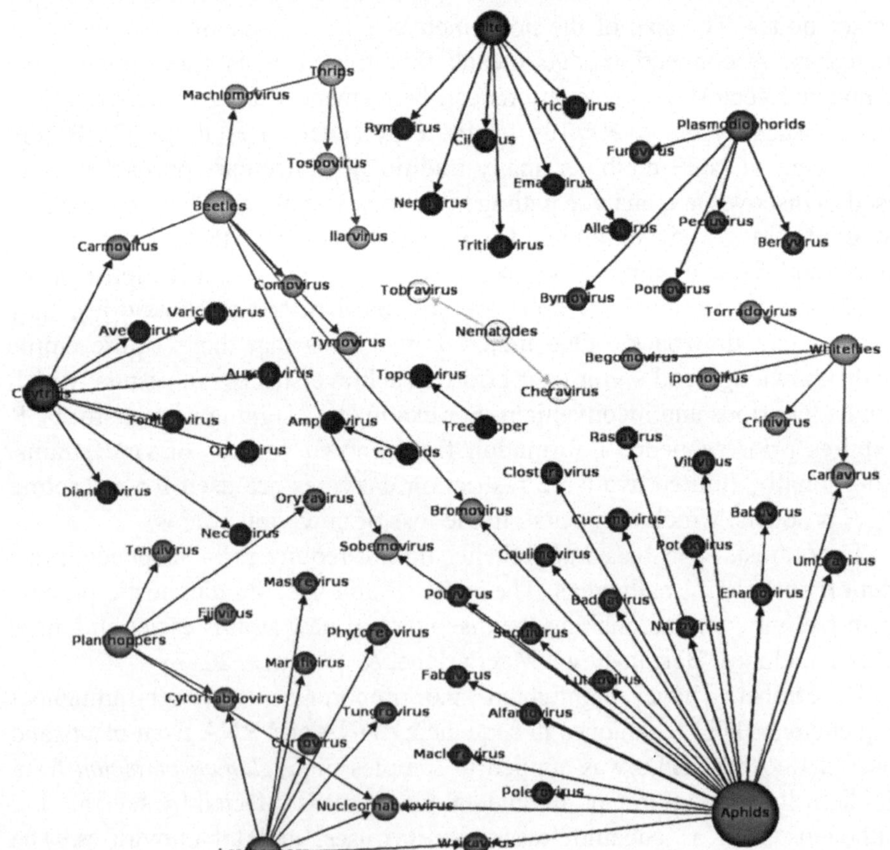

FIGURE 7.1 A network graph of the viruses and their vectors.

7.3 VIRAL DIAGNOSTICS METHODS

A detailed diagnosis and detection of viral strains have the potential to impact several fields, ranging from the environment and human health to global food security and crop production. The enzyme-linked immunosorbent assay (ELISA) has been extensively used for detecting virus-specific antibodies in blood serum and plant extracts. Another important use of ELISA is in quarantine protocols employed for screening viral infections in imported germplasm (Abd. El-Aziz, 2019).

The introduction of quantitative (q) PCR enabled speedy and accurate detection and diagnosis of plant viruses. However, the main problem lies in obtaining prior sequence information of viral pathogens, thus limiting the diagnosis to specific viral strains (Bester, Pepler, Burger, & Maree, 2014). As a result, unknown viruses remain unexplored due to the non-availability of particular diagnostic kits and procedures. The detection and discovery of the number and type of viral species become quite challenging in the absence of specific symptoms/indications linked with particular viral infections. Therefore, NGS tools have solved this problem by offering reliable diagnostic methods for the detection of mixed viral infections (Balasubramanian, 2015).

Deep sequencing of RNA extract collected from a vineyard through the Illumina Genome Analyzer platform revealed the presence of Grapevine rupestris stem pitting-associated virus, *Grapevine virus A,* and *Grapevine leafroll-associated virus 3* (GLRaV-3), along with a novel virus *Grapevine virus E* (Coetzee et al., 2010). The use of NGS tools has expedited the detection process of viruses, as well as played a significant role in reducing the costs of viral detection by enabling NGS-based large-scale viral diagnostic assays (Eichmeier, Komínková, Komínek, & Baránek, 2016; Stewart et al., 2014). In this aspect, sequencing of small RNA molecules has also proved to be a promising tool for diagnosing viral infections (Wu et al., 2010). Some novel strains of potyvirus containing 10,057 nucleotides (named *tomato necrotic stunt virus* (ToNSV)) were detected in tomato plants of Mexican origin by forming a *de novo* genome assembly of small RNA molecules enriched with viruses using the Illumina Genome Analyzer IIx platform (Li et al., 2012).

Recently, RNA-seq has been used for the identification of many novel viral species. Multiple viral infections, known as "pepper viromes," have been reported in *Capsicum* species using Illumina's HiSeq 2000 platform (Jo, Kim, Kim, Yoo, & Lee, 2017). The same research also reported a novel viral species, *pepper virus A* (PepVA), and further analysis identified it as a member of the *Betaflexiviridae* family. NGS platforms offer multiple solutions to researchers and are frequently available at a relatively low cost.

However, there are several issues that need to be addressed when using any of the aforementioned options, such as proper handling and assembly of sequenced data, extensive computing expertise and facilities, availability of bioinformatics software/tools, pruning of redundant/non-redundant sequences, and accurate mapping of NGS reads. Proper management of all NGS-based problems will play a crucial role in redefining the field of diagnostics and the discovery of novel viruses.

7.4 VIRAL METAGENOMICS IN DISCOVERY OF NOVEL VIRUSES IN PLANTS

Metagenomics covers the complete sequencing and recovery of genomic material collected from environmental samples. The whole procedure of developing metagenomics is known as metagenomics. One main advantage of metagenomics is that it allows the analysis of whole genomes of non-culturable microbes, such as viruses. Previously, large-scale sequencing of multiple samples collected from different resources was quite challenging using conventional methods like Sanger sequencing. However, NGS-based DNA/RNA sequencing tools have offered cost-effective approaches for studying metagenomics and microbial dynamics. Moreover, NGS tools have revolutionized metagenomic studies due to their ability to generate billions of sequencing reads, resulting in the production of a vast amount of data. The combination of metagenomic studies with recent advancements in NGS tools and easily accessible bioinformatics tools has led to the exploration of novel viruses and virus-like agents in plant and insect samples collected from various origins (Ho & Tzanetakis, 2014; Wu, Ding, Zhang, & Zhu, 2015).

Deep sequencing of RNAs produced through different processes of viral infections has led to the discovery of novel viruses. To date, a carlavirus is the only known virus infecting *Hevea brasiliensis*. However, a novel virus, *Hevea brasiliansis virus* (HBrV), was discovered during deep sequencing of siRNAs extracted from *H. brasiliensis* and its subsequent virome analysis (Fonseca et al., 2018). Metagenomics studies performed through NGS tools on infected samples of okra and tomato produced the whole genome sequence information, as well as identified some SNPs and satellite molecules linked with the viruses (Idris et al., 2014). In addition to several other advantages, effective plant protection strategies can also be designed through metagenomics. For instance, up or down-regulation of different genes in plants may be linked with some viral infections, but the suppression of these particular genes may not provide effective protection against that particular viral pathogen. Hence, only NGS tools are capable of detecting traces of any pathogen. Thus, NGS-based data can be exploited for devising effective and appropriate plant protection strategies used in plant protection programs.

7.5 DETECTION OF NON-CODING RNAS SPECIFIC TO VIRUSES

The small non-coding RNAs, such as microRNA (miRNA), comprising 21–28 nucleotides, play an important role in regulating gene expression and plant development, depending on the developmental stages of the plant. For example, upon infection of *Turnip yellow mosaic virus* in *Brassica*, tobacco, and *Arabidopsis*, several miRNAs, including miR158, miR164, miR156, and miR160, were induced (He, Fang, Feng, & Guo, 2008; Kasschau et al., 2003). In addition to miRNAs, viral infections also lead to the production of short interfering RNAs (siRNAs). By utilizing Illumina/Solexa platforms for deep sequencing of siRNAs, followed by subsequent assemblies, some known viruses such as *blackberry chlorotic ringspot virus* (BCRV), *Prunus necrotic ringspot virus* (PNRSV), and *apple stem grooving virus* (ASGV), as well as a novel virus tentatively named *rose leaf rosette-associated virus* (*RLRaV*), were detected, highlighting the importance of NGS tools (He et al., 2015). Similarly, miRNAs may exhibit altered expression upon viral infections, resulting in the up or down-regulation of gene expression. Therefore, a comprehensive understanding of the pathways involved in siRNAs, miRNAs, and other non-coding RNAs is crucial for a complete understanding of the dynamics of plant-virus interactions. High-throughput Illumina sequencing in *Vigna mungo* detected varying patterns of several known miRNAs, including miR159, miR156, miR398, miR2118, etc. *Vigna mungo* is completely susceptible to *Mung bean yellow mosaic India virus* (MYMIV), which contains single-stranded circular DNA and belongs to the *Begomovirus* genus (Kundu, Paul, Dey, & Pal, 2017). Additionally, they also demonstrated the presence of new strains of miRNAs, such as vmu-miRn8, vmu-miRn7, vmu-miRn14, and vmu-miRn13. In another study, NGS tools were employed to explore the miRNA profiles in tomato cultivars (Pusa Ruby and LA1777) upon infection of *tomato leaf curl virus* (ToLCV), which led to the detection of 91 known miRNAs and 15 novel homologs, along with 53 new miRNAs (Tripathi, Goswami, Tiwari, Mukherjee, & Sanan-Mishra, 2018).

Several other small RNA molecules, including transacting si(tasi)RNA, phased siRNA, and repeat-associated (rasi)RNA, also play a significant role in mediating plant-virus interactions. Comprehensive knowledge of these RNA molecules is essential for understanding plant-virus interactions more effectively. Additionally, long non-coding (lnc) RNAs have been involved in mediating plant-virus interactions, and these lnc RNAs could be of plant, viral, or mixed origin (Fortes & Morris, 2016). Unfortunately, the role of lncRNAs has not been explored yet; otherwise, better strategies could have been devised to protect plants from viral infections effectively. Furthermore, the detection of small RNA molecules could be challenging for mixed viral infections, requiring a complete analysis of the virome. The study of whole genomes of pathogenic viruses could be a cumbersome task without using NGS tools. Lastly, NGS tools coupled with a complete analysis of small RNA molecules may lead to the detection and discovery of several novel factors responsible for imparting viral resistance in plants. Recent efforts have focused on viromics and vectoromics of insects to explore the full diversity of viruses and find novel viral sequences of importance for plant health.

7.5.1 VIROMICS OF WHITEFLY

Whitefly (*Bemisia tabaci*) is a polyphagous plant pest and is considered the most detrimental insect vector, causing considerable economic losses to multiple important crops in different parts of the world. It poses a serious threat to food security (Figure 7.2).

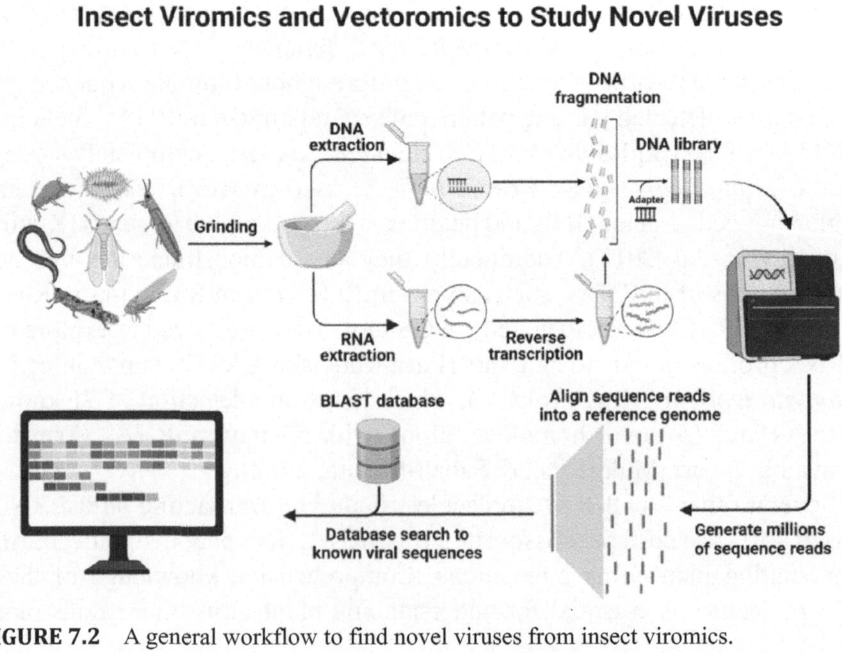

FIGURE 7.2 A general workflow to find novel viruses from insect viromics.

Whitefly damages crops directly by feeding on the phloem and indirectly by excreting honeydew, which promotes the growth of sooty mold, and transmitting pathogenic viruses (Navas-Castillo, Fiallo-Olivé, & Sánchez-Campos, 2011; Oliveira, Henneberry, & Anderson, 2001). Both DNA and RNA viruses are transmitted by whitefly; however, it is most renowned for the exclusive transmission of single-stranded DNA (ssDNA) viruses belonging to the genus *Begomovirus* – known to transmit 280 species of begomoviruses (Fiallo-Olivé, Pan, Liu, & Navas-Castillo, 2020). RNA viruses transmitted by whiteflies belong to the genera *Crinivirus, Carlavirus, Ipomovirus,* and *Torradovirus* (Navas-Castillo et al., 2011).

A metagenomic study conducted on whiteflies from Florida resulted in the detection of both DNA and RNA viruses. In this study, DNA and RNA were extracted and sequenced separately. Among the DNA viruses, 97.3% were identified as begomoviruses. RNA (cDNA) sequencing reads produced a smaller number of contigs (n = 64) compared to DNA sequencing reads (n = 259). The RNA contigs led to the identification of diverse viruses with similarities to at least five different viral families: *Betaflexiviridae, Closteroviridae, Bunyaviridae, Bromoviridae,* and *Virgaviridae*. Interestingly, viruses from the last three families had not been previously reported in whiteflies. However, the RNA contigs were predominantly associated with *cowpea mild mottle virus* (CpMMV) (genus *Carlavirus*). Since CpMMV had not been reported in Florida before, whiteflies carrying CpMMV were used to infect common bean plants to support the findings. All three plants exhibited mild mottling and were confirmed to be infected with CpMMV, suggesting that CpMMV is part of the insect vectorome (Rosario, Capobianco, Ng, Breitbart, & Polston, 2014).

In a similar study, whiteflies were collected from two different sites (460 km apart) in Florida, namely Citra (soybean and watermelon plants) and Homestead (tomato and squash plants, as well as other cucurbit crops such as cucumber, cantaloupe, pumpkin, and watermelon). The "Citra virome" was dominated by *cucurbit leaf crumple virus* (CuLCrV), while two sequences showed similarity to the DNA B of *Sida golden mosaic virus* (SiGMV). A novel sequence was also discovered, showing 81% similarity to *Desmodium leaf distortion virus* (DeLDV). The majority of sequences in the Homestead sample exhibited high similarity to known begomoviruses, with *tomato yellow leaf curl virus* (TYLCV) and SiGMV being the most dominant. Some of the sequences were novel, as they showed less than 88% similarity to any known sequences in the database and were tentatively named *Sida golden mosaic Linguanea virus* [USA:Florida:Homestead:2009:VEM].

Additionally, several novel virus-associated satellites were found, which were similar in size and genome organization to ToLCV *satellite* (Ng et al., 2011).

In a global study, whitefly was collected from Israel, Spain, Puerto Rico, Guatemala, and the USA (California) and subjected to metagenomics in the quest to explore viral diversity in different locations. Sequence analysis of contigs from all locations returned begomoviruses, showing the highest diversity in samples from Guatemala, with the identification of up to seven *Begomovirus* species from a single field. The lowest diversity was displayed by samples from California, with the identification of only *squash leaf curl virus* (SLCV). Both monopartite and bipartite begomoviruses were identified from all samples, except Spain, which contained only monopartite viruses. The novel viruses were observed in Guatemala, Puerto Rico, and Spain, based on less than 91% sequence similarity (Brown et al., 2015). One of the novel viruses discovered contained a nonanucleotide motif that varies at one nucleotide (CAATATTAC) instead of a bona fide TAATATTAC sequence (Heyraud, Matzeit, Schaefer, Schell, & Gronenborn, 1993). Samples collected from squash fields in Israel contained sequences showing high similarity to *hollyhock leaf crumple virus* (HLCrV) and *Cotton leaf curl Gezira virus genomes*, while the samples taken from Israeli tomato fields were highly enriched with TYLCV. Sweepoviruses, including *sweet potato leaf curl virus* (SPLCV), were recovered from samples from Puerto Rico and Spain. Three sequences from Spain were too divergent to be defined as a known species. They showed sequence similarity to *Tomato leaf curl Namakely virus* and *Tomato leaf curl Comoros virus* (Rosario et al., 2015). Very recently, an RNA virome of *B. tabici* recovered a new isolate of *potato virus S* (PVS) and 32 other novel viruses that belong to different insect/plant-associated viral order/families such as *Flaviviridae*, *Orthomyxoviridae*, *Negevirus*, *Picornavirales*, *Virgaviridae*, *Lispiviridae*, *Totiviridae*, and *Nidovirales* (Huang et al., 2021).

7.5.2 VIROME OF APHIDS

Aphids are serious sap-sucking insect pests that cause substantial damage to many crucial crops. Aphids, like whiteflies, can damage plants directly by feeding on them, and indirectly by acting as vectors for dangerous plant viruses. They also release honeydew, which promotes the growth of sooty molds (Will & van Bel, 2006). The aphid virome contains both DNA and RNA viruses, but the insect's vectorome mostly consists of RNA viruses. Most aphid-vectored plant viruses do not replicate in insects and can be transmitted in a persistent or non-persistent manner. However, some plant rhabdoviruses can also replicate in insect hosts. In addition to plant viruses, metagenomic efforts have discovered many novel insect-specific viruses belonging to *Dicistroviridae, Ifaviridae*, and a newly proposed group of insect-specific viruses known as negeviruses (Kondo et al., 2020).

Aphids from barley fields in Japan were collected, and total RNA was extracted before subjecting it to sequencing. The sequences recovered around 60 virus sequences, which were used for genome assembly. The contigs' sequence similarity was determined using the basic local alignment search tool (BLAST). The BLAST reflected similarity to different virus families such as *Kitaviridae*, *Phenuiviridae*, *Narnaviridae, Flaviviridae, Chrysoviridae, Tombusviridae, Partitiviridae, Rhabdoviridae*, and *Luteoviridae*. Three contigs exhibited high similarity to plant viruses such as *barley yellow dwarf virus* (BYDV) belonging to *Luteoviridae* (Kondo et al., 2020). In further exploration, RNA sequencing libraries were constructed from 10 different aphid species, resulting in the identification of a total of 18 novel RNA viruses, including 11 positive-stranded and seven negative-stranded viruses. These contigs were similar to different viral families such as *Iflaviridae*, *Picornaviridae*, *Dicistroviridae*, *Phenuiviridae*, *Mesoniviridae*, and *Phasmaviridae* however, most signicant order was *Bunyaviridales*. *Aphid bunyavirus 1* was found in 10 different aphid species and horizontal transmission to plants was detected in *Vicia faba* leaves (An et al., 2021).

Aphids have a significant impact on the growth of alfalfa and can transmit *alfalfa mosaic virus* (AMV) in a non-persistent manner (Garran & Gibbs, 1982; van Leur, Duric, George, & Boschma, 2019), as well as *Alfalfa leaf curl virus* (ALCV) in a circulative manner (Ryckebusch, Sauvion, Granier, Roumagnac, & Peterschmitt, 2020). RNA-sequencing of the aphids from the alfalfa fields and subsequent BLAST analysis of the assembled contigs identified several viruses including AMV, *Medicago sativa alphapartitivirus 1* (MsAPV1), *Medicago sativa deltapartitivirus 1* (MsDPV1), *Medicago sativa amalgavirus 1* (MsAV1), *Cymbidium mosaic virus* (CymMV), *bean yellow mosaic virus* (BYMV), *cauliflower mosaic virus* (CaMV) and ALCV. These viruses were also detected in the alfalfa transcriptome assembly and verified by RT-PCR (Li et al., 2021). Similarly, RNA sequencing was conducted on maize and aphids collected from maize fields, and the presence of viruses was compared between insects and maize. *Rhopalosiphum padi virus* (RhPV), *aphid lethal paralysis virus* (ALPV), and a novel *Big Sioux River virus* (BSRV)-like dicistrovirus were identified in the transcriptome and confirmed by RT-PCR (Wamonje et al., 2017).

7.5.3 VIROMICS OF OTHER INSECTS

Thrips are economically important insect pests that have been reported to transmit tospoviruses in plants (Ullman et al., 2002). Thrips are known for transmitting *tomato spotted wilt virus* (TSWS) and *groundnut bud necrosis virus* (GBNV) (Daimei et al., 2017; Keough et al., 2018). High-throughput small interfering RNA (siRNA) sequencing of thrips and *Arachis hypogaea* revealed the presence of siRNA for TSWS in both thrips and *A. hypogaea.* It was further observed that the processing of siRNA varied between the host and thrips (Fletcher et al., 2016). Similarly, RNA-sequencing of wheat and grasshoppers from wheat identified the presence of a novel virus, *Wheat yellow striate virus*, belonging to the *Nucleorhabdovirus* genus (Liu et al., 2018).

7.6 CONCLUSION

Viruses pose a significant threat to global crop production and food security. It is crucial to continuously monitor the occurrence, frequency, and types of viruses in different infections in order to develop strategies to mitigate virus-related risks. Traditionally, plant samples have been used for virus surveillance. However, plant vectors can also be utilized to detect plant viruses, offering valuable insights into transmission patterns and host ranges. Since most viruses are transmitted by vectors, it is essential to include virus vectors in our approach to identify new viruses and determine their hosts. In recent years, insect viromics has led to the discovery of several novel viruses and their associated hosts.

The high mutation rates and complex recombination patterns of viruses can give rise to new viral strains. These emerging viruses may evade conventional detection methods due to their unique characteristics. NGS methods provide unparalleled precision in detecting both known and unknown viruses, even in mixed infections, without the need for prior sequence information. Advancements such as long-read sequencing further enhance the capabilities of NGS methods, ensuring greater confidence in the final assembled genome sequence. Additionally, single-cell genomics of infected plants and viruliferous vectors offer insights into the cellular-scale interactions between plants and viruses, as well as insects and viruses.

KEYWORDS

- **evolution**
- **plant-vector-virus interaction**
- **vector host range**
- **vectorome**
- **vectors**
- **virome**
- **virus replication**
- **viruses**

REFERENCES

Abd El-Aziz, M. H., (2019). Three modern serological methods to detect plant viruses. *J. Plant. Sci. Phytopathol., 3*, 101–106.

Adams, I., & Fox, A., (2016). Diagnosis of plant viruses using next-generation sequencing and metagenomic analysis. In: Wang, A., & Zhou, X., (eds.), *Current Research Topics in Plant Virology* (pp. 323–335). Cham: Springer International Publishing.

An, X., Zhang, W., Ye, C., Smagghe, G., Wang, J. J., & Niu, J., (2021). Discovery of a widespread presence bunyavirus that may have symbiont-like relationships with different species of aphids. *Insect Science*. doi: Https://doi.org/10.1111/1744-7917.12989.

Antipov, D., Raiko, M., Lapidus, A., & Pevzner, P. A., (2020). Metaviral SPAdes: Assembly of viruses from metagenomic data. *Bioinformatics, 36*(14), 4126–4129. doi: 10.1093/bioinformatics/btaa490.

Balasubramanian, S., (2015). Solexa sequencing: Decoding genomes on a population scale. *Clinical Chemistry, 61*(1), 21–24. doi: 10.1373/clinchem.2014.221747.

Bester, R., Pepler, P. T., Burger, J. T., & Maree, H. J., (2014). Relative quantitation goes viral: An RT-qPCR assay for a grapevine virus. *Journal of Virological Methods, 210*, 67–75. doi: Https://doi.org/10.1016/j.jviromet.2014.09.022.

Blawid, R., Silva, J. M. F., & Nagata, T., (2017). Discovering and sequencing new plant viral genomes by next-generation sequencing: Description of a practical pipeline. *Annals of Applied Biology, 170*(3), 301–314. doi: https://doi.org/10.1111/aab.12345.

Bragard, C., Caciagli, P., Lemaire, O., Lopez-Moya, J. J., MacFarlane, S., Peters, D., & Torrance, L., (2013). Status and prospects of plant virus control through interference with vector transmission. *Annual Review of Phytopathology, 51*(1), 177–201. doi: 10.1146/annurev-phyto-082712-102346.

Brown, J. K., Zerbini, F. M., Navas-Castillo, J., Moriones, E., Ramos-Sobrinho, R., Silva, J. C. F., & Varsani, A., (2015). Revision of *Begomovirus* taxonomy based on pairwise sequence comparisons. *Archives of Virology, 160*(6), 1593–1619. doi: 10.1007/s00705-015-2398-y.

Buermans, H. P. J., & Den, D. J. T., (2014). Next generation sequencing technology: Advances and applications. *Biochimica et Biophysica Acta (BBA) – Molecular Basis of Disease, 1842*(10), 1932–1941. doi: https://doi.org/10.1016/j.bbadis.2014.06.015.

Coetzee, B., Freeborough, M. J., Maree, H. J., Celton, J. M., Rees, D. J. G., & Burger, J. T., (2010). Deep sequencing analysis of viruses infecting grapevines: Virome of a vineyard. *Virology, 400*(2), 157–163. doi: https://doi.org/10.1016/j.virol.2010.01.023.

Daimei, G., Raina, H. S., Devi, P. P., Saurav, G. K., Renukadevi, P., Malathi, V. G., & Rajagopal, R., (2017). Influence of groundnut bud necrosis virus on the life history traits and feeding preference of its vector, *Thrips palmi*. *Phytopathology, 107*(11), 1440–1445.

Edgar, R. C., Taylor, J., Lin, V., Altman, T., Barbera, P., Meleshko, D., & Babaian, A., (2022). Petabase-scale sequence alignment catalyzes viral discovery. *Nature, 602*(7895), 142–147. doi: 10.1038/s41586-021-04332-2.

Eichmeier, A., Komínková, M., Komínek, P., & Baránek, M., (2016). Comprehensive virus detection using next generation sequencing in grapevine vascular tissues of plants obtained from the wine regions of bohemia and Moravia (Czech Republic). *Plos One, 11*(12), e0167966. doi: 10.1371/journal.pone.0167966.

Elbeaino, T., Marais, A., Faure, C., Trioano, E., Candresse, T., & Parrella, G., (2018). High-throughput sequencing reveals *Cyclamen persicum* mill. as a natural host for *Fig mosaic virus*. *Viruses, 10*(12), 684.

Fereres, A., & Moreno, A., (2009). Behavioral aspects influencing plant virus transmission by homopteran insects. *Virus Research, 141*(2), 158–168. doi: https://doi.org/10.1016/j.virusres.2008.10.020.

Fereres, A., & Raccah, B., (2015). *Plant Virus Transmission by Insects eLS* (pp. 1–12).

Fiallo-Olivé, E., Pan, L. L., Liu, S. S., & Navas-Castillo, J., (2020). Transmission of *Begomoviruses* and other whitefly-borne viruses: Dependence on the vector species. *Phytopathology, 110*(1), 10–17.

Fletcher, S. J., Shrestha, A., Peters, J. R., Carroll, B. J., Srinivasan, R., Pappu, H. R., & Mitter, N., (2016). The *Tomato spotted wilt virus* genome is processed differentially in its plant host arachis hypogaea and its thrips vector *Frankliniella fusca*. *Frontiers in Plant Science, 7*. doi: 10.3389/fpls.2016.01349.

Fonseca, P. L. C., Badotti, F., De Oliveira, T. F. P., Fonseca, A., Vaz, A. B. M., Tomé, L. M. R., & Góes-Neto, A., (2018). Virome analyses of *Hevea brasiliensis* using small RNA deep sequencing and PCR techniques reveal the presence of a potential new virus. *Virology Journal, 15*(1), 184. doi: 10.1186/s12985-018-1095-3.

Fortes, P., & Morris, K. V., (2016). Long noncoding RNAs in viral infections. *Virus Research, 212*, 1–11. doi: https://doi.org/10.1016/j.virusres.2015.10.002.

Garran, J., & Gibbs, A., (1982). Studies on *Alfalfa mosaic virus* and alfalfa aphids. *Australian Journal of Agricultural Research, 33*(4), 657–664.

He, X. F., Fang, Y. Y., Feng, L., & Guo, H. S., (2008). Characterization of conserved and novel microRNAs and their targets, including a TuMV-induced TIR–NBS–LRR class R gene-derived novel miRNA in Brassica. *FEBS Letters, 582*(16), 2445–2452. doi: https://doi.org/10.1016/j.febslet.2008.06.011.

He, Y., Yang, Z., Hong, N., Wang, G., Ning, G., & Xu, W., (2015). Deep sequencing reveals a novel closterovirus associated with wild rose leaf rosette disease. *Molecular Plant Pathology, 16*(5), 449–458. doi: https://doi.org/10.1111/mpp.12202.

Heyraud, F., Matzeit, V., Schaefer, S., Schell, J., & Gronenborn, B., (1993). The conserved nonanucleotide motif of the geminivirus stem-loop sequence promotes replicational release of virus molecules from redundant copies. *Biochimie, 75*(7), 605–615. doi: 10.1016/0300–9084(93)90067-3.

Ho, T., & Tzanetakis, I. E., (2014). Development of a virus detection and discovery pipeline using next generation sequencing. *Virology, 471–473*, 54–60. doi: https://doi.org/10.1016/j.virol.2014.09.019.

Hohn, T., (2007). Plant virus transmission from the insect point of view. *Proceedings of the National Academy of Sciences, 104*(46), 17905–17906. doi: doi:10.1073/pnas.0709178104.

Huang, H. J., Ye, Z. X., Wang, X., Yan, X. T., Zhang, Y., He, Y. J., & Li, J. M., (2021). Diversity and infectivity of the RNA virome among different cryptic species of an agriculturally important insect vector: Whitefly *Bemisia tabaci. npj Biofilms and Microbiomes, 7*(1), 43. doi: 10.1038/s41522-021-00216-5.

Idris, A., Al-Saleh, M., Piatek, M. J., Al-Shahwan, I., Ali, S., & Brown, J. K., (2014). Viral metagenomics: Analysis of *Begomoviruses* by illumina high-throughput sequencing. *Viruses, 6*(3), 1219–1236.

Jo, Y. S., Kim, S. S., Kim, M. S., Yoo, N. J., & Lee, S. H., (2017). Frameshift mutation of FXR1 encoding a RNA-binding protein in gastric and colorectal cancers with microsatellite instability. *Pathology & Oncology Research, 23*(2), 453, 454. doi: 10.1007/s12253-016-0177-z.

Jones, S., Baizan-Edge, A., MacFarlane, S., & Torrance, L., (2017). Viral diagnostics in plants using next generation sequencing: Computational analysis in practice. *Frontiers in Plant Science, 8*. doi: 10.3389/fpls.2017.01770.

Kasschau, K. D., Xie, Z., Allen, E., Llave, C., Chapman, E. J., Krizan, K. A., & Carrington, J. C., (2003). P1/HC-Pro, a viral suppressor of RNA silencing, interferes with *Arabidopsis* development, and miRNA function. *Developmental Cell, 4*(2), 205–217. doi: https://doi.org/10.1016/S1534-5807(03)00025-X.

Kehoe, M. A., Coutts, B. A., Buirchell, B. J., & Jones, R. A. C., (2014). Plant virology and next generation sequencing: Experiences with a potyvirus. *PLOS One, 9*(8), e104580. doi: 10.1371/journal.pone.0104580.

Keough, S., Danielson, J., Marshall, J. M., Lagos-Kutz, D., Voegtlin, D. J., Srinivasan, R., & Nachappa, P., (2018). Factors affecting population dynamics of thrips vectors of *Soybean vein necrosis virus*. *Environmental Entomology, 47*(3), 734–740.

Kieft, K., Adams, A., Salamzade, R., Kalan, L., & Anantharaman, K., (2022). vRhyme enables binning of viral genomes from metagenomes. *Nucleic Acids Research*. doi: 10.1093/nar/gkac341.

Kondo, H., Fujita, M., Hisano, H., Hyodo, K., Andika, I. B., & Suzuki, N., (2020). Virome analysis of aphid populations that infest the barley field: The discovery of two novel groups of nege/kita-like viruses and other novel RNA viruses. *Frontiers in Microbiology, 11*. doi: 10.3389/fmicb.2020.00509.

Kundu, A., Paul, S., Dey, A., & Pal, A., (2017). High throughput sequencing reveals modulation of microRNAs in *Vigna mungo* upon *Mungbean yellow mosaic India Virus* inoculation highlighting stress regulation. *Plant Science, 257*, 96–105. doi: https://doi.org/10.1016/j.plantsci.2017.01.016.

Li, J., Gu, H., Liu, Y., Wei, S., Hu, G., Wang, X., & Ban, L., (2021). RNA-seq reveals plant virus composition and diversity in alfalfa, thrips, and aphids in Beijing, China. *Archives of Virology, 166*(6), 1711–1722. doi: 10.1007/s00705-021-05067-1.

Li, R., Gao, S., Hernandez, A. G., Wechter, W. P., Fei, Z., & Ling, K. S., (2012). Deep sequencing of small RNAs in *Tomato for virus* and viroid identification and strain differentiation. *Plos One, 7*(5), e37127. doi: 10.1371/journal.pone.0037127.

Liu, Y., Du, Z., Wang, H., Zhang, S., Cao, M., & Wang, X., (2018). Identification and characterization of *Wheat yellow striate virus,* a novel leafhopper-transmitted *Nucleorhabdovirus* infecting wheat. *Frontiers in Microbiology, 9*. doi: 10.3389/fmicb.2018.00468.

Massart, S., Olmos, A., Jijakli, H., & Candresse, T., (2014). Current impact and future directions of high throughput sequencing in plant virus diagnostics. *Virus Research, 188*, 90–96. doi: https://doi.org/10.1016/j.virusres.2014.03.029.

Navas-Castillo, J., Fiallo-Olivé, E., & Sánchez-Campos, S., (2011). Emerging virus diseases transmitted by whiteflies. *Annual Review of Phytopathology, 49*, 219–248.

Ng, T. F. F., Duffy, S., Polston, J. E., Bixby, E., Vallad, G. E., & Breitbart, M., (2011). Exploring the diversity of plant DNA viruses and their satellites using vector-enabled metagenomics on whiteflies. *PLOS One, 6*(4), e19050. doi: 10.1371/journal.pone.0019050.

Nouri, S., Matsumura, E. E., Kuo, Y. W., & Falk, B. W., (2018). Insect-specific viruses: From discovery to potential translational applications. *Current Opinion in Virology, 33*, 33–41. doi: https://doi.org/10.1016/j.coviro.2018.07.006.

Nurk, S., Meleshko, D., Korobeynikov, A., & Pevzner, P. A., (2017). metaSPAdes: A new versatile metagenomic assembler. *Genome Research, 27*(5), 824–834. doi: 10.1101/gr.213959.116.

Oliveira, M. R. V., Henneberry, T. J., & Anderson, P., (2001). History, current status, and collaborative research projects for *Bemisia tabaci. Crop Protection, 20*(9), 709–723. doi: https://doi.org/10.1016/S0261-2194(01)00108-9.

Pecman, A., Kutnjak, D., Gutiérrez-Aguirre, I., Adams, I., Fox, A., Boonham, N., & Ravnikar, M., (2017). Next generation sequencing for detection and discovery of plant viruses and viroids: Comparison of two approaches. *Frontiers in Microbiology, 8*, 1998–1998. doi: 10.3389/fmicb.2017.01998.

Rosario, K., Capobianco, H., Ng, T. F. F., Breitbart, M., & Polston, J. E., (2014). RNA viral metagenome of whiteflies leads to the discovery and characterization of a whitefly-transmitted *Carlavirus* in North America. *PLOS One, 9*(1), e86748. doi: 10.1371/journal.pone.0086748.

Rosario, K., Seah, Y. M., Marr, C., Varsani, A., Kraberger, S., Stainton, D., & Breitbart, M., (2015). Vector-enabled metagenomic (VEM) surveys using whiteflies (*Aleyrodidae*) reveal novel *Begomovirus* species in the new and old worlds. *Viruses, 7*(10), 5553–5570. doi: 10.3390/v7102895.

Ryckebusch, F., Sauvion, N., Granier, M., Roumagnac, P., & Peterschmitt, M., (2020). *Alfalfa leaf curl virus* is transmitted by *Aphis craccivora* in a highly specific circulative manner. *Virology, 546*, 98–108.

Stewart, L. R., Teplier, R., Todd, J. C., Jones, M. W., Cassone, B. J., Wijeratne, S., & Redinbaugh, M. G., (2014). Viruses in maize and johnsongrass in Southern Ohio. *Phytopathology®, 104*(12), 1360–1369. doi: 10.1094/PHYTO-08-13-0221-R.

Tripathi, A., Goswami, K., Tiwari, M., Mukherjee, S. K., & Sanan-Mishra, N., (2018). Identification and comparative analysis of microRNAs from tomato varieties showing contrasting response to ToLCV infections. *Physiology and Molecular Biology of Plants, 24*(2), 185–202. doi: 10.1007/s12298-017-0482-3.

Ullman, D. E., Meideros, R., Campbell, L. R., Whitfield, A. E., Sherwood, J. L., & German, T. L., (2002). Thrips as vectors of *Tospoviruses Advances in Botanical Research* (Vol. 36, pp. 113–140). Academic Press.

Van, L. J., Duric, Z., George, J., & Boschma, S., (2019). *Alfalfa mosaic virus* infects the tropical legume *Desmanthus virgatus* in Australia and the potential role of the cowpea aphid (*Aphis craccivora*) as the virus vector. *Australasian Plant Disease Notes, 14*(1), 1–4.

Wamonje, F. O., Michuki, G. N., Braidwood, L. A., Njuguna, J. N., Musembi, M. J., Djikeng, A., & Carr, J. P., (2017). Viral metagenomics of aphids present in bean and maize plots on mixed-use farms in Kenya reveals the presence of three dicistroviruses including a novel Big Sioux River virus-like dicistrovirus. *Virology Journal, 14*(1), 188. doi: 10.1186/s12985-017-0854-x.

Wei, T., & Li, Y., (2016). Rice reoviruses in insect vectors. *Annual Review of Phytopathology, 54*(1), 99–120. doi: 10.1146/annurev-phyto-080615-095900.

Will, T., & Van, B. A. J. E., (2006). Physical and chemical interactions between aphids and plants. *Journal of Experimental Botany, 57*(4), 729–737. doi: 10.1093/jxb/erj089.

Wu, Q., Ding, S. W., Zhang, Y., & Zhu, S., (2015). Identification of viruses and viroids by next-generation sequencing and homology-dependent and homology-independent algorithms. *Annual Review of Phytopathology, 53*(1), 425–444. doi: 10.1146/annurev-phyto-080614-120030.

Wu, Q., Luo, Y., Lu, R., Lau, N., Lai, E. C., Li, W. X., & Ding, S. W., (2010). Virus discovery by deep sequencing and assembly of virus-derived small silencing RNAs. *Proceedings of the National Academy of Sciences, 107*(4), 1606–1611. doi: 10.1073/pnas.0911353107.

CHAPTER 8

Weeds and Wild Relatives as Undetermined Plant Virus Sources: Detection and Diagnosis

ALANGAR ISHWARA BHAT and ARJUNAN JEEVALATHA

ICAR–Indian Institute of Spices Research, Kozhikode, Kerala, India

ABSTRACT

Diseases caused by viruses result in significant crop losses worldwide. These diseases are systemic and cannot be eliminated through the use of chemicals. In the absence of the main crops, viruses can survive in wild or volunteer plants of the same species, as well as in several weed species that grow in and around the main crop, often without showing any symptoms. These host plants serve as a reservoir for viruses and act as the primary source for virus transmission to the main crop through vectors. Additionally, some viruses can survive in seeds, tissue culture materials, and vegetatively propagated materials without showing any symptoms, further contributing to the spread of viruses in the field. Therefore, it is crucial to accurately detect and diagnose viruses in all the aforementioned hosts in order to establish successful virus-free crops. Various methods, such as biological, physicochemical, protein-based, and nucleic acid-based methods, are available for virus detection and diagnosis. Among these methods, serological (protein) and nucleic acid-based methods, such as PCR, rolling circle amplification (RCA), and next-generation sequencing (NGS), provide quick and reliable results for identifying the virus.

New and Emerging Plant Viruses: The Threat to Food Security.
Alireza Golnaraghi and Rajarshi Kumar Gaur (Eds.)

8.1 INTRODUCTION

Viral diseases are serious constraints on crop productivity, and severe epidemics of newly emerging viruses can affect a wide range of crops. The emergence of new viruses or virus variants is influenced by various factors, including the development of virus variants, changes in vector biology and cropping systems, the introduction of new crops or varieties, the movement of infected planting materials, the introduction of new crop species and genotypes, and the exchange of germplasm carrying host susceptibility genes (Hull, 2014). In the absence of main crops, viruses can survive in wild or volunteer plants of the same species, as well as in several weed species that grow in and around the main crop, often without showing symptoms. These hosts act as reservoirs of viruses and serve as the primary source for virus spread through vectors to the main crop. Many crops are susceptible to infection by multiple viruses, each of which can cause significant economic losses. Infected plant material may also be unsuitable for export. Some viruses in crops do not exhibit symptoms, and the virus can remain latent in infected plants, leading to substantial yield reduction when exposed to different environments. These symptomless plants can serve as a source for virus spread by insect vectors. Many virus-infected vegetative planting materials, including seeds, may not show symptoms. Therefore, proper identification of viruses associated with planting units is necessary to prevent the introduction of viruses into the environment. This chapter provides a brief overview of biological, serological, and nucleic acid-based methods that have been reported for the diagnosis of unknown plant viruses.

8.2 DETECTION AND DIAGNOSIS OF PLANT VIRUSES

8.2.1 BIOLOGICAL AND PHYSICAL METHODS

Initially, viral diseases could be diagnosed based on biological properties such as symptoms, host range, and transmission. Symptoms have traditionally been the primary means of diagnosing and naming viral diseases, both in the past and today (Bhat & Rao, 2020). The symptoms caused by a specific virus in a host can vary and may even remain dormant depending on environmental conditions and seasons. Therefore, the symptoms observed in a range of indicator hosts provide initial indications about the identity of a virus in the field or laboratory. Additionally, the transmission of the test virus by vectors (such as aphids, leafhoppers, whiteflies, thrips, mealybugs, beetles, motes, fungi, and nematodes) can also shed light on the identity of the virus.

The host range of the test virus, determined through controlled conditions using sap, vectors, or graft inoculation, provides further information about the identity of the test virus (Hull, 2014). Biological tests are important for diagnosing plant viruses, but they are more time-consuming and labor-intensive compared to laboratory assays. Viruses can also be diagnosed based on physicochemical characteristics, such as particle shape and dimensions determined through electron microscopy, and coat protein molecular weight determined through SDS-polyacrylamide gel electrophoresis (Zechmann & Zellnig, 2009; Bhat & Rao, 2020). However, these tests are less sensitive and cannot identify viruses at the species level.

8.2.2 SEROLOGICAL METHODS

Based on the primary clue obtained from biological and other assays, such as particle morphology and coat protein molecular weight, further diagnosis of the test virus can be done using serological assays. This involves subjecting the test sample to ELISA or other serological assays using antisera to different viruses to further narrow down the identity of the test virus, probably to the genus level or, in some cases, even up to the species level. Among various serological methods, ELISA and its variants (double antibody sandwich (DAS)-ELISA, direct antigen-coated (DAC)-ELISA, $F(ab)_2$-ELISA, plate trapped ELISA) are used in the detection and diagnosis of plant viruses (Clark & Adams, 1977; Barbara & Clark, 1982; Clark et al., 1986). The serological reaction is based on the specific reaction of the antigen (viral coat protein) with antibodies raised against the coat protein of the virus. The main advantages of ELISA are that the assay can be completed in a few hours, it is amenable for large-scale testing, it is more sensitive than biological assays, crude samples can be directly used in the assay, no sophisticated equipment is needed, and results are not affected by changes in the environment (Bhat & Rao, 2020). However, this assay requires virus-specific antiserum and labeling of antibody molecules with enzymes to increase the sensitivity of the assay. Another serological assay used in the diagnosis is immunosorbent electron microscopy (ISEM), which involves the use of an electron microscope. Electro-blot immunoassay (EBIA) utilizes sodium dodecyl sulfate-polyacrylamide gel electrophoresis (SDS-PAGE) to separate viral proteins, followed by transfer to a membrane and detection using labeled antibodies (Derrick, 1973; O'Donell et al., 1982). Among all serological assays, lateral flow immunoassay (LFIA) can be performed onsite by anyone without any special skills.

8.2.3 POLYMERASE CHAIN REACTION (PCR)-BASED METHODS

Once genus-level identification is completed based on biological and serological assays, further identification of the test virus can be achieved through polymerase chain reaction (PCR) using primers designed for the conserved region (CR) in all known species within that genus. If the expected product is obtained during PCR, the PCR product can be cloned, and its nucleotide sequence determined using Sanger's chain termination method. The obtained sequence is then analyzed using the BLAST on the NCBI website (https://blast.ncbi.nlm.nih.gov/Blast.cgi?PROGRAM=blastn&PAGE_TYPE=BlastSearch&LINK_LOC=blasthome) and other bioinformatics analyzes including phylogenetic relationship to identify the exact taxonomic status of the test virus (Bhat & Rao, 2020). PCR-based assays are more sensitive than serological assays as they involve prior amplification of the viral genomic segments before detection. Before setting up PCR reactions, isolation of template nucleic acid from the test sample and synthesizing specific primers for the target virus are needed. If the test virus has an RNA genome, then total RNA isolated from the test plant should be first subjected to cDNA using either virus-specific or random primers. The product of cDNA is then subjected to PCR using both forward and reverse primers designed for the CR. If an antibody against the test virus is available, one can use immunocapture (IC) PCR to amplify the viral genomic segments. In such cases, PCR tubes may be first incubated with virus-specific antibodies to capture the antibody molecules on the inner walls of the PCR tube (Mumford & Seal, 1997). After washing the tube, crude sap extracted from the test sample is added and incubated. After the incubation period, the PCR tube is washed to remove the excess sap, added with sterile water needed for the PCR reaction, and subjected to a freeze-thaw cycle to disrupt the coat protein and expose the virus genome captured in the PCR tube. This is followed by the addition of other PCR components, including primers, and performing PCR. If the test plant has inhibitory substances against PCR or the test virus concentration is very low, in such cases subjecting the extracted nucleic acid to nested PCR may efficiently amplify the test virus. The nested PCR uses two rounds of PCRs, the first round to increase the template concentration uses degenerate primers. A portion of the first-round reaction is used as the template for the second round using primers that anneal within the first amplicon (Vincelli & Tisserat, 2008).

8.2.4 ROLLING CIRCLE AMPLIFICATION (RCA)

Rolling circle amplification (RCA) is used for the amplification of circular DNA molecules. This technique is useful for identifying novel circular DNA viruses. The total DNA isolated from the test plant is denatured and incubated with a random/specific/combination of both primers (Inoue-Nagata et al., 2004). The primer extension occurs through DNA ϕ polymerase with strand displacement activity, allowing the newly synthesized DNA molecule to displace the previously generated DNA molecule, releasing single-stranded DNA (ssDNA). This process generates a long ssDNA containing a repeated sequence that is complementary to the circular DNA template. After the reaction, the product is subjected to restriction using an enzyme that has a unique restriction site for the test virus. When identifying a novel virus, RCA products may be subjected to restriction using different enzymes to identify the enzyme with a unique restriction site. The restricted product is then run on an agarose gel, and the band of interest can be eluted, cloned, sequenced, and analyzed to determine the identity of the test virus, as described in the PCR-based methods section (Lau & Botella, 2017). This method has been used for the identification of many novel begomo-, nano-, and badnaviruses.

8.2.5 NEXT-GENERATION SEQUENCING (NGS) FOR THE DIAGNOSIS OF VIRUSES

When all the techniques discussed above fail to identify the virus associated with a disease, the next-generation sequencing (NGS) approach may be used. It is an unbiased method for plant viral disease diagnosis that requires no prior knowledge of the host or the pathogen. The total DNA (if the test virus is a DNA virus)/RNA/dsRNA/small RNA extracted from the test plant is first subjected to cDNA synthesis, fragmented, immobilized on a solid surface, amplified, and then subjected to sequencing using an appropriate platform (Adam et al., 2009; Kreuze et al., 2009; Pecman et al., 2017; Bhat & Rao, 2020). The sequence data obtained from the NGS is analyzed for quality and used to develop contigs using an appropriate program. The contigs are then analyzed using BLAST against NCBI's plant virus database to identify the contigs related to viruses. The contigs that match with viruses are assembled and mapped against the reference virus genome, and further analysis is conducted to determine the identity and taxonomy of the test virus (Bhat & Rao, 2020). The test virus identified through NGS can be confirmed by subjecting nucleic acid isolated from the test plant to PCR using primers designed to target contigs that match viruses, followed by Sanger sequencing of the PCR product. Once the identity and sequence of the test virus are known, assays such as PCR, real-time PCR, loop-mediated isothermal amplification (LAMP), and recombinase polymerase amplification (RPA) can be developed for routine detection of the virus in plants. NGS technology has been successfully used to identify viruses associated with diseases of unknown etiology in various crops (Adam et al., 2009; Kreuze et al., 2009; AIRwahnih et al., 2011; Pecman et al., 2017; Rott et al., 2017; Bhat et al., 2020a, b). Recently, Hasiów-Jaroszewska et al. (2021) reviewed the application of NGS for the detection of viruses in weeds and wild plants.

8.3 DETECTION OF VIRUSES IN WEEDS AND WILD RELATIVES

Here, we present a concise overview of the detection and diagnosis of both known and unknown viruses in weeds and wild relatives, which act as reservoirs for major crops.

8.3.1 BEGOMOVIRUSES

Detection of many begomoviruses, including those infecting various weed hosts, has been studied using monoclonal antibodies raised against the Indian cassava mosaic virus (ICMV) (Muniyappa et al., 1987; Harrison et al., 1991; Nateshan et al., 1996; Swanson et al., 1998; Ramappa et al., 1998; Garg et al., 2001; Muniyappa et al., 2003; Devaraja et al., 2005) (Table 8.1). PCR assay with degenerate primers, followed by southern hybridization tests, was used to identify geminivirus infection in *Launaea acaulis* (Raj et al., 2000). Kashina et al. (2003) used dot blot hybridization assay to identify tomato yellow leaf curl Tanzania virus infection in weed hosts such as *Ageratum conyzoides* and *Sida acuta.* Natural infection of ToLCV was observed in symptomless plants of Parthenium *hystrophorus* and *Sonchus asper* in Uttar Pradesh, India. The vector, *Bemisia tabaci*, could transmit the virus from these weeds to tomato plants (Ansari & Tiwari, 2005). Viruses that predominantly infect only uncultivated plant species may pose a potential risk to cultivated plant species. For example, the maize streak virus strain, MSV-A, which causes severe disease in maize, is the recombinant progeny of ancestral MSV-B and MSV-G/-F variants adapted to grasses (Varsani et al., 2008). Universal begomovirus and betasatellite primers were used to clone the viral components in *Digera arvensis,* a weed host affected by yellow vein disease. The study found the presence of multiple and recombinant betasatellites, which could potentially affect crop plants as an important source of begomovirus components (Mubin et al., 2009).

A total of 122 dicotyledonous plant species from 25 families were screened using TAS-ELISA, conventional PCR assay, and real-time Taqman assay. Among these techniques, real-time PCR was found to be the most sensitive, enabling the detection of TYLCV in 461 samples from 49 different species belonging to various families including Amaranthaceae, Chenopodiaceae, Compositae, Convolvulaceae, Cruciferae, Euphorbiaceae, Geraniaceae, Leguminosae, Malvaceae, Orobanchaceae, Plantaginaceae, Primulaceae, Solanaceae, Umbelliferae, and Urticaceae (Papayiannis et al., 2011).

Tahir et al. (2015) identified *Ageratum enation virus* (AEV) from *Ageratum conyzoides, Sonchus oleraceus,* and turnip using PCR followed by sequencing of viral components. The study revealed that this virus has the capacity to infect crop plants.

In a recent study by Zubair et al. (2020), DAS-ELISA and PCR assay with tomato leaf curl New Delhi virus (ToLCNDV) specific primers were used to detect ToLCNDV in *Physalis minima*, a common weed plant found in and around luffa fields. *Combretum hispidum* and *Euphorbia hirta* were found in both wet and dry seasons in cassava fields in Nigeria. *E. hirta* was found to be an alternative host to ACMV, EACMV, and their co-infection (Badamasi, 2020). Nightshade (*Solanum nigrum*) may act as a reservoir of ToLCNDV and is involved in the development of epidemics in cucurbit species. Nightshade serves as an intermediate host between tomatoes and cucurbits, and the strain of ToLCNDV has likely adapted from solanaceous to cucurbitaceous hosts (Ansar et al., 2021). Kil et al. (2021) detected TYLCV through PCR in 28 species, including *Eleusine indica, Digitaria ciliaris, Echinochloa crus-galli, Panicum dichotomiflorum*, and *Setaria faberi*, belonging to the family Poaceae. Two new species of begomovirus, malvastrum yellow vein Baoshan virus (MaYVBsV) and malvastrum yellow vein Honghe virus (MaYVHhV), were found in the weed host *Malvastrum coromandelianum* through PCR followed by sequencing of genomic components (Wang et al., 2021).

RCA has tremendously accelerated the discovery of novel circular ssDNA viruses from various environmental sources. RCA was used to identify and characterize a highly divergent geminivirus, *Euphorbia* caput-medusae latent virus (EcmLV), from wild spurge, *Euphorbia caputmedusae*, which did not show any symptoms in its natural host wild spurge. However, the cloned components produced severe symptoms in tomato and *Nicotiana benthamiana* (Bernardo et al., 2013). Similarly, Varsani et al. (2009) isolated *Eragrostis curvula* streak virus (ECSV) from an uncultivated African grass species, *Eragrostis curvula*, using RCA. *Abutilon* golden yellow mosaic virus (AbGYMV) was found to cause striking golden/yellow mosaic symptoms in the malvaceous weed, *Abutilon* sp. In this study, RCA products were cloned and sequenced, and further phylogenetic analysis confirmed the presence of AbGYMV, a new bipartite begomovirus in *Abutilon* sp. (Maliano et al., 2021). Using next-generation technology, the causal agent of citrus chlorotic dwarf disease (CCDD), a virus-like disorder of citrus, was identified as citrus chlorotic dwarf-associated virus (CCDaV) (Loconsole et al., 2012). Similarly, by adapting the Illumina MiSeq platform, Nhlapo et al. (2018) showed mixed infections of DNA (begomovirus) and RNA (potyvirus and crinivirus) in sweet potato virus disease (SPVD)-affected symptomatic and asymptomatic sweet potato plants. In this study, RCA and deep sequencing were used to identify begomovirus. Tomato severe rugose virus (ToSRV) was found to infect the weed host *Physalis angulate* using the RCA method, followed by nanopore sequencing (Duarte et al., 2020). High-throughput sequencing (HTS) and genome assembly were used to identify a new divergent geminivirus, *Opuntia* virus 1 (OpV1), in Cactus species, and the infectious clone of OpV1 could replicate in *Nicotiana benthamiana* (Fontenele et al., 2020). Malvaceous weeds *Malachra* sp. and Abutilon sp., tested with degenerate primers for begomovirus infection, showed tobacco leaf curl Cuba virus (TbLCuCV) and AbGYMV infection, respectively, in the Caribbean Basin (Maliano et al., 2021).

8.3.2 CUCUMOVIRUSES

CMV may also easily spread in agroecosystems due to its ability to infect many plant hosts (over 1,200 species), including overwintering weeds or plants for which seed transmission of this virus was confirmed. Bioassays, serological tests, and dsRNA analysis were used to investigate the presence of cucumber mosaic virus (CMV) and associated satRNAs in 71 weed species collected from tomato fields with an 80% incidence of tomato lethal necrosis in Croatia. CMV was found in eight weed species including *Anagallis arvensis*, *Solanum nigrum*, and *Sonchus* sp. (Skoric et al., 2000). NGS coupled with metagenomic analysis was applied to *Gomphrena globosa* infected with an unknown pathogen originally isolated from the flowering plant *Liatris spicata* and was found to contain a new cucumovirus, 'gayfeather mild mottle virus (GMMV)' (Adams et al., 2009). *Datura stramonium, D. metal, Portulaca oleracea, Cyprus rotundus,* and *Trianthema pentandra* were found to be weed hosts for CMV by mechanical inoculation followed by DAC-ELISA in Pakistan (Iqbal et al., 2011). RT-PCR assay using CMV specific primers targeting RNA 1 detected new strains of CMV in naturally infected *Basella alba, Telfairia occidentalis*, and *Talinum fruticosum* in Nigeria (Adediji, 2019) (Table 8.1). NGS to sequence virus-derived small RNAs combined with RT-PCR was used to detect viruses including CMV in wild plants in common bean fields in Tanzania (Mwaipopo et al., 2021).

8.3.3 ILARVIRUSES

Lavanya et al. (2005) reported several weed species, including *Trianthema portulacastrum, Priva leptostachya, Digeria arvensis, Clitoria ternatea, Solanum nigrum, Vernonia cineraria,* and *Trichodesma indicum,* as alternate hosts of tobacco streak virus (TSV), causing necrosis disease in sunflower. Based on DAS-ELISA and biological assay methods, Sokmen et al. (2005) reported the occurrence of apple mosaic virus in eight weed species in hazelnut orchards in Turkey. Sharman et al. (2015) reported several weed species, including *Parthenium hysterophorus* and *Verbesina encelioides,* as sources of TSV. In a metagenomics study using HTS, Ma et al. (2020) reported the occurrence of a novel ilarvirus (*Solanum nigrum* ilarvirus 1 (SnIV 1)) in tomato and European black nightshade (*Solanum nigrum*) (Table 8.1).

8.3.4 ORTHOTOSPOVIRUSES

Of the 1,446 samples of weeds and wild plants belonging to 22 families tested, 14 samples of the species *Portulaca oleracea, Salvia pratensis, Solanum nigrum,* and *Stellaria media* were infected with tomato spotted wilt virus (TSWV) (van Os et al., 1993). Of the 26 species of weeds associated with chrysanthemum, 16 were positive for the TSWV (Daniel et al., 1996). Through ELISA testing, Wilson (1998) showed that TSWV was present in a range of weed species, among which *Arctotheca calendula, Sonchus oleraceus, Malva sylvestris, Brassica rapa* ssp. *silvestris,* and *Erodium moschatum* and *Trifolium* sp. are the most important reservoir host species. Mertelík & Mokrá (1999) reported that ornamental plant species (*Dendranthema orifolium* and *Zantedeschia* sp.) and the weed hosts (*Stellaria media*

TABLE 8.1 Viruses Recorded on Different Weeds and Wild Plants from Different Regions

Cultivated Host	Weeds/Wild Plants/Plants Other Than Cultivated Hosts	Viruses Recorded	Region/ Country	References
Rice	*Cynodon dactylon*, *Digitaria sanguinalis*, *Echinocloa colonum*, *E. crusgalli*, *Leptochloa chinensis*, *Panicum repens*, *Cyperus rotundus*, *Fimbrystylis miliaceae*, *Dinebra aratica*, *Brachiaria ramosa*, *Paspalum dilatum*, *Leersia hexandra*, and *Paspalum hydrophilum*	Rice tungro virus	India	Patel et al. (2018)
	Digitaria sanguinalis, *Monochoria vaginalis*, *Eleusine indica*, *Alopecurus aequalis*	Rice stripe mosaic virus	China	Wang et al. (2021)
Wheat, corn, barley, rye, foxtail millet, pearl millet, Sorghum, oats	*Bromus tectorum*, *Aegilops cylindricae*, *Avena fatua*, *Elymus repens*, *Bromus secalinus*	Wheat streak mosaic virus	Montana, US	Ito et al. (2012)
	Anthoxanthum odoratum (*sweet vernal grass*), *Arrhenatherum elatius* (*false oat-grass*), *Lolium multiflorum* (*Italian rye-grass*), *Bromus japonicus* (*Japanese chess*), *Echinochloa crus-galli* (*barnyard grass*), *Holcus lanatus* (*meadow soft grass) and Holcus mollis* (*creeping soft grass*)	-do-	Czech Republic	Chalupniková et al. (2017)
Wheat	*Bromus tectorum*, *Setaria viridis*, *Hordeum jubatum*	Barley yellow dwarf virus	Idaho	Rashidi et al. (2021)
	Achnatherum lettermanii, *Achnatherum occidentale*, *Achnatherum thurberianum*, *Danthonia intermedia*, *Poa fendleriana*, *Sporobolus airoides*, and *Sporobolus cryptandrus*	-do-	US Pacific Northwest	Ingwell & Bosque-Pérez (2015)
Maize	*Pennisetum purpureum*, *Digitaria abyssinica*, *Cyperus rotundus*, *Amaranthus spinosus*, *Commelina benghalensis and Eleusine indica*	Maize chlorotic mottle virus	Uganda	Mudde et al. (2019)
	Digitaria sanguinalis, *Phalaris paradoxa*, *Oplismenus hirtellus*, *Echinocloa colona*, *Cynodon nlemfuensis*, *Pennisetum purpureum*, *Cyperus cyperoids*	-do-	Ethiopia	Regassa et al. (2021)
	Sorghum	Maize dwarf mosaic virus	Spain	Clemente-Orta et al. (2020)
	Halepense, *Setaria verticillate*, *Digitaria sanguinalis*	Sugarcane mosaic virus		
	Sorghum			
	Halepense, Setaria verticillate			

TABLE 8.1 *(Continued)*

Cultivated Host	Weeds/Wild Plants/Plants Other Than Cultivated Hosts	Viruses Recorded	Region/ Country	References
Tomato	*Ageratum conyzoides*, *Sida acuta*	Tomato yellow leaf curl Tanzania virus	Tanzania	Kashina et al. (2003)
	28 species including *Eleusine indica*, *Digitaria ciliaris*, *Echinochloa crus-galli*, *Panicum dichotomiflorum*, and *Setaria faberi*	Tomato yellow leaf curl virus	Korea	Kil et al. (2021)
	Abutilon indicum, *Ageratum conyzoides*, *Croton bonplandianum*, Euphorbia *hirta*, *E. geniculate*, *Nicandra physaloides*, *Nicotiana plumbaginifolia*, *Parthenium hysterophorus*, *Rumex dentatus*, *Salvia plebia*, *Sida mysoriansis*, *Sida rhombadifolia*, *Solanum nigrum*, *Sonchus arvensis*, *Veronia cineria*, *Xanthium strumarium*	Tomato leaf curl virus	India	Pandey et al. (2017)
–	*Physalis minima*	Tomato leaf curl New Delhi virus	Pakistan	Zubair et al. (2020)
	Solanum nigrum	-do-	India	Ansar et al. (2021)
	Amaranthus sp., *Spergularia* sp., *Atriplex* sp., *Chenopodium ambrosioides* L., *Chenopodium* sp., *Halogetum sativus*; *Senebiera didyma*; *Malva* sp.; *Polygonum* sp.; *Nicotiana glauca*, and *Solanum nigrum*	Tomato torrado virus	Spain	Alfaro-Fernández et al. (2008)
	Conyza canadensis, *Erigeron annuus*, *Sonchus asper*, *Youngia japonica*, *Trigonotis peduncularis*, *Cardamine flexuosa*, *Cerastium glomeratum*, *Stellaria media*, *Chenopodium album*, *Ipomoea hederacea*, *Quamoclit coccinea*, *Vicia angustifolia var. segetilis*, *V. tetrasperma*, *Phytolacca americana*, *Mazus pumilus*, *Solanum americanum* and *S. nigrum*	Tomato chlorosis virus	Korea	Kil et al. (2015)
	Amaranthus hybridus, *Solanum nigrum*, *Tagetes minuta*, *Datura stramonium*	Tomato spotted wilt virus	Kenya	Macharia et al. (2016)
	Chenopodium album, *Calotropis procera*, *Heliotropium indicum*, *Cornopus didymus*, *Convolvulus arvensis*, *Melilotus indica*, *Galium aparine*, *Nicotiana plumbaginifolia*, and *Withania somnifera*	Cucumber mosaic virus Sub-group I	Pakistan	Akhtar et al. (2019)
Bhendi	*Ageratum conyzoides*	Bhendi yellow vein mosaic virus	Punjab	Cheema (1991)

TABLE 8.1 *(Continued)*

Cultivated Host	Weeds/Wild Plants/Plants Other Than Cultivated Hosts	Viruses Recorded	Region/ Country	References
Brinjal or eggplant	*Euphorbia* spp. and *Parthenium hysterophorus*	Cucumber mosaic virus and Tomato leaf curl. New Delhi virus	Tamil Nadu, India	Abirami et al. (2022)
–	*Solanum seaforthianum*	Eggplant mosaic virus	Venezuela	Debrat et al. (1997)
Cucurbits	*Moluccella laevis*, *Withania somnifera*, *Amaranthus graecizans*, *A. muricatus*, *Ecballium elaterium*, and *Chrozophora tinctoria*	Cucumber green mottle mosaic virus	Israel	Shargil et al. (2017)
–	*Alhagi camelorum*	Zucchini yellow mosaic virus	Iran	Safaeizadeh (2008)
–	*Cucumis* spp.	Cucumber mosaic virus	Florida	Goyal et al. (2012)
	Melothria pendula	Papaya ringspot virus type W, Zucchini yellow mosaic virus		
–	*Momordica charantia*	Squash vein yellowing virus	Florida	Shrestha et al. (2016)
–	*Solanum nigrum*	Tomato leaf curl New Delhi virus	India	Ansar et al. (2021)
Pepper	*Leonurus sibiricus*	Tomato spotted wilt virus, pepper mild mottle virus, beet western yellow virus	Korea	Kwon et al. (2016)
Cassava	*Euphorbia hirta*	ACMV, EACMV	Nigeria	Badamasi (2020)
	Senna occidentalis, *Leucana leucocephala*, *Combretum confertum*, *Manihot glaziovii*	-do-	-do-	Alabi et al. (2008)
	Centrosema pubescens, *Chromolaena odorata*, *Senna alata*	-do-	-do-	Eni et al. (2020)
–	*Datura stramonium* and 38 species of *Nicotiana*	SLCMV	India	Jose et al. (2008)
–	*Albizia zygia*, *Senna obtusifolia*, *Pupalia lappacea*, *Strophanthus hispidus*, *Manihot glaziovii*	EACMV	Togo	Agjata et al. (2014)

TABLE 8.1 *(Continued)*

Cultivated Host	Weeds/Wild Plants/Plants Other Than Cultivated Hosts	Viruses Recorded	Region/ Country	References
–	*Zanha africana*, *Trichodesma zeylanicum*	Cassava brown streak virus	Mozambique	Amisse et al. (2019)
Potato	*Amaranthus viridis*, *Chenopodium album*, *Datura stramonium*, *Physalis minima*, *Solanum nigrum*, and *Withania somnifera*	Potato leaf roll virus	India	Singh (2016)
	Sonchus arvensis	-do-	Russia	Sobko et al. (2021)
–	*Pisum sativum* and *Solanum nigrum*	Potato virus X	India	Shah et al. (2020)
–	*Erodium cicutarium*, *Geranium pusillum*, *Lactuca serriola*, and *Lamium purpureum*	Potato virus Y	Europe	Kaliciak & Syller (2009)
	Parthernium hysterophorus, *Sonchus oleraceus*, *Chenopodium murale*, *Pisum sativum*, *Capsicum annuum*, *Solanum lycopersicum*, *S. melongena*, and *S. nigrum*	-do-	India	Shah et al. (2020)
	Sonchus arvensis, *Trifolium pratense*, *Chenopodium album*, *Plantago major*, *Barbarea vulgaris*, *Ambrosia artemisiifolia*	-do-	Russia	Sobko et al. (2021)
	Solanum nigrum, *Physalis* sp., *Nicotiana tabacum*	-do-	Syria	Chikh Ali et al. (2008)
Sweet potato	*Convolvulus arvensis*, *Euphorbia heterophylla*, and *Corchorus olitorius*	Sweet potato feathery mottle virus (SPFMV) and sweet potato virus G (SPVG)	Egypt	Mazyad et al. (2021)
–	*Amaranthus retroflexus* L., *Atriplex hastate* L., *Brassica oleracea* var. capitata., *Calendula officinalis* L., *Capsicum annum* L., *Chenopodium* sp., *Chenopodium hybridum* L., *Chenopodium polyspermum* L., *Chenopodium vulvaria* L., *Convolvulus arvensis* L., *Cucurbit maxima* Dush., *Heliotropium europaeum* L., *Ipomoea violacea* L., *Malva neglecta* Wallr., *Mentha* arvensis L., *Phaseolus lanatus* L., *Sinapis arvensis* L., *Spinacea oleracea* L., and *Traxacum officinale*	Sweet potato feathery mottle virus	Syria	Akel et al. (2010)
	22 *Ipomoea* spp., *Hewittia sublobata*, and *Lepistemon owariensis*	-do-	Uganda	Tugume et al. (2008)

TABLE 8.1 *(Continued)*

Cultivated Host	Weeds/Wild Plants/Plants Other Than Cultivated Hosts	Viruses Recorded	Region/ Country	References
–	38 species of *Ipomoea*	*Sweet potato leaf curl virus*	US	Ling et al. (2011)
Sugarbeet, tobacco, potato	*Amaranthus palmeri*, *Chenopodium album*	Tobacco rattle virus	Florida	Goyal et al. (2012)
Pigeonpea	*N. benthamiana*, *N. clevelandii*, *Phaseolus vulgaris*, and *Chrozophora rottleri*	Pigeonpea sterility mosaic virus	India	Kumar et al. (2004)
–	*Nicotiana benthamiana*, *Chrozophora rottleri*, wild *Cajanus* species – *Cajanus platycarpus*, *Cajanus scarabaeoides*	Pigeonpea sterility mosaic virus-I (PPSMV-I) and pigeonpea sterility mosaic virus-II (PPSMV-II)	India	Sayiprathap et al. (2022)
–	*Chrozophora rottleri*	Pigeonpea sterility mosaic virus	India	Kulkarni et al. (2003)
Cajanus cajan, *Glycine max*, *Phaseolus vulgaris*, *Vigna mungo*, *Vigna radiata*, and *Vigna unguiculata*	*Ageratum conizoides*, *Vigna trilobata*, *Corchorus olitorius*, and *Alternanthera sessilis*	Mungbean yellow mosaic India virus	India	Marabi et al. (2017)
Sesame	*Amaranthus hybridus*, *Arachis hypogaea*, *Crotalaria incana*, *Crotalaria juncea*, *Crotalaria spectabilis*, and *Vigna unguiculata*	Cowpea aphid-borne mosaic virus	Paraguay	González-Segnana et al. (2013)

and *Galinsoga parviflora*) are the most common sources of TSWV in the Czech Republic. Of the 5,543 plants representing 45 weed species tested through ELISA, 59 samples belonging to 16 different species were found positive for TSWV. Among these, the highest virus infections were observed in *Arctotheca calendula* and *Sonchus asper* (Latham & Jones, 1997). Chatzivassiliou et al. (2001) reported differential infection of weeds in the greenhouse and fields with TSWV, which act as weed host reservoirs of the virus. Reservoirs of TSWV include diseased planting stock, volunteer plants of vegetatively propagated crops, and weeds. Kahn et al. (2005) identified weeds *Ipomoea purpurea, I. hederacea, Mollugo verticillata, Amaranthus palmeri, Cassia obtusifolia,* and *Richardia scabra* as potential reservoirs of TSWV in North Carolina. Ghotbi et al. (2005) reported different weed host reservoirs of TSWV, Impatiens Necrotic Spot Virus (INSV), Tomato Varamin Virus (ToVV), tomato chlorotic spot virus (TCSV), and iris yellow spot virus (IYSV) through ELISA tests in Iran. Gopal et al. (2011) listed several weed hosts as a source of peanut bud necrosis virus (PBNV) in peanut crops in India.

8.3.5 POLEROVIRUSES AND LUTEOVIRUSES

Around 12 weed species from five families, including new weed species such as *Cardamine oligosperma, Matricaria maritima, Polygonum lapathifolium,* and *Raphanus raphanistrum*, were infected with Beet Western Yellows Virus (BWYV). Potato leaf roll virus (PLRV) was found in *Capsella bursa-pastoris* and *Solanum nigrum* (Ellis, 1992). Thomas & Hassan (2002) reported 22 new hosts of PLRV. Weed plants such as *Amaranthus viridis, Chenopodium album, Datura stramonium, Physalis minima, Solanum nigrum,* and *Withania somnifera* were found infected with PLRV by DAS-ELISA and RT-PCR. Among them, *C. album* and *S. nigrum* were abundant and served as a source of PLRV inoculum in Punjab (Singh, 2016). BYDV-PAV was found in *Echinochloa crus-galii, Setaria pumila,* and *Phalaris canariensis* plants in the Czech Republic through DAS-ELISA, which serve as a major source of BYDV-PAV for winter cereals (Pokorny, 2006). Yellow Dwarf Viruses (BYDV-PAV and BYDV-MAV), which mainly infect cereal crops, were found in dicotyledonous crops such as *Geranium dissectum* and *Juncus compressus* in Turkey through DAS-ELISA and RT-PCR methods (Ilbagi et al., 2019). Rashidi et al. (2021) reported corn (*Zea mays*) and three commonly found grassy weeds, downy brome (*Bromus tectorum* L.), green foxtail (*Setaria viridis*), and foxtail barley (*Hordeum jubatum*), as a potential source of virus inoculum in winter wheat through aphid transmission studies.

8.3.6 POTEXVIRUSES

Plantago asiatica mosaic virus (PlAMV) has been isolated from a variety of plants, including lilies (monocots) and the eudicots *Plantago asiatica, Primula sieboldii,* and *Nandina domestica* (Komatsu et al., 2017). Pepino mosaic virus (PepMV) was found in 20 weed species in the families of Amaranthaceae, Chenopodiaceae, Compositae, Convolvulaceae, Malvaceae, Plantaginaceae, and Solanaceae in Cyprus with DAS-ELISA and real-time RT-PCR (Papayiannis, 2012). Furthermore, PepMV was detected in *Amaranthus* sp., *M. parviflora, N. glauca, Solanum nigrum* and *Sonchus oleraceus* which were asymptomatic but upon mechanical inoculation produced symptoms in tomato plants (Jordá et al., 2001). *Pisum sativum* and *Solanum nigrum* are the known hosts of potato virus x (PVX) and the status of six plant species viz., *A. spinosus, L. siceraria, Melilotus indicus, T. foenum-graecum, R. obtusifolius, Rosa indica* as host of PVX is not known (Shah et al., 2020).

8.3.7 POTYVIRUSES

Veronica hederifolia, Capsella bursa-pastoris, Lactuca serriola, Lythospermum arvervensis, Rumex crispus, and Erigeron annuus were identified as possible hosts of plum pox virus (PPV) through DAS-ELISA of which, *E. annuus* could proliferate aphid and spread of the virus (Milusheva & Rankova, 2002; Kil et al., 2021). Using DAS-ELISA with specific polyclonal antibodies and RT-PCR assay, the zucchini yellow mosaic virus (ZYMV) was detected in the weed *Alhagi camelorum*, which is a common weed host in the cucurbits field in Iran (Safaeizadeh, 2008). *Erodium cicutarium, Geranium pusillum, Lactuca serriola,* and *Lamium purpureum* were found as new hosts of potato virus Y (PVY) in Europe by experimental inoculation followed by testing through ELISA (Kaliciak & Syller, 2009). Around 19 species were found infected with sweet potato feathery mottle virus (SPFMV) in Syria using tissue blot immunoassay (Akel et al., 2010). Baghel et al. (2011) identified potyvirus infection in *Amaranthus gracilis, Withania somnifera,* and *Oxalis corniculate* in India based on symptomatology and virus transmission studies through sap, aphid inoculation experiments, and electron microscopy studies. Yam mosaic virus (YMV) was the most widely distributed virus in weeds (*Ludwigia abyssinica, Amaranthus* spp., *Eclipta Prostrata, Hibiscus esculentus,* and *Euphorbia heterophylla*) of yam fields in Nigeria, as determined through DAS-ELISA (Asala et al., 2014). RT-PCR analysis detected the presence of papaya ringspot virus (PRSV) in *Abutilon abutilastrum, Anoda cristata,* and *Solanum rostratum* associated with papaya crops in Mexico (Bermúdez-Guzmán et al., 2017). Cassava Brown Streak Virus (CBSV) was detected in two non-cassava perennial wild plant species: Zanha africana and Trichodesma zeylanicum in Mozambique through RT-PCR, which was also confirmed through biological assays via mechanical inoculation of CBSV to *T. zeylanicum* (Amisse et al., 2019). Johnsongrass was found to harbor maize dwarf mosaic virus and sugarcane mosaic virus (SCMV) through DAS-ELISA (Clemente-Orta et al., 2020). PVY was confirmed through DAS-ELISA in *Parthenium hysterophorus, Sonchus oleraceus, Chenopodium murale, Pisum sativum, Capsicum annuum, Solanum lycopersicum, S. melongena*, and *S. nigrum* collected from in and around potato fields in India (Shah et al., 2020). SPFMV and sweet potato virus G (SPVG) infection in weeds including *Convolvulus arvensis, Euphorbia heterophylla,* and *Corchorus olitorius* were confirmed through Indirect ELISA, DBIA, TBIA, and RT-PCR assay (Mazyad et al., 2021). NGS was used to characterize and detect the potyviruses from the cultivated and weed hosts (Kehoe et al., 2014; Kwon et al., 2016; Glasa et al., 2021).

8.3.8 OTHER VIRUSES

Six weed species, namely *Solanum nigrum, Datura stramonium, Chicorium intybus, Heliotropium eurapaeum, Plantago major,* and *Polygonum aviculare,* were collected from sugar beet fields in Turkey. These species were determined as alternative hosts for beet necrotic yellow vein virus (BNYVV) using DAS-ELISA (Yanar et al., 2006). In Korea, *Alternanthera philoxeroides, Conyza canadensis, Erigeron annuus, Sonchus asper, Youngia japonica, Trigonotis peduncularis, Cardamine flexuosa, Cerastium glomeratum, Stellaria media, Chenopodium album, Ipomoea hederacea, Quamoclit coccinea, Vicia angustifolia var. segetilis, V. tetrasperma, Phytolacca americana, Mazus pumilus, Solanum americanum,* and *S. nigrum* were identified as natural weed hosts of tomato chlorosis virus (ToCV) through RT-PCR analysis of root samples (Kil et al., 2015). Additionally, *Alternanthera philoxeroides* was found to be infected with cucurbit chlorotic yellows virus in China (Tang et al., 2017). Thapa et al. (2015) reported the occurrence of many unknown viruses associated with asymptomatic weeds and wild plants from the USA using NGS approach. Similarly, Bernardo et al. (2018) reported 94 unknown viruses in cultivated and uncultivated plants from South Africa and France using the NGS approach. Small RNA sequencing from *Plantago lanceolate* revealed the presence of novel viruses belonging to the *Caulimovirus, Betapartitivirus, Enamovirus,* and *Closterovirus* genera in Finland (Susi et al., 2019). Richet et al. (2019) documented the occurrence of five novel circular DNA viruses infecting *Arctopus echinatus, Lolium perenne, Panicum ecklonii*, and *Stipagrostis* sp. in South Africa and New Zealand using the NGS approach. The first-time occurrence of clover yellow mosaic virus and melandrium yellow fleck virus associated with *Verbena officinalis, Silene latifolia, Rorippa x prostrata,* and *Robinia pseudoaccacia* was reported in Poland (Minicka et al., 2020).

8.4 CONCLUSION

Weeds and wild plants often grow in and around cultivated crops, playing an important role in plant virus epidemics. They serve as potential reservoirs for viruses and their vectors, which can spread to the cultivated crops. Many times, these wild plants and weeds remain asymptomatic despite virus infection, evading growers' attention for eradication. Mixed infections by multiple viruses in these hosts can lead to recombination among viruses, resulting in the emergence of new variants that may pose a threat to cultivated crops. Although the occurrence of viruses in weeds and wild hosts has been reported, the actual biodiversity of viruses in such reservoirs is not yet fully known. Most viral pathogens that threaten cultivated plants have broad host ranges and naturally infect many species from various families. Therefore, they can thrive in alternative plant hosts, such as weeds, when annual crops are absent in the fields or when resistant cultivars are grown. Recent studies have also shown the emergence of novel viruses from weeds and wild plants, posing a threat to cultivated crops. Hence, studies on viruses and their biodiversity in wild plants and weeds are essential for understanding the epidemiology, spread, and potential threat to cultivated crops. Various methods based on biological, physical, serological, and nucleic acid approaches are available for the identification and characterization of viruses infecting cultivated and non-cultivated hosts. Among these methods, PCR, RCA, and NGS-based methods can identify viruses up to the species level. Additionally, RNA-sequencing, small RNA sequencing, and dsRNA sequencing of suspected weed hosts using the NGS approach can identify both known and novel viruses, thus aiding in obtaining a complete biodiversity of viruses present in tested plants. Once the identification of the virus infecting weeds and wild plants is known, routine diagnostics can be developed and used for the detection of the virus in potential sources. This will enable appropriate management practices to be implemented to contain the spread of the virus into cultivated crops. Future studies should include the identification of viruses in reservoir hosts, as well as the development and validation of specific and sensitive diagnostic assays for the quick detection of the virus *in situ.* This will allow infected plants that act as the primary source of the virus for cultivated crops to be completely eradicated, preventing disease epidemics.

KEYWORDS

- **diagnosis**
- **host range**
- **isolation**
- **next-generation sequencing**
- **nucleic acid**
- **protein**
- **serology**
- **symptoms**
- **vector**

REFERENCES

Abirami, R., Manoranjitham, S. K., Rajasree, V., Mohankumar, S., & Karthikeyan, G., (2022). Weeds as reservoirs for viruses infecting brinjal in its ecosystem. *J. of Curr. Crop Sci. Technol.* https://doi.org/10.29321/MAJ.10.000589.

Adams, I. P., Glover, R. H., Monger, W. A., Mumford, R., Jackeviciene, E., Navalinskiene, M., Samuitiene, M., & Boonham, N., (2009). Next-generation sequencing and metagenomic analysis: A universal diagnostic tool in plant virology. *Molecular Plant Pathology, 10*(4), 537–545.

Adediji, A. O., (2019). Molecular detection of *Cucumber mosaic virus* from *Basella alba, Telfairia occidentalis* and *Talinum fruticosum* in Nigeria. *Journal of Plant Protection Research, 59*(2), 177–184.
Adjata, K. D., Woegan, A. Y., Mivedor, A. S., Muller, E., Peterschmitt, M., Sanda, K., & Gumedzoe, Y. M. D., (2014). Wild plants as sources of the permanency of viruses infecting cultivated plants: Case of cassava *Begomoviruses* in Togo. *American Journal of Biochemistry and Molecular Biology, 4*, 136–142.
AIRwahnih, M., Daubert, S., U'rbez-Torres, J. R., Cordero, F., & Rowhani, A., (2011). Deep sequencing evidence from single grapevine plants reveals a virome dominated by mycoviruses. *Arch. Virol., 156*, 397–403.
Alabi, O. J., Ogbe, F. O., Bandyopadhyay, R., Lava, K. P., Dixon, A. G., Hughes, J. D., & Naidu, R. A., (2008). Alternate hosts of *African cassava mosaic virus* and East African *Cassava mosaic Cameroon virus* in Nigeria. *Arch. Virol., 153*(9), 1743–1747.
Alfaro-Fernández, A., Córdoba-Sellés, C., Cebrián, M. C., Herrera-Vásquez, J. A., Sánchez-Navarro, J. A., Juárez, M., Espino, A., et al., (2008). First report of *Tomato torrado virus* on weed hosts in Spain. *Plant Dis., 92*(5), 831.
Anitha, J., Makeshkumar, T., & Edison, S., (2008). Host range of *Sri Lankan cassava mosaic virus. Journal of Root Crops, 34*(1), 21–25.
Ansar, M., Agnihotri, A. K., Ranjan, T., Karn, M., Srinivasaraghavan, A., Kumar, R. R., & Bhagat, A. P., (2021). Nightshade (*Solanum nigrum*), an intermediate host between tomato and cucurbits of *Tomato leaf curl New Delhi virus. Phytopathologia Mediterranea, 60*(3), 409–419.
Ansari, N. A., & Tiwari, J. P., (2005). Hitherto unrecorded weed reservoirs of *Tomato leaf curl virus* (TLCV) in Eastern U.P. and its role in epidemiology of the disease. *Journal of the Living World, 12*, 22–26.
Asala, S., Alegbejo, M. D., Kashina, B. D., Banwo, O. O., & Shinngu, C. P., (2014). Viruses in weeds in *Dioscorea* yam fields in Nigeria. *African Crop Science Journal, 22*(2), 109–115.
Badamasi, H., Alegbejo, M. D., Kashina, B. D., & Banwo, O. O., (2020). Alternative hosts of cassava viruses in Kaduna and Sokoto states, Nigeria. Science World Journal, 15, 51–55.
Baghel, G., Afreen, B., Khan, A. A., & Naqvi, Q. A., (2011). Weed plants harboring potyvirus in western UP- A TEM based study. *National Academy Science Letters, 34*(9), 333–336.
Barbara, D. J., & Clark, M. F., (1982). A simple indirect ELISA using F(ab')$_2$ fragments of immunoglobulin. *J. Gen. Virol., 58*, 315–322.
Bermúdez-Guzmán, M. J., Guzmán-González, S., Lara-Reyna, J., Palmeros-Suárez, P. A., López-Muraira, I. G., & Gómez-Leyva, J. F., (2017). Presence of *Papaya ringspot virus* (PRSV) in weed associated with *Carica papaya* in Colima, Mexico. *Revista Mexicana de Fitopatología 36*, 1–15.
Bernardo, P., Charles-Dominique, T., Barakat, M., Ortet, P., Fernandez, E., Filloux, D., Hartnady, P., Rebelo, T. A., Cousins, S. R., Mesleard, F., et al., (2018). Geometagenomics illuminates the impact of agriculture on the distribution and prevalence of plant viruses at the ecosystem scale. *ISME J., 12*, 173–184.
Bernardo, P., Golden, M., Akram, M., Naimuddin, Nadarajan, N., Fernandez, E., Granier, M., et al., (2013). Identification and characterization of a highly divergent geminivirus: Evolutionary and taxonomic implications. *Virus Res., 177*, 35–45.
Bhat, A. I., & Rao, G. P., (2020). Characterization of plant viruses: Methods and protocols. *Springer Protocols Handbook Series* (p. 510). Humana Press.

Chatzivassiliou, E. K., Boubourakas, I., Drossos, E., Eleftherohorinos, I., Jenser, G., Peters, D., & Katis, N. I., (2001). Weeds in greenhouses and tobacco fields are differentially infected by *Tomato spotted wilt virus* and infested by its vector species. *Plant Dis., 85*(1), 40–46.

Cheema, S. S., (1991). *Ageratum conyzoides* Linn. – a new host for *Yellow vein mosaic virus* of bhindi in the Punjab. *Plant Disease Research, 6*, 104. https://eurekamag.com/research/002/294/002294438.php (accessed on 02 March 2024).

Chikh, A. M., Katayama, K., Maoka, T., & Natsuaki, K. T., (2008). Significance of weed hosts for *Potato virus Y* protection in Syria. *Bulletin OEPP/EPPO Bulletin, 38*(2), 226–232.

Clark, M. F., & Adams, A. N., (1977). Characteristics of the microtitre plate method of enzyme-linked immunosorbent assay for the detection of plant viruses. *J. Gen. Virol., 34*, 475–483.

Clark, M. F., Lister, R. M., & Bar-Joseph, M., (1986). ELISA techniques. *Methods Enzymol., 118*, 742–766.

Clemente-Orta, G., Albajes, R., & Achon, M. A., (2020). Early planting, management of edges and non-crop habitats reduce potyvirus infection in maize. *Agron. Sustain. Dev., 40*, 21.

Daniel, M., Ochoa, L., Zavaleta-Mejía, E., Johansen, N. R. M., Herrera, G. A., & Cárdenas, S. E., (1996). *Tospoviruses*, weeds and thrips associated with chrysanthemum (*Dendranthema grandiflora* Tzvelev cv. Polaris). *International J. Pest Management, 42*, 157–159.

Debrat, E. A., Lastra, R., & Deuzeategui, R. C., (1977). *Solanum seaforthianum*, A weed host of *Eggplant mosaic virus* in Venezuela. *Plant Dis. Reporter, 61*, 628–631.

Derrick, K. S., (1973). Quantitative assay for plant viruses using serologically specific electron microscopy. *Virology, 56*, 652–653.

Devaraja, Narayanaswamy, K., Savithri, H. S., & Muniyappa, V., (2005). Purification of *Tomato leaf curl Bangalore virus* and production of polyclonal antibodies. *Curr. Sci., 89*, 181–183.

Duarte, M. F., Fonseca, M. E., Boiteux, L. S., Costa, H., Ribeiro, B. M., Melo, F. L., & Pereira-Carvalho, R. C., (2020). Identification of *Physalis angulata* (*Solanaceae*) as a natural alternative weed host of *Tomato severe rugose virus* in Brazil. *Plant Dis., 104*, 600.

Ellis, P. J., (1992). Weed hosts of *Beet western yellows virus* and *Potato leafroll virus* in British Columbia. *Plant Disease, 76*(11), 1137–1139.

Eni, A. O., Efekemo, O. P., Onile-ere, O. A., & Pita, J. S., (2021). South West and North Central Nigeria: Assessment of cassava mosaic disease and field status of *African cassava mosaic virus* and East African *Cassava mosaic virus*. *Ann. Appl. Biol., 178*, 466–479.

Fontenele, R. S., Salywon, A. M., Majure, L. C., Cobb, I. N., Bhaskara, A., Avalos-Calleros, J. A., Argüello-Astorga, G. R., et al., (2020). A novel divergent geminivirus identified in asymptomatic new world *Cactaceae* plants. *Viruses, 12*(4), 398.

Garg, I. D., Khurana, S. M. P., Kumar, S., & Lakra, B. S., (2001). Association of a geminivirus with potato apical leaf curl in India and its immuno-electron microscopic detection. *Potato J., 28*, 227–232.

Ghotbi, T., Shahraeen, N., & Winter, S., (2005). Occurrence of *Tospoviruses* in ornamental and weed species in Markazi and Tehran Provinces in Iran. *Plant Dis., 89*(4), 425–429.

Glasa, M., Hanˇcinský, R., Šoltys, K., Predaj ˇna, L., Tomašechová, J., Hauptvogel, P., Mrkvová, M., et al., (2021). Molecular characterization of *Potato Virus Y* (PVY) using high-throughput sequencing: Constraints on full genome reconstructions imposed by mixed infection involving recombinant PVY strains. *Plants, 10*, 753.

Gopal, K., Muniyappa, V., & Jagadeeshwar, R., (2011). Weed and crop plants as reservoirs of peanut bud necrosis *Tospovirus* and its occurrence in South India. *Archives of Phytopathology and Plant Protection, 44*(12), 1213–1224.

Harrison, B. D., Muniyappa, V., Swanson, M. M., Roberts, I. M., & Robinson, D. J., (1991). Recognition and differentiation of seven whitefly-transmitted geminiviruses from India, and their relationships to *African cassava mosaic* and Thailand *Mungbean yellow mosaic viruses*. *Ann. Appl. Biol., 18*, 299–308.

Hasiów-Jaroszewska, B., Boezen, D., & Zwart, M. P., (2021). Metagenomic studies of viruses in weeds and wild plants: A powerful approach to characterize variable virus communities. *Viruses, 13*, 1939.

Hull, R., (2014). *Matthews' Plant Virology* (5th edn.). Academic Press, San Diego, California, USA.

Ilbaği, H., Çitir, A., Kara, A., Uysal, M., & Olden, F. A., (2019). First report of *Barley yellow dwarf viruses* (BYDVs) on dicotyledonous weed hosts in Turkey. *Cereal Research Communications Cereal Research Communications, 47*(2), 292–303.

Ingwell, L. L., & Bosque-Perez, N. A., (2015). New experimental hosts of *Barley yellow dwarf virus* among wild grasses, with implications for grassland habitats. *Plant Pathol., 64*, 1300–1307.

Inoue-Nagata, A. K., Albuquerque, L. C., Rocha, W. B., & Nagata, T., (2004). A simple method for cloning the complete *Begomovirus* genome using the bacteriophage 29 DNA polymerase. *J. Virol. Methods, 116*, 209–211.

Ito, D., Miller, Z., Menalled, F., Moffet, M., & Burrows, M., (2012). Relative susceptibility among alternative host species pre valent in the great plains to *Wheat streak mosaic virus*. *Plant Dis., 96*, 1185–1192.

Jamisse, J. G. A., Joseph, N., Fred, T., Laura, M. B., Monica, A. K., Nurbibi, C., Elijah, A., Chrissie, R., & Peter, S., (2019). First report of *Cassava brown streak viruses* on wild plant species in Mozambique. *Physiological and Molecular Plant Pathology, 105*, 88–95.

Jana, C., Jiban, K. K., Khushwant, S., Pavla, B., & Eva, B., (2017). *Wheat streak mosaic virus*: Incidence in field crops, potential reservoir within grass species and uptake in winter wheat cultivars. *Journal of Integrative Agriculture, 16*(3), 523–531. https://doi.org/10.1016/S2095-3119(16)61486-7.

Jordá, C., Lázaro, P. A., Martínez, P. V., & Lacasa, A., (2001). First report of *Pepino mosaic virus* on natural hosts. *Plant Disease, 85*, 1292.

Kahn, N. D., Walgenbach, J. F., & Kennedy, G. G., (2005). Summer weeds as hosts for *Frankliniella occidentalis* and *Frankliniella fusca* (*Thysanoptera: Thripidae*) and as reservoirs for *Tomato spotted wilt Tospovirus* in North Carolina. *Journal of Economic Entomology, 98*, 1810–1815.

Kai-Shu, L., Ling, K., Howard, F. H., Harrison, H. F., Alvin, M. S., Simmons, A. M., Shuo, C. Z., et al., (2011). Experimental host range and natural reservoir of *Sweet potato leaf curl virus* in the United States. *Crop Protection, 30*, 1055–1062. doi: 10.1016/j.cropro.2011.03.009.

Kaliciak & Syller, (2009). New hosts of *Potato virus Y* (PVY) among common wild plants in Europe. *European Journal of Plant Pathology, 124*(4), 707–713.

Kashina, B. D., Mabagala, R. B., & Mpunami, A. A., (2003). First report of *Ageratum conyzoides* L. and *Sida acuta* Burm F. as new weed hosts of *Tomato yellow leaf curl Tanzania virus*. *Plant Protect. Sci., 39*, 18–22.

Kehoe, M. A., Coutts, B. A., Buirchell, B. J., & Jones, R. A. C., (2014). Plant virology and next generation sequencing: Experiences with a potyvirus. *PLoS One, 9*(8), e104580.

Khalid, P. A., Pervaiz, A. K., Marium, A., Anwer, M., Muhammad, Y. S., Yussouf, S. M., Sumaira, Y., Yousaf, S., et al., (2019). Identification of natural weed hosts of *Cucumber mosaic virus* subgroup-I and the absence of seed transmission in weed hosts in Pakistan. *Journal of Horticultural Science & Biotechnology, 94*, 468–474. doi: 10.1080/14620316.2019.1565947.

Kil, E. J., Byun, H. S., Hwang, H., Lee, K. Y., Choi, H. S., Kim, C. S., & Lee, S., (2021). *Tomato yellow leaf curl virus* infection in a monocotyledonous weed (*Eleusine indica*). *The Plant Pathology Journal, 37*(6), 641–651.

Kil, E. J., Ho, P. T., Fadhila, C., et al., (2021). *Plum pox virus*: Diagnosis and spread inhibition by weed control. *J. Plant Dis. Prot., 128*, 1091–1099.

Kil, E. J., Lee, Y. J., Cho, S., et al., (2015). Identification of natural weed hosts of *Tomato chlorosis virus* in Korea by RT-PCR with root tissues. *Eur. J. Plant Pathol., 142*, 419–426.

Komatsu, K., Yamashita, K., Sugawara, K., Verbeek, M., Fujita, N., Hanada, K., Uehara-Ichiki, T., & Fuji, S. I., (2017). Complete genome sequences of two highly divergent Japanese isolates of *Plantago asiatica mosaic virus*. *Arch. Virol., 162*(2), 581–584.

Kreuze, J. F., Perez, A., Untiveros, M., Quispe, D., Fuentes, S., Barker, I., & Simon, R., (2009). Complete viral genome sequence and discovery of novel viruses by deep sequencing of small RNAs: A generic method for diagnosis, discovery, and sequencing of viruses. *Virology, 388*, 1–7.

Kulkarni, N. K., & Kumar, P. L., & Muniyappa, V., & Jones, A. T., (2003). Studies on the natural and experimental host range of *Pigeonpea sterility mosaic virus*. *Journal of Mycology and Plant Pathology, 33*(1). 141–145.

Kumar, P. L., Jones, A. T., & Waliyar, F., (2004). Biology, etiology, and management of pigeonpea sterility mosaic disease. *Annu. Rev. Plant Pathol., 3*, 77–100.

Kwon, S. J., Choi, G. S., Yoon, J. Y., Seo, J. K., & Choi, H. S., (2016). Identification of *Leonurus sibiricus* as a weed reservoir for three pepper-infecting viruses. *The Plant Pathology Journal, 32*, 65–69. Korean Society of Plant Pathology. https://doi.org/10.5423/ppj.nt.07.2015.0138.

Latham, L. J., & Jones, R. A. C., (1997). Occurrence of tomato spotted wilt *Tospovirus* in native flora, weeds, and horticultural crops. *Australian Journal of Agricultural Research, 48*(3), 359–369.

Lau, H. Y., & Botella, J. R., (2017). Advanced DNA-based point of care diagnostic methods for plant diseases detection. *Plant Sci., 8*, 2016.

Lavanya, N., Ramiah, M., Sankaralingam, M., Renukadevi, & Velazhahan, R., (2005). Identification of hosts for ilarvirus associated with sunflower necrosis disease. *Acta Phytopathologica et Entomologica Hungarica, 40*, 31–34.

Loconsole, G., Saldarelli, P., Doddapaneni, H., Savino, V., Martelli, G. P., & Saponari, M., (2012). Identification of a single-stranded DNA virus associated with citrus chlorotic dwarf disease, a new member in the family *Geminiviridae*. *Virology, 432*(1), 162–172.

Luis, R. G.-S., Arnaldo, E. F., Diego, D. G. et al., (2013). Alternative hosts of *Cowpea aphid-borne mosaic virus* (CABMV) in sesame (*Sesamum indicum*) crops grown in Paraguay. *Trop. Plant Pathol., 38*(6), 539–542. doi: 10.1590/S1982-56762013000600010.

Ma, Y., Marais, A., Lefebvre, M., Faure, C., & Candresse, T., (2020). Metagenomic analysis of virome cross-talk between cultivated *Solanum lycopersicum* and wild *Solanum nigrum*. *Virology, 540*, 38–44.

Maliano, M. R., Macedo, M. A., Rojas, M. R., & Gilbertson, R. L., (2021). Weed-infecting viruses in a tropical agroecosystem present different threats to crops and evolutionary histories. *PLoS One, 16*(4), e0250066.

Marabi, R. S., Deepti, B. S., Das, S. B., Tripathi, N., & Noda, H., (2017). Molecular identification of mungbean yellow mosaic India virus (MYMIV) from alternate Weed and Crop hosts. *Annals of Plant Protection Sciences*, *25*(1), 152–155.

Mazyad, A., El-Attar, A., Amer, W., Ahmed, A., & Ismail, M., (2021). Investigations on the prevalence of two sweet potato viruses and their potential weed reservoirs in Egypt. *Egyptian Journal of Botany*, *61*(1), 105–125.

Mertelík, J., & Mokrá, V., (1999). Occurrence of tomato spotted wilt *Tospovirus* in horticultural crops and weeds in the Czech Republic in 1992/1997· OEPP/EPPO. *Bulletin O€PP/€PPO Bulletin 29*, 81–83.

Milusheva, S., & Rankova, Z., (2002). Plum pox potyvirus detection in weed species under field conditions. *Acta Hortic., 577*, 283–287.

Minicka, J., Zarzy´nska-Nowak, A., Budzy´ nska, D., Borodynko-Filas, N., & Hasiów-Jaroszewska, B., (2020). High-throughput sequencing facilitates discovery of new plant viruses in Poland. *Plants, 9*, 820.

Mubin, M., Briddon, R. W., & Mansoor, S., (2009). Diverse and recombinant DNA betasatellites are associated with a *Begomovirus* disease complex of *Digera arvensis*, a weed host. *Virus Res., 142*(1, 2), 208–212.

Mudde, B., Kilalo, D. C., Olubayo, F. M., Asea, G., Kiggundu, A., Kwemoi, D. B., & Miano, D. W., (2019). Role of prevalent weeds and cultivated crops in the epidemiology of maize lethal necrosis disease in major maize growing agroecological zones of Uganda. *Annual Research & Review in Biology*, *32*(3), 1–17. https://doi.org/10.9734/arrb/2019/v32i330084.

Muhammad, Z., Zubair, M., Raja, A. A. K., Asad, A. K. R., Asad, A., Ali, A., Najeeb, U., et al., (2020). First report of *Tomato leaf curl New Delhi virus* in *Physalis minima* in Pakistan. *Plant Disease, 104*, 1878.

Mumford, R. A., & Seal, S. E., (1997). Rapid single-tube immunocapture RT-PCR for the detection of two yam potyviruses. *J. Virol. Methods, 69*, 73–79.

Muniyappa, V., Rajeshwari, R., Bharathan, N., Reddy, D. V. R., & Nolt, B. L., (1987). Isolation and characterization of a geminivirus causing yellow mosaic disease of horsegram (*Macrotyloma uniflorum* [Lam.] Verdc.) in India. *J. Phytopath., 119*, 81–87.

Mwaipopo, B., Rajamäki, M. L., Ngowi, N., Nchimbi-Msolla, S., Njau, P. J. R., Valkonen, J. P. T., & Mbanzibwa, D. R., (2021). Next-generation sequencing-based detection of common bean viruses in wild plants from Tanzania and their mechanical transmission to common bean plants. *Plant Dis., 105*(9), 2541–2550.

Nateshan, H. M., Muniyappa, V., Swanson, M. M., & Harrison, B. D., (1996). Host range, vector relations and serological relationships of *Cotton leaf curl virus* from Southern India. *Ann. Appl. Biol., 128*, 233–244.

Nhlapo, T. F., Rees, D. J., Odeny, D. A., Mulabisana, J. M., & Rey, M. E., (2018). Viral metagenomics reveals sweet potato virus diversity in the Eastern and Western Cape provinces of South Africa. *S. Afr. J. Bot., 117*, 256–267.

O'Donell, I. J., Shukla, D. D., & Gough, K. H., (1982). Electroblot immunoassay of virus-infected plant sap- a powerful technique for detecting plant viruses. *J. Virol. Methods, 4*, 19–26.

Papayiannis, L. C., Katis, N. I., Idris, A. M., & Brown, J. K., (2011). Identification of weed hosts of *Tomato yellow leaf curl virus* in *Cyprus*. *Plant Dis., 95*, 120–125.

Papayiannis, L. C., Kokkinos, C. D., & Alfaro-Fernández, A., (2012). Detection, characterization, and host range studies of *Pepino mosaic virus* in Cyprus. *Eur. J. Plant Pathol., 132*, 1–7.

Pecman, A., Kutnjak, D., Gutiérrez-Aguirre, I., Adams, I., Fox, A., Boonham, N., & Ravnikar, M., (2017). Next-generation sequencing for detection and discovery of plant viruses and viroids: Comparison of two approaches. *Front. Microbiol., 8*, 1998.

Pokorný, R., (2006). Occurrence of viruses of the family *Luteoviridae* on maize and some annual weed grasses in the Czech Republic. *Cereal Research Communications, 34*, 1087–1092.

Raj, S. K., Chandra, G., & Singh, B. P., (2000). Detection of a whitefly-transmitted geminivirus in *Launaea* spp. in India. Plant virology, National Botanical Research Institute, Lucknow 226 001, India. *Plant Dis., 84*, 593.

Ramappa, H. K., Muniyappa, V., & Colvin, J., (1998). The contribution of tomato and alternative host plants to *Tomato leaf curl virus* inoculum pressure in different areas of South India. *Ann. Appl. Biol., 133*, 187–198.

Rashidi, M., Cruzado, R. K., Hutchinson, P. J. S., Bosque-Pérez, N. A., Marshall, J. M., & Rashed, A., (2021). Grassy weeds and corn as potential sources of *Barley yellow dwarf virus* spread into winter wheat. *Plant Dis., 105*(2), 444–449.

Ravindra, K. P., Nafees, A. A., & Tewari, J. P., (2017). Studies on weed reservoirs of Eastern U.P., (India), and it's role in epidemiology of disease. *International Journal of Current Research, 9*, (6), 53410–53412.

Richet, C., Kraberger, S., Filloux, D., Bernardo, P., Harkins, G. W., Martin, D. P., Roumagnac, P., & Varsani, A., (2019). Novel circular DNA viruses associated with *Apiaceae* and *Poaceae* from South Africa and New Zealand. *Arch. Virol., 164*, 237–242.

Rott, M., Xiang, Y., Boyes, I., Belton, M., Saeed, H., & Kesanakurti, P., (2017). Application of next-generation sequencing for diagnostic testing of tree fruit viruses and viroids. *Plant Dis., 101*, 1489–1499.

Safaeizadeh, M., (2008). Reservoir weed host for *Zucchini yellow mosaic virus* in Iran. *Phytopathol. Mediterr., 47*(1), 77, 78.

Sayiprathap, B. R., Patibanda, A. K., Prasanna, K. V., Jayalalitha, K., Ramappa, H. K., Rajeswari, E., Karthiba, L., et al., (2022). Salient findings on host range, resistance screening, and molecular studies on sterility mosaic disease of pigeonpea induced by *Pigeonpea sterility mosaic viruses* (*PPSMV-I* and *PPSMV-II*). *Front Microbiol., 13*, 838047. doi: 10.3389/fmicb.2022.838047. PMID: 35432270; PMCID: PMC9012581.

Shah, M. A., Ravinder, K., & Priyanka, K. S. S., (2020). Prevalence of natural infection of potato viruses in weeds and other crops. *Mycol. Pl. Pathol., 50*(1), 105–109.

Shargil, D., Smith, E., Lachman, O., et al., (2017). New weed hosts for *Cucumber green mottle mosaic virus* in wild Mediterranean vegetation. *Eur. J. Plant Pathol., 148*, 473–480.

Sharman, M., Thomas, J., & Persley, D., (2015). Natural host range, thrips, and seed transmission of distinct *Tobacco streak virus* strains in Queensland, Australia. *Ann. Appl. Biol., 167*, 197–207.

Shomaila, I., Muhammad, A., & Hussain, S., (2011). Biological characterization of Pakistani isolates of cucumber mosaic *Cucumovirus* (CMV). *Pak. J. Bot., 43*(6), 3041–3047.

Shrestha, D., McAuslane, H. J., Adkins, S. T., Smith, H. A., Dufault, N., & Webb, S. E., (2016). Transmission of *Squash vein yellowing virus* to and from cucurbit weeds and effects on sweetpotato whitefly (*Hemiptera: Aleyrodidae*) behavior. *Environmental Entomology, 45*, 967–973. doi: http://dx.doi.org/10.1093/ee/nvw086.

Singh, B., (2016). Survey and indexing of weeds growing around potato fields for their role as an inoculum source for *Potato leafroll virus* (PLRV). *Biotechnology Journal International, 16*(1), 1–8.

Škorić, D., Krajacic, M., Ćurković-Perica, M., Edyta, Đ., Topić, J., & Igrc-Barčić, J., (2000). *Cucumber mosaic virus* (*Cucumovirus*) and associated satRNA in weed species under the natural epidemic conditions of tomato lethal necrosis in Croatia. *Zeitschrift fur Pflanzenkrankheiten und Pflanzenschutz, 107*, 304–309.

Sobko, O. A., Matsishina, N. V., Fisenko, P. V., Kim, I. V., Didora, A. S., Boginskay, N. G., & Volkov, D. I., (2021). Viruses in the agrobiocenosis of the potato fields. *IOP Conf. Ser.: Earth Environ. Sci., 677*, 052093.

Sokmen, M. A., Yilmaz, N. D. K., Mennan, H., & Sevik, M. A., (2005). Natural weed hosts of *Apple mosaic virus* in hazelnut orchards in Turkey. *Journal of Plant Pathology, 87*(3), 239–242.

Susi, H., Filloux, D., Frilander, M. J., Roumagnac, P., & Laine, A. L., (2019). Diverse and variable virus communities in wild plant populations revealed by metagenomic tools. *PeerJ, 7*, e6140.

Swanson, M. M., Valand, G. B., Muniyappa, V., & Harrison, B. D., (1998). Serological detection and antigenic variation of two whitefly-transmitted geminiviruses: Tobacco leaf curl and *Croton yellow vein mosaic viruses*. *Ann. Appl. Biol., 132*, 427–443.

Tahir, M., Amin, I., Haider, M. S., Mansoor, S., & Briddon, R. W., (2015). Ageratum enation virus-a *Begomovirus* of weeds with the potential to infect crops. *Viruses, 7*(2), 647–665.

Tang, X., Shi, X., Zhang, D., et al., (2017). Detection and epidemic dynamic of ToCV and CCYV with *Bemisia tabaci* and weed in Hainan of China. *Virol. J., 14*, 169.

Thapa, V., McGlinn, D. J., Melcher, U., Palmer, M. W., & Roossinck, M. J., (2015). Determinants of taxonomic composition of plant viruses at the nature conservancy's tallgrass prairie preserve, Oklahoma. *Virus Evol., 1*, vev007.

Thomas, P. E., & Hassan, S., (2002). First report of twenty-two new hosts of *Potato leafroll virus*. *Plant Dis., 86*(5), 561.

Tugume, A. K., Mukasa, S. B., & Valkonen, J. P. T., (2008). Natural wild hosts of *Sweet potato feathery mottle virus* show spatial differences in virus incidence and virus-like diseases in Uganda. *Phytopathology, 98*, 640–652.

Van, O. B., Stancanelli, G., Mela, L., & Lisa, V., (1993). Role of wild plants and weeds in the epidemiology of *Tospovirus* in Liguria. *Informatore Fitopatologico, 43*, 40–44.

Varsani, A., Shepherd, D. N., Dent, K., Monjane, A. L., Rybicki, E. P., & Martin, D. P., (2009). A highly divergent South African geminivirus species illuminates the ancient evolutionary history of this family. *Virol. J., 6*, 36.

Varsani, A., Shepherd, D. N., Monjane, A. L., Owor, B. E., Erdmann, J. B., Rybicki, E. P., Peterschmitt, M., et al., (2008). Recombination, decreased host specificity and increased mobility may have driven the emergence of *Maize streak virus* as an agricultural pathogen. *Journal of General Virology, 89*(Pt. 9), 2063–2074.

Vincelli, P., & Tisserat, N., (2008). Nucleic acid-based pathogen detection *in* applied *plant pathology*. *Plant Dis., 92*, 660–669.

Wang, Y., Zhang, X., Hu, T., et al., (2021). Identification of two distinct *Begomoviruses* infecting *Malvastrum coromandelianum*. *Phytopathol Res., 3*, 8.

Wang, Z., Chen, B., Zhang, T., Zhou, G., & Yang, X., (2021). Rice stripe mosaic disease: Characteristics and control strategies. *Front. Microbiol., 12*, 715223. doi: 10.3389/fmicb.2021.715223.

Wilson, C. R., (1998). Incidence of weed reservoirs and vectors of tomato spotted wilt *Tospovirus* on southern Tasmanian lettuce farms. *Plant Pathology, 47*, 171–176.

Yanar, Y., Kutluk, N. D., & Erkan, S., (2006). Alternative weed hosts of *Beet necrotic yellow vein virus* and *Beet soil borne virus* in North East of Turkey. *International Journal of Virology, 2,* 50–54.

Zechmann, B., & Zellnig, G., (2009). Rapid diagnosis of plant virus diseases by transmission electron microscopy. *Jour. Virol. Methods, 162*, 163–169.

PART II
Plant Virus-Host Interaction/Evolution

CHAPTER 9

Implications of Mixed Infection on Disease Pathogenicity and Epidemiology

PUNAM RANJAN,[1] BHAVIN BHATT,[2] and ACHUIT KUMAR SINGH[3]

[1]*Department of Botany, Patna Science College, Patna University, Bihar, India*

[2]*Shree Ramkrishna Institute of Computer Education and Applied Sciences, Faculty of Science, Sarvajanik University, Surat, Gujarat, India*

[3]*Crop Improvement Division, ICAR–Indian Institute of Vegetable Research, Jakhini (Shahanshahpur), Varanasi, Uttar Pradesh, India*

ABSTRACT

Mixed virus infections play an important role in the evolution of new viruses. Studies of related and unrelated viruses that infect a plant simultaneously indicate that they may have biochemical interrelated mechanisms to manifest pathogenicity. The understanding of the emergence of new virus species and factors influencing their disease epidemiology will help us to develop control strategies. In this chapter, we first look into the trends in mixed infection with special reference to Gemini viruses. Second, we understand the molecular interactions happening between virus populations present in a host and how they contribute to the disease progression. Third, we review whether mixed infection can provide an opportunity for the evolution of new viruses. Finally, we lead up to the details of molecular epidemiology which can be targeted to devise effective control and management strategies.

New and Emerging Plant Viruses: The Threat to Food Security.
Alireza Golnaraghi and Rajarshi Kumar Gaur (Eds.)

9.1 INTRODUCTION

The enormous diversity of plant viruses has been overlooked for decades because the viruses of economic importance have been given much attention and studied in detail. These diversified viruses parasitize almost all groups of plant kingdom, and they belong to numerous families (almost 30). Many of them caused considerable economic loss but designing control strategies for them have extraordinary challenges. The frequent emergence of new viruses, breaking of resistance and broadening host range are few of the major challenges. International trade, climate change and rapid evolution ability of viruses are also adding to the complexity in devising virus control strategies.

The mixed infection and unknown wild host reservoirs of viruses provide ample opportunity for evolution and adaptability of viruses. Two or more viruses infecting together on single plant is known as the mixed infection of viruses. In broad sense mixed infection includes not only two viruses but more than one associated molecule (like satellite molecules) too. It is a common situation in nature, but it makes the disease epidemiology even more complicated to study and it is one of the major causes for origin of new variants and species. One of the major causes for mixed infection is the polyphagous (organisms associated with multiple hosts) nature of vectors. Plant virus vectors which feed on multiple plants cause their mixing in host and able to even make them proliferate while transmitting. These days, the metagenomics approaches using high-throughput next-generation sequencing (NGS) techniques are being applied to analyze the extraordinary diversity of virome in crops and wild plants. The mixed virus infections can be even more frequent than considered earlier. Their interactions can be very complex and depends on the characteristics of virus, plant and vectors all cross-interacting together. Therefore, it becomes important to study different aspects of mixed infection that affects pathogenicity, host resistance, emergence, and evolution of pathogen. To estimate the dimensions of virome interactions in the host and understand the different aspects of virus control is the focus of this chapter, as it is an area of instant attention.

"Nevertheless, the new concept of the virome is finally gaining interest in the research community (including funding agencies). Logically, plant virologists are intensifying their efforts to describe "what is out there"" (Moreno & López-Moya, 2020).

9.2 TRENDS IN MIXED INFECTION

Mixed infection leads to a great variety of host-virus-vector interactions. Some may result in the generation of variants showing novel genetic features, thus changing the genetic structure of the viral population. Mixed infection can be classified into two types based on the difference in time intervals of pathogen infection.

9.2.1 CO-INFECTION

In co-infection, two or more viruses infect the host simultaneously or successively. This occurs during the early stage of an epidemic when a large number of uninfected hosts are available, and the viral density is low. During this stage, viruses have the opportunity to randomly infect separate hosts due to a lower environmental viral load and a larger number of hosts being available. When two or more homologous viruses enter the same host cell, a struggle for the environmental niche occurs, favoring the relatively fitter virus.

9.2.2 SUPER-INFECTION

In super-infection, different viruses infect the same host at different times. In advanced epidemic conditions, a large number of hosts become infected, and the viral density is high. This provides an opportunity for a new virus variant or another species to infect plants that are already infected. The primary viruses have the advantage of utilizing the host resources early, which becomes a limiting factor for the incoming new virus. In this scenario, the secondary viruses may benefit from easier invasion due to the weakened host defense caused by the primary virus.

9.3 TYPES OF INTERACTION DURING MIXED INFECTION

During mixed viral infections, different kinds of molecular interactions occur between plants and viruses. These interactions can be synergistic, antagonistic, or neutral. The stress caused on the host due to infection and environmental stress together play a determinative role in deciding the impact of mixed infection. It is even beyond general accessing and analyzing techniques to fully derive all dimensions of impact in a mixed infection condition. Several examples of mixed plant virus infections, their nature of interaction, and impact on epidemiology are listed in Table 9.1. The following section provides brief information on the different statuses of interaction during mixed infection and suggests that these interactions cannot be completely demarcated from each other during a live interaction, but it depends upon their effect on a virus.

TABLE 9.1 Mixed Plant Virus Infection, Their Nature of Interaction, and Impact on Epidemiology

Virus 1	Virus 2	Type of Interaction	Host Plant	Impact on Disease Epidemiology	References
African cassava mosaic virus (ACMV) (isolate ACMV-CM; Begomovirus)	East African cassava mosaic Cameroon virus (EACMCV; Begomovirus)	Synergism	*N. benthamiana* and tobacco (*N. tabacum*) protoplasts	Causes increase in symptom severity and viral accumulation of both viruses, involves post-transcriptional gene silencing (PTGS).	Vanitharani et al. (2004)
PVX (Potexvirus)	PVY (Potyvirus)	Synergism	Tobacco (*N. tabacum*)	Causes increase of infectious PVX particles.	Rochow & Ross (1955)
WSMV (Tritimovirus)	Triticum mosaic virus (TriMV; Poacevirus)	Synergism	Wheat (*T. aestivum*)	Cultivar-specific synergism and influence of different temperatures, impact on endogenous and viral small RNA profiles.	Tatineni et al. (2010)
Sweet potato chlorotic stunt virus (SPCSV; Crinivirus).	Sweet potato feathery mottle virus (SPFMV; Potyvirus).	Synergism	Sweet potato (Ipomoea batatas)	Cause the sweet potato virus disease, a synergistic interaction with a 600-fold increase of SPFMV and severe symptoms.	Mingot et al. (2000)
Tomato chlorosis virus (ToCV; Crinivirus)	Tomato spotted wilt virus (TSWV; Tospovirus)	Synergism	Tomato (*Solanum lycopersicum*)	Cause increase in ToCV only breakdown of resistance against TSWV in plants pre infected with ToCV, in contrast to coinfection.	Gracio-Cano et al. (2006)
Groundnut rosette virus (GRV) and its satellite (sat RNA)	Groundnut rosette assistor virus (GRAV)	Helper dependence synergism	Groundnut (*Arachis hypogea*)	GRV and it sat RNA require CP of GRAV for encapsidation and transmission.	Murant (1990)
Potyvirus	Potexvirus – Potato aucuba mosaic virus	Helper dependence	Potato (*Solanum tuberosum*)	Trans complementation that helps in transmission by aphids.	Baulcombe et al. (2001)
Tomato leaf curl New Delhi virus (ToLCNDV)	Tomato leaf curl Gujarat virus (ToLCGV)	Asymmetric synergism	Tomato (*Solanum lycopersicum*), and protoplast assay.	Accumulation of ToLCNDV DNA-A and ToLCGV DNA-B is increased.	Chakraborty et al. (2008)

TABLE 9.1 *(Continued)*

Virus 1	Virus 2	Type of Interaction	Host Plant	Impact on Disease Epidemiology	References
Tomato torrado virus	Pepino mosaic virus (PepMV)	Antagonism	Tomato (*Solanum lycopersicum*)	ToTV-induced symptoms was unaffected but virus load of ToTV is increased in early phase; however, PepMV viral load remained reduced throughout.	Gomez et al. (2010)
Papaya ringspot virus (PRSV)	Papaya mosaic virus (PapMV)	Cross protection	Papaya (*Carissa papaya*)	Prior infection of PapMV leads to antago-nistic effect for PRSV.	Chavez-Calvillo et al. (2016)
Tomato chlorosis virus (ToCV)	Tomato infectious chlorosis virus (TICV)	Cross protection	*Nicotiana benthamiana*	Experimental, TICV titers increased and ToCV titers decreased.	Wintermantel et al. (2014)
Citrus tristeza virus (CTV) mild strain	Citrus tristeza virus (CTV) virulent strain	Cross protection	Citrus plants	Mild strain protect infection by virulent strain.	Ziebell & Carr (2010)
Between three strains of Barley yellow dwarf virus (BYDV).	–	Mutual exclusion	Barley (*Hordeum vulgare*)	BYDV strains inoculated together during early stage of growth, after appearance of mild symptoms later on plant gets recovered.	Jedlinks & Brown (1965)

Source: Modified from: Moreno & Moya (2020).

9.3.1 *SYNERGISTIC APPROACH IN MIXED INFECTION*

In a synergistic interaction during mixed infection, at least one of the viruses benefits due to the presence of others when compared with single infection. The titer of the virus, disease severity, or dispersal efficacy of vectors can be enhanced due to synergism. In comparison to single infection (Figure 9.1), disease symptoms are more severe in mixed infection that involves synergistic interactions (Figure 9.2). This occurs because of the individual or combined effect of cellular events and molecular interactions that either facilitate each other or facilitate uni-directionally. It may lead to a suppressed host response that benefits the co-infecting viruses. Synergistic interactions can be prolonged throughout the lifespan of both virus partners, and it may involve a brief period of interaction in which only one partner benefits.

FIGURE 9.1 Symptoms on tomato plant infected with ToLCNDV (left) and plant infected with ToLCGV (right).

FIGURE 9.2 Symptoms on tomato plant infected with ToLCNDV and ToLCGV both.

9.3.1.1 LONG-TERM SYNERGISM

Synergistic interactions do not always lead to different disease symptoms, but sometimes additive symptoms may appear. In a synergistic interaction, it is not necessary that all viruses benefit, but at least one does. Generally, an increase in viral titer, pathogenicity, and enhanced dissemination is observed. The severity of symptoms can be increased due to replication, leading to the invasion of new plant tissue and the suppression of host defenses. A classic example of synergism that causes a high economic impact on crops is studied in Cassava. Severe crop loss occurs due to mixed infection of isolates of *African cassava mosaic virus* and isolates of the Uganda strain of *East African cassava mosaic virus* (Legg & Fauquet, 2004; Vanitharani, Chellappan, Pita, & Fauquet, 2004).

Another example of synergism between plant viruses is reported in potyviruses (Genus Potyvirus, Family Potyviridae). In tobacco plants (*Nicotiana tabacum*), the interaction between *potato virus Y* (PVY) and *potato virus X* (PVX) resulted in increased disease symptoms and a 10-fold increase in the titer of PVX compared to single infection. A similar synergistic interaction is not observed when they infect *Nicotiana benthamiana* together (Rochow & Ross, 1955). Therefore, synergistic interactions are host-dependent. The host cultivar can also affect the synergy pattern, as reported in three cultivars of wheat during coinfection with *wheat streak mosaic virus* (WSMV) and *Triticum mosaic virus* (TriMV) (Tatineni, Graybosch, Hein, Wegulo, & French, 2010).

In case of potyvirus combination with a virus from another group, the second partner (nonpotyvirus) benefits from synergistic interaction. The RNA silencing suppressor, helper-component proteinase (HC-Pro), of the potyvirus assists the second associated virus in pathogenicity. However, in the case of the combination of SPFMV and *sweet potato chlorotic stunt virus* (SPCSV; cirnivirus), it is the Potyviruses partner SPFMV that benefits, as the RNA silencing function has shifted to a different gene (Mingot et al., 2016).

HC-Pro mediates synergistic interaction by suppressing PTGS. In *turnip mosaic virus* (TuMV), HC-Pro interferes with micro-RNA. In addition to HC-Pro, other proteins encoded by different viruses also interfere with PTGS and affect synergism. The P25 protein of PVX degrades Ago protein of the host via the proteasome pathway (Kasschau et al., 2003; Chiu, Chen, Baulcombe, & Tsai, 2010).

Epidemiology is greatly affected by synergism. In tomato plants, resistance breaking is caused by co-infection of *tomato spotted wilt virus*

(TSWV) and ToCV (cirnivirus) (García-Cano, Resende, Fernández-Muñoz, & Moriones, 2006). Epidemiological variation, as a result of synergism, may lead to resistance breaking in the host. The disease synergism during mixed infection and its impact on vector transmission led to serious ecological and epidemiological consequences.

In neutral synergism, interactions among different viruses do not significantly impact the accumulation of viral DNA (vDNA) and/or mature virus particles *in vivo* (Alves-Júnior et al., 2009). Such co-infection will result in viral coexistence. This coexistence, over a long period of time, may undergo sequence divergence and produce a novel gene pool, which ultimately results in a diverse host range and appearance of symptoms (Barboza, Blanco-Meneses, Esker, Moriones, & Inoue-Nagata, 2017). In a study conducted by Minor et al. (2022), it was observed that Tomato plants exhibited more severe and early symptoms when co-inoculated with all three viruses, namely *tomato yellow mottle virus* (ToYMoV), *tomato leaf curl Sinaloa virus* (ToLCSiV), and *tomato yellow leaf curl virus* (TYLCV). The results clearly indicate additive synergistic interactions.

9.3.1.2 HELPER DEPENDENCE SYNERGISM OR SHORT-TERM SYNERGISM

In helper dependence synergism, one virus is defective and is supported by a helper/facilitator virus. For example, groundnut rosette disease (GRD) is caused by the interaction between three partners: *groundnut rosette virus* (GRV), its satellite (sat RNA), and *groundnut rosette assistor virus* (GRAV) (Murant, 1990). In this case, GRV and it sat RNA require the capsid protein (CP) of GRAV for encapsidation and transmission. This is an example of Umbravirus, Luteoviridae co-infection, where the vector transmission of the first virus is dependent on the second one because umbravirids lack CP. Protein HC-Pro mediated helper dependence is reported in cases of co-infection, where a HC-Pro deficient non-transmissible isolate of the same virus is facilitated by a HC-Pro efficient one, thus called HC-transcomplementation. In transcomplementation or heterologous complementation synergistic interaction, some non-structural protein of one virus helps the partner for vector transmission. This is observed in cases where potyvirus mixed infection with potexvirus – Potato aucuba mosaic virus (PAMV) helps the latter one to be transmitted by aphids (Baulcombe et al., 1993; Froissart, Doumayrou, Vuillaume, Alizon, & Michalakis, 2010).

In a multi-infected cell, one virus member can transcomplement cell-to-cell or long-distance movement. As an example of asymmetric synergism, *tomato leaf curl New Delhi virus* (ToLCNDV) can trans replicate DNA-B of *tomato leaf curl Gujarat virus* (ToLCGV). Their mixed infection in nature is repeatedly reported (Chakraborty, Vanitharani, Chattopadhyay, & Fauquet, 2008; Zaidi, Martin, Amin, Farooq, & Mansoor, 2017).

9.3.2 ANTAGONISM

In an antagonistic interaction, the presence of one or more viruses is detrimental to at least one of them. In an interaction where one partner may have an advantage due to synergism, the other one may face antagonism. For example, antagonism is observed in the interaction between PVX and PVY in tobacco. Another example of antagonism is seen in mixed infection of *tomato torrado virus* (ToTV) and *Pepino mosaic virus* (PepMV), where the severity of ToTV-induced symptoms remains unaffected, but the virus load of ToTV increases in the early phase. However, the viral load of PepMV remains reduced throughout. This interaction affects the epidemiology of the virus (Gómez, Sempere, Amari, Gómez-Aix, & Aranda, 2010).

9.3.2.1 CROSS PROTECTION, HOMOLOGOUS INTERFERENCE, OR SUPERINFECTION EXCLUSION

A virus infection may protect the plant from another homologous virus infection due to their antagonistic interaction. It serves as a tool for cross-protection that occurs due to homologous interference.

The condition of superinfection and co-infection in mixed infection can also affect the interaction, as observed in the case of PRSV and *papaya mosaic virus* (PapMV). Prior infection of PapMV has an antagonistic effect on PRSV. However, if PRSV infects earlier or both viruses co-infect, it leads to synergism (Chávez-Calvillo et al., 2016).

In *N. benthamiana* plants, mixed infection of ToCV and *Tomato infectious chlorosis virus* (TICV) resulted in increased TICV titers and decreased ToCV titers (Wintermantel, Cortez, Anchieta, Gulati-Sakhuja, & Hladky, 2008). Therefore, the same combination of viruses can interact synergistically or antagonistically depending on the sequence of infection in the host or the nature of the hosts themselves.

Another example of cross protection is seen with *citrus tristeza virus* (CTV), where mild isolates prevent infection from virulent strains (Ziebell & Carr, 2010). Thus, cross protection can be applied as a preventive technique, with varying degrees of success, to protect cultivated plants. The evolutionary significance of superinfection exclusion is that it promotes the dissemination of newly produced viral variants by favoring their entry into uninfected plants.

Transient antagonism involves competition for resources required for mature virus particle production from host plants (Alves-Júnior et al., 2009; Syller, 2012; Syller & Grupa, 2016). Such competition allows the host plant to trigger and initiate a robust anti-viral defense system during mixed infection, leading to the eradication of the infection or the appearance of low levels of symptoms. Studies on PRSV and PapMV have shown that they produce different post-infection symptoms depending on the inoculation sequence. When papaya plants are first agroinfected with PapMV and later inoculated with PRSV, a striking antagonism develops, resulting in significantly low accumulation of PRSV. The opposite situation occurs when the inoculation sequence is reversed. When papaya plants are infected first with PRSV or coinfected at the same time with PRSV and PapMV, a synergistic effect occurs, and the virus titer for both viruses is high (Chávez-Calvillo et al., 2016). This study clearly indicates that the order of the infection cycle is also an important parameter in determining symptom development and interactions among plant viruses.

Thus, an actively replicating virus antagonizes the infectivity and accumulation of other viruses by activating host defense pathways such as imbalance in translation machinery, activation of virus DNA methylation and host gene silencing, and inappropriate recruitment of cellular replicative machinery, among others. In a study conducted by Torchetti et al. (2016), coinoculation of *tomato yellow leaf curl Sardinia virus* (TYLCSV) and *potato spindle tuber viroid* (PSTVd) on tomato plants significantly reduces the accumulation of TYLCSV. Molecular dissection reveals the activation of the DNA methylation machinery of the host plant by PSTVd. These results suggest that different geminiviruses interact differently with host DNA methylation pathways, resulting in an antagonistic relationship with each other.

9.3.2.2 MUTUAL EXCLUSION

When there is a mixed infection, partners act as mutual competitive inhibitors of each other. In oat plants, if three strains of *barley yellow dwarf virus*

(BYDV) are inoculated together during the early stage of growth, the plant later recovers after the appearance of mild symptoms (Jedlinski & Brown, 1965). In the study of CMV, it is found that two strains, when inoculated together, do not colonize the same cell, thus observing a kind of spatial exclusion.

9.4 ROLE OF PATHOGEN IN MIXED INFECTION

Mixed infection of plant viruses is a triad between the host, multiple strains of viruses, and their respective vectors. Mixed infection is a result of a complex interaction between these three cross-interacting factors and leads to more disastrous symptoms in plants. Plant viruses depend on the host cell machinery for their life cycle. During their life cycle, viruses enter the plant cell's genome and initiate replication and other molecular processes. In the case of mixed infection, viruses may have the opportunity to modify their genome through gene reassortments, recombination, or duplication. These gene modifications can give rise to new recombinant viruses. Viruses are obligate parasites, so they exist in continuous synchronization with their hosts. The coexistence of different viruses creates a potential for mixed infection and recombination. Additionally, various epigenetic and environmental factors favor the occurrence of mixed infections, especially in weed plants, which can serve as a source for the evolution of newly recombined viruses with a broad host range. Furthermore, virus transmission vectors are polyphagous and associated with multiple hosts for their nutritional resources. This dependence also contributes to mixed viral infection in plants. Understanding the biology and epidemiology of mixed infection is crucial for comprehending the evolution of plant viruses and their altered host range. Therefore, there is a need to develop robust, specific, and field-testing protocols to detect virus titer in mixed infections. Early detection of mixed infection will aid in formulating better disease management strategies.

9.5 IMPACT OF MIXED INFECTION ON EVOLUTION OF NEW VIRUSES

All viruses replicate in the host cell using their genome as a template. This process presents a significant opportunity for the generation of novel virus molecules by switching templates during replication when multiple genomes are present in the same cell. The replicase protein of begomovirus can give rise to chimeric genomes, which may later evolve into new viruses with either impaired or improved virulence. Even if virulence is affected, it contributes to the diversity that can lead to better host adaptation, increased host range, and differences in vector transmission. The probability of evolution is determined by various factors such as nutrient levels in the host plant, temperature, other stresses, age of the plant, and the different partners involved in a co-infection model. Recent years have seen significant advancements in investigative research that aim to elucidate the dynamics of virus evolution. While studies have been published on chimeric molecules and their probable recombination pathways for many viruses, here we mention a few examples of begomovirus genomes to explain this concept without delving into the complexities of evolution, which are even more intricate than commonly considered.

Plant viruses have the ability to undergo rapid evolution, and mixed infection provides additional opportunities for this process. ToLCNDV has been reported to co-occur with many other begomoviruses, as well as viruses from other groups. It exhibits pseudo recombination with several associated viruses, such as *tomato leaf curl Palampur virus* (ToLCPaV) and ToLCGV (Zaidi, Martin, Amin, Farooq, & Mansoor, 2017). This co-occurrence allows for the evolution of chimeric molecules. A similar phenomenon has been studied in the case of *tomato leaf curl Ranchi virus* (ToLCRnV), which exhibits recombination in the IR region, rep, and coat proteins regions. Trans-encapsidation and trans-replication are common occurrences in mixed infections. Trans-encapsidation can alter vector specificity, thereby enabling differential transmission. Trans-replication is one of the major causes of the generation of recombinant molecules, which can later evolve into novel virus strains.

9.6 DISEASE MANAGEMENT STRATEGIES

Plant viruses cause severe economic losses and pose a threat to sustainable agriculture due to the indirect use of insecticides and chemical control methods to maintain productivity. The frequent emergence of new viral diseases is mainly attributed to international trade, climate change, and the ability of viruses to rapidly evolve, especially in cases of mixed infections. Plant viral disease control relies on two broad strategies:

- Immunization (genetic resistance obtained through plant breeding, plant transformation, cross-protection, or other methods); and
- Prophylaxis to limit virus spread (quarantine, certification, removal of infected plants, control of natural vectors, or other procedures).

Accurate diagnosis of the causal viral sample is essential for implementing proper control measures. Sharma et al. (2009) described diagnostic specificity and accuracy as statistical measures performed through binary classification (Sharma, Yadav, & Sharma, 2009). The ratio of true positive samples to the total samples analyzed provides an idea of the positive predictive value, while the ratio of true negative samples to the total samples analyzed provides an idea of the negative predictive value (Table 9.2). Measuring these values ensures the accuracy and diagnostic significance of the test for management strategies. However, the sensitivity of diagnostic tests may be limited by low virus titers or early-stage infections, which can result in false positive results.

TABLE 9.2 Measurement of Specificity and Sensitivity of Diagnostic Tests

Symptom Appearance	Predicted Outcome	Number of Viral Titer Present	Number of Viral Titer Absent	Total Numbers of Plants Diagnosed	Conclusion
Yes	Test positive	X (True positive)	Y (False positive)	X + Y	Positive predictive value = X/X + Y
No	Test negative	A (False negative)	B (True negative)	A + B	Negative predictive value = A/A + B

Note: Sensitivity of diagnostic test employed = X/X + A.
Specificity of diagnostic test employed = Y/Y + B.
X, Y, A, and B are total number of plants samples, respectively.

Looking at the dynamic nature of viruses and their easy adaptability with epigenetic factors, the eradication of virus diseases poses a significant challenge (Acosta-Leal, Duffy, Xiong, Hammond, & Elena, 2011). The genetic diversity, broad host range, polyphagous nature of vectors, and evolution of virus populations add complexity to disease management. It is noteworthy to quote Ramakrishnan (1961) here, "Diseases exact a heavy toll on crop yields and cause considerable damage year after year." Therefore, better and improved methods should be employed to reduce the economic losses caused by plant diseases. The following are disease management strategies that can be broadly classified into three categories (Rubio, Galipienso, & Ferriol, 2020):

i. Immunization measures;
ii. Prophylactic measures; and
iii. Quarantine measures.

9.6.1 IMMUNIZATION MEASURES

Immunization methods aim to develop innate resistance in the plant cultivar against targeted pathogens. Backcrossing and plant breeding methods are employed to confer resistance in a cultivated variety through the integration of a cascade of R genes (resistance genes) from another variety of plant that is resistant to infection. The introduction of resistance genes establishes systemic resistance in plants, preventing the establishment of systemic infection by the pathogen. Various molecular cascades may be triggered upon virus entry into plants where R genes have been integrated. These molecular events halt viral integration and the completion of the life cycle in the plant, preventing infection. This process is driven by resistance proteins.

There are two types of resistance events that are initiated: (i) active resistance, which is a dominant trait encoded by resistance proteins. It determines specific conformational patterns of viral gene products (e.g., coat protein, spike protein, etc.) and results in a hypersensitive response that leads to induced cell death of the infected cell (De Ronde, Butterbach, & Kormelink, 2014); and (ii) passive resistance, which is a recessive trait and encodes various critical host factors such as translation initiation factors. Passive resistance genes interfere with the recruitment of host replication and translation machinery for viral particle replication and protein production, leading to eradication by inhibiting the formation of mature particles (Truniger & Aranda, 2009; Hashimoto, Neriya, Yamaji, & Namba, 2016).

Although the establishment of resistance genes through breeding is the most plausible technique for developing resistance against viruses, it is unsuitable for many cultivated crops due to the following reasons: (i) quick modulation and alteration of host range by virus particles; and (ii) Unavailability of resistance genes in genetically compatible plant varieties. In cases where complete active or passive resistance genes are unavailable, alternative approaches such as reduction in virus particle accumulation (partial resistance) or reduction of tissue damage and symptom subsidence (tolerance resistance) can be implemented.

Another alternative is the adaptive radiation and artificial evolution of resistance genes against novel viruses or mixed infections. This involves random mutation in resistance genes followed by the selection of mutants that confer resistance, which is a suitable approach for developing broad spectrum resistance against plant pathogens (Farnham & Baulcombe, 2006; Harris, Slootweg, Goverse, & Baulcombe, 2013). However, a major limitation of this approach is that most of the generated mutations are non-functional, and the process of screening and selecting mutants is a time-consuming and laborious task.

Genome editing techniques, such as the CRISPR-Cas9 system, ZFN, or TALENs, can be used to directly edit and introduce mutations in host resistance genes (Piron et al., 2010; Chandrasekaran et al., 2016; Pyott, Sheehan, & Molnar, 2016). It is possible that mutated genes play important roles in the physiology or developmental biology of plants, which may result in adverse or unexpected physiological and/or reproductive events in mutated cultivars.

Another strategy is to engineer the plant genome to express artificial miRNAs (amiRNAs) that target viral sequences (Niu et al., 2006; Qu, Ye, & Fang, 2007; Liu, Zhou, Hu, Wei, & Zhang, 2017), or small interfering RNAs (syn-tasiRNAs). These miRNAs or siRNAs have short sequences that target conserved regions (CRs) in the viral genome, making them suitable for targeting multiple sites in the viral genome and can be used in mixed infections (Lin et al., 2009; Lafforgue et al., 2011). This approach provides long-term and broad-spectrum resistance. By expressing multiple miRNAs that target multiple regions in the viral genome, it is possible to enhance resistance against a wide range of viruses (Kung et al., 2012; Lafforgue, Martínez, Niu, Chua, Daròs, & Elena, 2013; Kis et al., 2016).

The RNA silencing mechanism is a natural antiviral defense mechanism in plants. It serves as one strategy to combat viral infections. Virus-encoded RNA silencing suppressors can downregulate the RNA silencing mechanism by directly interacting with key components of the mechanism. The cumulative expression of silencing suppressors during a synergistic infection can lead to increased viral load and more severe necrotic symptoms in plants (Syller, 2012). Resistance breakdown against the silencing mechanism can occur due to spontaneous mutation or selection pressure. Additionally, in mixed infections, the presence of other viruses can amplify the resistance-breaking mechanism in the host plant. Engineering plants with multiple silencing constructs can minimize the possibility of resistance breakdown during mixed infections. Clay nanoparticles and small peptides can be utilized to stabilize dsRNA in the silencing mechanism, thereby enhancing the degree of resistance (Numata, Ohtani, Yoshizumi, Demura, & Kodama, 2014; Mitter et al., 2017).

Another method of immunization involves "vaccinating" plants with mild or attenuated strains of viruses. The mechanism by which resistance is conferred through this method is not well understood. Vaccination immunization can inhibit various stages of the viral life cycle, including the recognition of peptide signal sequences on host cells, the entry of the virus' genome into host cells, the recruitment of replication machinery, the modulation of host cell homeostasis, and the induction of RNA silencing. All of these processes result in the degradation of the viral genome or interfere with the production of mature viral particles. The cross-protection technique is used to confer resistance in tomato and citrus crops (Pechinger, Chooi, MacDiarmid, Harper, & Ziebell, 2019).

The immunization approach for conferring resistance in plants is often based on host genetic variability, but virus genetic variability is overlooked. The genome variability of viruses should also be considered to minimize resistance breakdown in mixed infections. Active resistance mechanisms are conferred by dominant plant alleles that target essential functions of the virus life cycle. Thus, it is difficult for the virus to overcome and break down resistance, even in mixed infections.

Nevertheless, there is no assurance that any of the techniques employed provide enduring resistance against all virus species, as more epigenetic factors and the polymorphic nature of the virus can subside the resistance mechanism in host plants.

9.6.2 PROPHYLACTIC MEASURES

Prophylaxis involves the avoidance of introducing pathogens to the host plant. Spatial and temporal studies on genetic variation in the virus genome can provide valuable insights into the introduction and migratory behavior of plant viruses. Such studies are instrumental in designing effective prophylactic measures for managing mixed infections in plants.

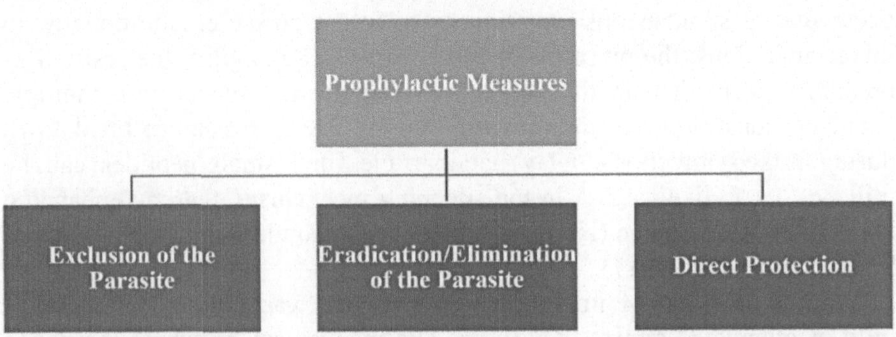

1. **Exclusion of the Parasite:** Quarantine strategies and certification measures are required to restrict the entry of diseased plants into the

field. These measures will prevent the introduction of pathogens into the field area and successfully control the spread of infection.

2. **Eradication or Elimination of the Parasite:** Once the virus enters the field and establishes a successful infection, efforts should be made to eradicate or eliminate it. This can be done through the following methods:

   i. **Elimination of Transmission Vectors:** Plant virus vectors are polyphagous in nature and are of utmost importance in virus dispersal and disease spread. The regulated use of insecticide in the field will destroy the vector population, thus controlling virus spread in the field.

   ii. **Removal or Destruction of Affected Parts (Roguing):** Regular practice involves the destruction of affected plants from the fields by burning or uprooting to eliminate the spread of infection in the field.

   iii. **Destruction of Wild Host and Weeds:** Weeds are considered prospective storehouses for viruses. Various recombination events are initiated in the virus genome during their presence in weed plants, resulting in mixed infection in nearby field crops. Eliminating such wild hosts and weeds will minimize the chances of mixed infection in field crops.

   iv. **Crop Rotation:** It is the most widely used sustainable practice for the elimination of infection. If a suitable host is not available for two to three cropping seasons, the load of pathogens will decrease from the field.

   v. **Improved Cropping Practices:** Using healthy and disease-free seeds for sowing will reduce the chances of infection. Providing proper manure, considering climatic conditions, and maintaining soil hygiene will enhance the innate immunity in plants to combat infection. Implementing mixed or alternate cropping and changing the planting season for annual crops will also result in a significant reduction of viruses.

3. **Direct Protection:** This strategy includes the use of antibiological substances or biological interactions for the elimination of pathogens. It includes the following measures:

i. **Biological Control Measures:** Various biological interactions such as commensalism, antagonism, competition, parasitism, etc., can be used in mixed infections to control the spread of viruses.

ii. **Use of Antibiological Substances:** Various fumigants, antimicrobial substances, and toxicants are used to reduce the spread of infection. Soil sterility and burning of plants are also used for the complete removal of crop biota. Various phytohormones, plant-derived chemicals, and pheromones are studied for their antiviral efficacies.

9.6.3 QUARANTINE MEASURES

Quarantine of infected samples restricts the spread of pathogens over long distances, while sanitary certification after proper molecular diagnosis certifies virus-free germplasm. The sanitized plants must be diagnosed with recent techniques like qRT-PCR or multiplex PCR to evaluate the chances of mixed infection. These are probably the first line of defense to limit the introduction, emergence, and spread of new viruses in a geographical location.

The introduction of viral pathogens on field crops will cause considerable economic loss. Therefore, much effort in the area of plant pathogen management is directed towards the management of viral diseases. The correct and early identification of viruses is important for designing proper sanitization measures and controlling the spread of infection. However, the modular nature of the genome and the adaptability of viruses with high recombination rates pose a major difficulty in virus detection. Although such constraints exist, various methods, either standalone or in combination, are practiced eradicating the spread of viral pathogens on plants.

9.7 CONCLUSIONS

Viral diseases in crop plants cause significant damage, leading to reduced yield of plant products. Viruses are nucleoproteins and obligate parasites, relying on the host's cellular machinery for replication, movement, and transmission. The small genome size of viruses allows for dynamic and diverse particles, facilitating recombination and re-assortment of genomic segments. Viral propagation is entirely dependent on the host, while environmental factors and the polyphagous nature of transmission vectors greatly influence virus acquisition, transmission, and infection establishment. The pervasiveness of viral infections can result in reduced production or complete loss, depending on the severity and spread of the infection. Mixed viral infections can lead to synergistic, antagonistic, or mutually exclusive processes. All of these lead to either an increased susceptibility of the host plant to virus infection or an expansion of the host range for virus transmission and establishment of infection. Hence, detection strategies must be efficiently designed to effectively analyze virus titer in the early stages of infection. Various detection methods such as immunoassays, NGS sequencing, qRT-PCR, molecular hybridization, RT-LAMP, etc., are commonly customized to detect the presence of specific viral antigens in host plants. These molecular detection methods provide new opportunities for designing efficient management strategies through genetic engineering and genome editing.

KEYWORDS

- **genome editing**
- **immunization**
- **mixed virus infection**
- **molecular interaction**
- **pathogen**
- **plant virus**
- **plant virus evolution**
- **viral disease epidemiology**

REFERENCES

Acosta-Leal, R., Duffy, S., Xiong, Z., Hammond, R. W., & Elena, S. F., (2011). Advances in plant virus evolution: Translating evolutionary insights into better disease management. *Phytopathology, 101*, 1136–1148.

Alves-Júnior, M., Alfenas-Zerbini, P., Andrade, E. C., Esposito, D. A., Silva, F. N., Da Cruz, A. C. F., Ventrella, M. C., et al., (2009). Synergism and negative interference during co-infection of tomato and *Nicotiana benthamiana* with two bipartite *Begomoviruses*. *Virology, 387*, 257–266.

Barboza, N., Blanco-Meneses, M., Esker, P., Moriones, E., & Inoue-Nagata, A., (2017). Distribution and diversity of *Begomoviruses* in tomato and sweet pepper plants in Costa Rica. *Annals of Applied Biology*, 172.

Baulcombe, D. C., Lloyd, J., Manoussopoulos, I. N., Roberts, I. M., & Harrison, B. D., (1993). Signal for potyvirus-dependent aphid transmission of *Potato aucuba mosaic virus* and the effect of its transfer to *Potato virus X*. *J. Gen. Virol., 74*, 1245–1253.

Chakraborty, S., Vanitharani, R., Chattopadhyay, B., & Fauquet, C. M., (2008). Supervirulent pseudorecombination and asymmetric synergism between genomic components of two distinct species of *Begomovirus* associated with severe tomato leaf curl disease in India. *J. Gen. Virol., 89*, 818–828.

Chandrasekaran, J., Brumin, M., Wolf, D., Leibman, D., Klap, C., Pearlsman, M., et al., (2016). Development of broad virus resistance in non-transgenic cucumber using CRISPR/Cas9 technology. *Mol. Plant Pathol., 17*, 1140–1153.

Chávez-Calvillo, G., Contreras-Paredes, C. A., Mora-Macias, J., Noa-Carrazana, J. C., Serrano-Rubio, A. A., Dinkova, T. D., Carrillo-Tripp, M., & Silva-Rosales, L., (2016). Antagonism or synergism between *Papaya ringspot virus* and *Papaya mosaic virus* in *Carica papaya* is determined by their order of infection. *Virology, 489*, 179–191.

Chiu, M. H., Chen, I. H., Baulcombe, D. C., & Tsai, C. H., (2010). The silencing suppressor p25 of *Potato virus X* interacts with Argonaute1 and mediates its degradation through the proteasome pathway. *Mol. Plant Pathol., 11*, 641–649.

De Ronde, D., Butterbach, P., & Kormelink, R., (2014). Dominant resistance against plant viruses. *Front. Plant Sci., 5*, 307. 10.3389/fpls.2014.00307.

Farnham, G., & Baulcombe, D. C., (2006). Artificial evolution extends the spectrum of viruses that are targeted by a disease–resistance gene from potato. *Proc. Natl. Acad. Sci. U. S. A., 103*, 18828–18833.

Froissart, R., Doumayrou, J., Vuillaume, F., Alizon, S., & Michalakis, Y., (2010). The virulence–transmission trade-off in vector-borne plant viruses: A review of (non-)existing studies. *Phil. Trans. R. Soc. B, 365*, 1907–1918.

García-Cano, E., Resende, R. O., Fernández-Muñoz, R., & Moriones, E., (2006). Synergistic interaction between *Tomato chlorosis virus* and *Tomato spotted wilt virus* results in breakdown of resistance in tomato. *Phytopathology, 96*, 1263–1269.

Gómez, P., Sempere, R. N., Amari, K., Gómez-Aix, C., & Aranda, M. A., (2010). Epidemics of *Tomato torrado virus*, *Pepino mosaic virus* and *Tomato chlorosis virus* in tomato crops: Do mixed infections contribute to torrado disease epidemiology? *Ann. Appl. Biol., 156*, 401–410.

Harris, C. J., Slootweg, E. J., Goverse, A., & Baulcombe, D. C., (2013). Stepwise artificial evolution of a plant disease resistance gene. *Proc. Natl. Acad. Sci. U. S. A., 110*, 21189–21194. 10.1073.

Hashimoto, M., Neriya, Y., Yamaji, Y., & Namba, S., (2016). Recessive resistance to plant viruses: Potential resistance genes beyond translation initiation factors. *Front. Microbiol., 7*, 1695. 10.3389.

Jedlinski, H., & Brown, C. M., (1965). Cross protection and mutual exclusion by three strains of *Barley yellow dwarf virus* in *Avena sativa* L. *Virology, 26,* 613–621.

Kasschau, K. D., Xie, Z., Allen, E., Llave, C., Chapman, E. J., Krizan, K. A., & Carrington, J. C., (2003). P1/HC-Pro, a viral suppressor of RNA silencing, interferes with *Arabidopsis* development and miRNA function. *Dev. Cell, 4*, 205–217.

Kis, A., Tholt, G., Ivanics, M., Varallyay, E., Jenes, B., & Havelda, Z., (2016). Polycistronic artificial miRNA-mediated resistance to *Wheat dwarf virus* in barley is highly efficient at low temperature. *Mol. Plant Pathol., 17*, 427–437.

Kung, Y. J., Lin, S. S., Huang, Y. L., Chen, T. C., Harish, S. S., Chua, N. H., et al., (2012). Multiple artificial microRNAs targeting conserved motifs of the replicase gene confer robust transgenic resistance to negative-sense single-stranded RNA plant virus. *Mol. Plant Pathol., 13*, 303–317.

Lafforgue, G., Martínez, F., Niu, Q., Chua, N., Daròs, J., & Elena, S. F., (2013). Improving the effectiveness of artificial microRNA (amiR)-mediated resistance against *Turnip mosaic virus* by combining two amiRs or by targeting highly conserved viral genomic regions. *J. Virol., 87*, 8254–8256.

Lafforgue, G., Martínez, F., Sardanyes, J., De La Iglesia, F., Niu, Q. W., Lin, S. S., et al., (2011). Tempo and mode of plant RNA virus escape from RNA interference-mediated resistance. *J. Virol., 85,* 9686–9695.

Legg, J. P., & Fauquet, C. M., (2004). Cassava mosaic geminiviruses in Africa. *Plant Mol. Biol., 56*, 585–599.

Lin, S. S., Wu, H. W., Elena, S. F., Chen, K. C., Niu, Q. W., Yeh, S. D., et al., (2009). Molecular evolution of a viral non-coding sequence under the selective pressure of amiRNA–mediated silencing. *PLoS Pathog., 5*, e1000312. 10.1371.

Liu, S., Zhou, J., Hu, C., Wei, C., & Zhang, J., (2017). MicroRNA–mediated gene silencing in plant defense and viral counter–defense. *Front. Microb., 8*, 1801. 10.3389.

Mingot, A., Valli, A., Rodamilans, B., San León, D., Baulcombe, D. C., García, J. A., & López-Moya, J. J., (2016). The P1N-PISPO trans-frame gene of sweet potato feathery mottle potyvirus is produced during virus infection and functions as an RNA silencing suppressor. *J. Virol., 90*, 3543–3557.

Minor, R. M., Maria, R. R., Monica, A. M., Natalia, B., & Robert, L. G., (2022). The invasion biology of tomato *Begomoviruses* in Costa Rica reveals neutral synergism that may lead to increased disease pressure and economic loss. *Virus Research, 317*.

Mitter, N., Worrall, E. A., Robinson, K. E., Li, P., Jain, R. G., Taochy, C., et al., (2017). Clay nanosheets for topical delivery of RNAi for sustained protection against plant viruses. *Nat. Plants, 3*, 16207.

Moreno, A. B., & López-Moya, J. J., (2020). When viruses play team sports: Mixed infections in plants. *Phytopathology, 110*(1), 29–48.

Murant, A. F., (1990). Dependence of groundnut rosette virus on its satellite RNA as well as on groundnut rosette assistor luteovirus for transmission by *Aphis craccivora*. *J. Gen. Virol. 71*, 2163–2166.

Niu, Q. W., Lin, S. S., Reyes, J. L., Chen, K. C., Wu, H. W., Yeh, S. D., et al., (2006). Expression of artificial microRNAs in transgenic *Arabidopsis thaliana* confers virus resistance. *Nat. Biotechnol., 24*, 1420–1428.

Numata, K., Ohtani, M., Yoshizumi, T., Demura, T., & Kodama, Y., (2014). Local gene silencing in plants via synthetic dsRNA and carrier peptide. *Plant Biotechnol. J., 12*, 1027–1034.

Pechinger, K., Chooi, K. M., MacDiarmid, R. M., Harper, S. J., & Ziebell, H., (2019). A new era for mild strain cross- protection. *Viruses, 11*, 670.

Piron, F., Nicolaï, M., Minoïa, S., Piednoir, E., Moretti, A., Salgues, A., et al., (2010). An induced mutation in tomato eIF4E leads to immunity to two potyviruses. *PLoS One, 5*, e11313. 10.1371.

Pyott, D. E., Sheehan, E., & Molnar, A., (2016). Engineering of CRISPR/Cas9-mediated potyvirus resistance in transgene-free *Arabidopsis* plants. *Mol. Plant Pathol., 17*, 1276–1288.

Qu, J., Ye, J., & Fang, R., (2007). Artificial microRNA–mediated virus resistance in plants. *J. Virol., 81*, 6690–6699.

Ramakrishnan, T. S., & Radhakrishna, P. P. N., (1961). Abnormal leaf fall disease of rubber caused by *Phytophthora palmivora* (Butl.). *Rubber Board Bull., 5*, 11–20.

Rochow, W. F., & Ross, A. F., (1955). Virus multiplication in plants doubly infected by *Potato viruses X and Y. Virology, 1*, 10–27.

Rubio, L., Galipienso, L., & Ferriol, I., (2020). Detection of plant viruses and disease management: Relevance of genetic diversity and evolution. *Front Plant Sci., 11*, 1092.

Sharma, D., Yadav, U., & Sharma, P., (2009). The concept of sensitivity and specificity in relation to two types of errors and its application in medical research. *J. Reliab. Stat. Stud., 2*, 53–58.

Syller, J., & Grupa, A., (2016). Antagonistic within-host interactions between plant viruses: Molecular basis and impact on viral and host fitness. *Mol. Plant Pathol., 17*, 769–782.

Syller, J., (2012). Facilitative and antagonistic interactions between plant viruses in mixed infections. *Mol. Plant Pathol., 13*(2), 204–216.

Tatineni, S., Graybosch, R. A., Hein, G. L., Wegulo, S. N., & French, R., (2010). Wheat cultivar-specific disease synergism and alteration of virus accumulation during co-infection with *Wheat streak mosaic virus* and *Triticum mosaic virus*. *Phytopathology, 100*, 230–238.

Torchetti, E. M., Pegoraro, M., Navarro, B., Catoni, M., Di Serio, F., & Noris, E., (2016). A nuclear-replicating viroid antagonizes infectivity and accumulation of a geminivirus by upregulating methylation-related genes and inducing hypermethylation of viral DNA. *Sci. Rep., 6*, 35101.

Truniger, V., & Aranda, M. A., (2009). Recessive resistance to plant viruses. *Adv. Virus Res., 75*, 119–159.

Vanitharani, R., Chellappan, P., Pita, J., & Fauquet, C., (2004). Differential roles of AC2 and AC4 of cassava geminiviruses in mediating synergism and suppression of posttranscriptional gene silencing. *J. Virol., 78*, 9487–9498.

Wintermantel, W. M., Cortez, A. A., Anchieta, A. G., Gulati-Sakhuja, A., & Hladky, L. L., (2008). Co-infection by two criniviruses alters accumulation of each virus in a host-specific manner and influences efficiency of virus transmission. *Phytopathology, 98*, 1340–1345.

Zaidi, S. S., Martin, D. P., Amin, I., Farooq, M., & Mansoor, S., (2017). *Tomato leaf curl New Delhi virus*: A widespread bipartite *Begomovirus* in the territory of monopartite *Begomoviruses*. *Mol. Plant Pathol., 18*(7), 901–911.

Ziebell, H., & Carr, J. P., (2010). Cross-protection: A century of mystery. *Adv. Virus Res., 76*, 211–264.

CHAPTER 10

Understanding Begomovirus and Its Host Relationship

NEHA RAUNIYAR and DEEPA SRIVASTAVA

Department of Botany, Deen Dayal Upadhyaya Gorakhpur University, Gorakhpur, Uttar Pradesh, India

ABSTRACT

Infectious agents, such as viruses, are known to be obligate parasites whose survival and ability to replicate depend entirely on their host, where genome replication takes place. The outcome of this process is that the virus is influenced by its host in every aspect during its life cycle. Over the years, numerous surveys and studies have attempted to express comprehensive knowledge about the interaction between plant hosts and viruses, dating back to ancient times. However, recent advances have shed light on the impact of hosts on the evolution of viruses in relation to plants.

This chapter focuses on how certain viruses induce abnormalities in their host, both histologically and cytologically, which are the result of abnormalities in cell division in response to host-virus interaction. Although there was initially a surge of research on the relationship between hosts and viruses, nature has witnessed a surprising evolution over time, from being parasitic to now, to some extent, being mutualistic. The present topic highlights the research interest in recent years, contributing to the evolution and improvement of viral management mechanisms in plants. This helps provide comparative information about both the host and the virus.

10.1 INTRODUCTION

Major strides have been made in the field of molecular biology as a result of the struggle between the virus and its host. This has helped us to better

New and Emerging Plant Viruses: The Threat to Food Security.
Alireza Golnaraghi and Rajarshi Kumar Gaur (Eds.)

grasp the relationship between the plant virus's genome and its replication mechanism within the host (Zailtin & Hull, 1987). The modest to large genomes of RNA/DNA plant viruses have produced compelling data that demonstrates the virus's profound dependence on the machinery of the host to carry out all of its cellular processes, as well as its replication within the host. Viruses currently affect all forms of life, including humans, plants, and animals, directly or indirectly. The begomovirus, the largest genus in the Geminiviridae family, is transmitted by whiteflies and severely damages various essential fresh crops (Rojas et al., 2018). There are two types of begomovirus species: monopartite (just one genetic component) and bipartite (dual genomic components, known as DNA-A and DNA-B) (Zerbini et al., 2017; Kumar, 2019). Both the nuclear shuttle protein (NSP) and the movement protein (MP) are encoded by B-DNA in bipartite (PM) begomovirus. In the case of systemic infection, the MP and the NSP cooperate to mediate the intra- and intercellular trafficking of vDNA (Lazarowitz & Beachy, 1999; Gafni & Epel, 2002; Sanderfoot et al., 1996). However, PM is mostly responsible for facilitating the movement of vDNA from cell to cell through plasmodesmata. Researchers have discovered that the MPs of the viruses "bean dwarf mosaic virus" (BDMV) and "squash leaf curl virus" (SLCV) facilitate the nucleocytoplasmic relocation of vDNA through nuclear pores (Noueiry et al., 1994; Sanderfoot & Lazarowitz, 1995, 1996).

A range of hosts of different species of Begomovirus is determined by the molecular connections that allow those viruses to hijack cellular processes and avoid detection by innate immune systems (Kumar, 2019). The virus's ability to replicate on a specific host can be restricted or hampered by the failure of any one of these phases. Both pro-viral and antiviral interactions that affect host susceptibility to begomovirus are mediated by NSP. Based on identified NSP-interacting proteins, the NSP-interacting immunological hub has been described as one of the primary information transmitters of the host immune system and the virulence tactics of the begomovirus.

10.2 PROVIRAL FUNCTIONS OF NSP-INTERACTING PROTEINS

As begomovirus proliferates in the nuclei of infected cells, vDNA is produced, which is then transferred from the nucleus to the cytosol through the nuclear pores, across the cytoplasm along the cell periphery, and finally near uninfected cells via plasmodesmata. Similar to other plant viruses, begomovirus B-DNA encodes a MP that facilitates this typical cellular motion, enlarges the plasmodesmata size exclusion limit (SEL), and mediates the translocation of viral nucleic acid into surrounding cells (Noueiry et al., 1994). Less is known about how vDNA exits the nucleus and moves through the cytoplasm towards the cell's edge. It is widely accepted that NSP accelerates the nuclear exit of newly duplicated vDNA through nuclear pores, although the interactions between the viral protein and the host nuclear transport system have not been thoroughly investigated. Although SLVC NHP only binds to *in vitro* SSDNA, BDMV NHP and *Abutilon mosaic virus* (AbMV) interact with both ds-vDNA and ss-vDNA, apparently in the nucleus of unhealthy cells, which is consistent with the NSP-mediated nucleocytoplasmic transfer of ss-vDNA (Pascal et al., 1994; Rojas et al., 1998; Hehnle et al., 2004). Systemically diseased leaves were immunogold labeled to identify the SLCV NSP, and the presence of an epitope-tagged AbMV NSP in Nicotiana benthamiana cells confirmed the NSP's nuclear location (Kleinow et al., 2009). When co-expressed MP is present, NSP from diverse begomoviruses is shown to be transported to the cell boundary from the nuclei of cells that are only expressing NSP fusions in transient tests (Sanderfoot & Lazarowitz, 1995; Sanderfoot et al., 1996; Lazarowitz & Beachy, 1999; Zhang et al., 2001). The discovery of an NSP-networking GTPase (NIG), interacting with the NSP on the cytosolic side of the nuclear pore complex and sharing transport capacities with the human protein interacting with Rev. (hRIP), has shed light on the mechanism for the release of the NSP-vDNA complex from the nuclear boundary into the cytoplasm (Carvalho et al., 2008a, b). Interaction of NSP with PM in the cytoplasm may facilitate translocation of the virus to the cell surface (Carvalho et al., 2008a; Noueiry et al., 1994; Sanderfoot & Lazarowitz, 1995; Frischmuth et al., 2007).

NIG interacts with various begomovirus NSPs *in vivo* and *in vitro*, enhances the translocation of NSP from the nucleus to the cytoplasm, and displays proviral motion, all of which are consistent with a role in the transportation of vDNA-NSP complex (Carvalho et al., 2008a). NIG and hRIP are important Rev cofactors that facilitate the release of HIV-1 RNA from the nuclear export complex into the cytosol (Sánchez-Velar et al., 2004). Both share structural characteristics and transport properties. Therefore, the function of begomovirus NSPs in plants may be comparable to that of HIV Rev in living beings. Both hRIP and NIG contain an N-terminal ArfGAP domain, which is involved in signaling, vesicular trafficking, and intracellular protein localization (Chavrier & Goud, 1999; Turner et al., 2001; Sabe et al., 2006; Bist et al., 2017). Based on its influence on NSP function and transport properties, NIG either acts as a promoter for the disassembly of nuclear export complexes on the cytoplasmic side and/or targets these nuclear export proteins to specific sites within the cytoplasm. It can act as a positive regulator (Carvalho et al., 2008a). However, the cytosolic target of NIG has not yet been identified, and reverse genetics studies have not focused on its proviral role, so we cannot fully understand the mechanisms underlying NIG's action. The only known partner of NIG at this time is AtWWP1 (*Arabidopsis thaliana* WW domain-containing protein 1), a nuclear body (NB)-forming protein that accumulates during virus-related infection and encases NIG in nuclear bodies (Calil et al., 2018). Nuclear shuttle protein interactor (AtNSI), a nuclear acetyltransferase that may regulate the nuclear export of ss-vDNA-NSP by acetylating CP and histones, has been shown to interact with the nucleocapsid protein (NSP) of the cabbage leaf curl virus (CaLCuV) (McGarry et al., 2003; Carvalho & Lazarowitz, 2004). Plant proteins involved in differentiation and/or plant protection are acetylated by highly active multimeric enzyme complexes formed by AtNSI (Carvalho et al., 2006). Due to NSP binding, AtNSI cannot form these highly active enzyme complexes and is recruited by NSP to acetylate specific CP in newly synthesized ss-vDNA. This impairs vDNA-CP interaction, promoting the formation of vDNA-NSP complexes and subsequent transfer of vDNA to the cytoplasm (Carvalho & Lazarowitz, 2004; Carvalho et al., 2006). Additionally, AtNSI acetylates histone H3, which is involved in the formation of minichromosomes and nucleosome synthesis in vDNA. This alters viral transcription and replication (McGarry et al., 2003). However, it has been demonstrated that H3 is a factor of the vDNA-protein composite that moves between and within cells. It also networks with NSP and MP from the BDMV (Zhou et al., 2011). The ability of histones to package themselves may explain why H3 is found in an NSP and MP-comprehending complex that drives competence. This helps the viral genome to pass through the nucleoporin complex and then through plasmodesma. It is not yet known whether the putative proviral activity of H3 is required for infection, as attempts to use gene silencing to determine the direct role of H3 in begomovirus infection were unsuccessful. However, it is possible that certain host proteins that interact with NSP and have proviral properties are somehow engaged in maintaining the stability of the ss-vDNA-NSP complex or dynamically supporting the viral protein's nucleocytoplasmic transportation activity.

According to the standard model of intracellular trafficking of vDNA, NSP binds to vDNA and H3 in the nucleus before exporting the NSP-H3-vDNA complex to the cytoplasm. While some elements of the begomoviral DNA transport mechanism may be species-specific, this model remains valid. NIG facilitates the release of the trimeric complex from the cytosolic side. Subsequently, NSP attaches to MP, which also binds to H3. Through plasmodesmata at the cell periphery, the MP-NSP-H3-vDNA complex is transported to nearby uninfected cells. In these neighboring cells, NSPs transport the viral genome to the nucleus, initiating a new round of viral genome replication through a rolling circle process (Krichevsky et al., 2006; Hanley-Bowdoin et al., 2013).

Using AbMV and BDMV MPs and NSPs, we have demonstrated that recombinant mastoviral replicons from the Mastovirus genus of the Geminiviridae family can move between cells, consistent with the normal intracellular movement properties of vDNA (Diamos et al., 2019). The size of the replicon and the precise characteristics of MP and NSP are crucial to maintain the lack of species specificity in the activities of begomoviral MP and NSP. This minimizes the risk of allergic reactions and interference by viral replicons during viral replication and expression.

10.3 PLANT DEFENSE SUPPRESSOR: NSP

NSP-interacting kinases (NIKs) belonging to the transmembrane leucine-rich repeat receptor-like kinase (LRR-RLK) superfamily have been shown to suppress host defense (Fontes et al., 2004; Mariano et al., 2004). The NSP-binding site on Arabidopsis NIK1 is an 80-amino acid region bounded by amino acids 422–502, containing an activation loop (subdomain VII-DFGAk/rx and subdomain VIII-GtxGyiaPEY) and together with a Ser/Thr kinase (subdomain VIb–HrDvKssNxLLD). The kinase's stimulation loop has a threonine residue at position 474 that corresponds to the stimulation site of NIK1 (Fontes et al., 2004; Santos et al., 2009). To facilitate trans-phosphorylation and sequential kinase activation, NIK1 is a single-by-skip transmembrane receptor kinase that may dimerize or multimerize with different receptors or with itself (Santos et al., 2010). NSP binds to NIK1 and prevents phosphorylation of the vital threonine-474 residue, which would otherwise cause NIK1 to activate and send an antiviral signal in response to begomovirus infection. Neither the virus nor the host has a specific role in the development of the NSP-NIK complex (Fontes et al., 2004; Mariano et al., 2004). NSPs of various begomoviruses, such as CaLCuV, tomato golden mosaic virus (TGMV), tomato crinkle yellow leaf virus (TCrYLV), tomato yellow spot virus (ToYSV), and NIK tomato in various plant species such as soybean, and Arabidopsis thaliana All adhere to the NIK in a similar manner (Fontes et al., 2004; Mariano et al., 2004; Sakamoto et al.,

2012). The NIK1 immune receptor-like kinase was chosen as a target as a result of this discovery in order to create broad begomovirus resistance (Brustolini et al., 2015). The NIK1-T474D gain-of-function variation, which decreased translation extensively and enriched resistance to multiple types of tomato-infecting begomoviruses, was produced in tomato transgenic lines with constitutive activation of NIK1. Despite a 25% drop in translation, the transgenic lines under greenhouse-optimized growth conditions were phenotypically identical to untransformed tomato plants. On the other hand, the fundamental initiation of NIK1-arbitrated antiviral signaling restricted the growth of Arabidopsis (Zorzatto et al., 2015). The finding that NIK1 has a negative impact on antibacterial immunity is yet another barrier to the development of widespread begomovirus resistance in crops (Li et al., 2019). It has been shown that Arabidopsis overexpressing NIK1 makes the plant more susceptible to bacterial contaminations, which is undesirable in arena settings because plants are regularly exposed to contaminations by multiple pathogens at the same time (Li et al., 2019).

Our understanding of NIK1-mediated signaling pathways was aided by the characterization of downstream components, including ribosomal protein L10 (RPL10) and transcriptional repressor L10-interacting Myb domain-containing protein (LIMYB) (Carvalho et al., 2008c; Rocha et al., 2008; Zorzatto et al., 2015). More recently, it has been demonstrated that nucleic acids derived from begomoviruses trigger NIK1-mediated antiviral signaling. These nucleic acids most likely act as PAMPs to induce or stabilize NIK1 oligomerization, which is a crucial initial event that initiates signaling and transduction from a receptor (Teixeira et al., 2019).

A mechanistic model of NIK1 activation and defense suggests that begomovirus-derived nucleic acids act as PAMPs in response to viral infection, and that NIK1 itself, its paralog NIK2, or another unidentified transmembrane receptor known as PAMP, act as PAMPs. NIK1 mediates dimerization with the body (Teixeira et al., 2019). The critical threonine residue at position 474 in the NIK1 kinase domain is phosphorylated as a result of NIK1 oligomerization (Machado et al., 2015, 2017). The downstream component RPL10 is phosphorylated by this phosphorylation-induced activation of NIK1 and translocates from the cytoplasm to the nucleus, where it interacts with LIMYB. Genes associated with the translational machinery, including those encoding initiation and elongation factors for translation and ribosomal proteins, are suppressed by the RPL10-LYMIB complex (Gouveia et al., 2017). Global translation is suppressed when the translational machinery in plant cells is downregulated over time (Brustolini et al., 2015; Zorzatto et al., 2015). The translational regulatory system of plant cells prevents viral mRNAs from bypassing it, compromising the infection (Brustolini et al., 2015). During interactions between the host and the begomovirus, NSP from the begomovirus binds to the NIK1 kinase domain, inhibiting NIK1-mediated antiviral signaling and creating a more favorable environment for virus infection (Santos et al., 2009). Thus, the virus has evolved to withstand this antiviral signaling system, which serves as a defense mechanism. Two other receptor-like kinases, Brassinosteroid-insensitive 1-associated kinase 1 (BAK1) and a PERK-like kinase called NSP-associated kinase (NsAK), have been shown to interact with NSPs (Florentino et al., 2006; Li et al., 2019). While the formation of the NSP-NsAK complex promotes viral contamination, the interaction of NSP with BAK1 may be linked to a host defense-suppressing function. BAK1 acts as a co-receptor for several PRRs, including FLS2 and EFR, which detect specific PAMPs and activate or enhance PTI (PAMP-triggered immunity), the initial layer of the plant's innate immune system (Zipfel et al., 2006; Chinchilla et al., 2007). There are numerous pieces of data that suggest NSP may inhibit the role of BAK1. First, there is a structural connection between NIK1 and BAK1; both kinases have conserved locations for the activation site and are members of the same subfamily II of LRR-RLKs (Sakamoto et al., 2012). Second, there is a strong conservation between the NSP binding site on NIK1 and the homologous categorization on BAK1 (Santos et al., 2010). Third, while having antiviral agent properties, NIK1 has been demonstrated to directly engage with the co-receptor BAK1 and the PRR FLS2 to block BAK1-mediated PTI (Li et al., 2019). The NIK1-BAK1-FLS2 compound establishment may be hampered by the viral NSP suppressor of NIK1. The Envelope protein (CP) of plum pox virus (PPV; Nicaise & Candresse, 2017), the mobility protein (MP) of cucumber mosaic virus (CMV; Kong et al., 2018), and the P6 virus of *cauliflower mosaic virus* (CaMV) have been found to directly inhibit PTI when present (Zvereva et al., 2016). PPV CPs also function as viral effectors (Avr factors), selectively recognizing intracellular host receptors known as resistance R proteins, thereby promoting the second layer of the innate immune system, effector-induced immunity (ETI) (Decroocq et al., 2009). This study suggests that viral suppressors may link PTI suppression to ETI activation, which is consistent with the zigzag evolutionary model of plant innate immunity (Jones & Dangl, 2006). It has also been shown that the BDMV NSP of the resistant kidney bean genotype acts as an Avr factor for ETI activation, linking ETI to cell surface primary antiviral immunity (NIK1 and BAK1) (Garrido-Ramirez et al., 2000).

Additionally, ASYMMETRIC LEAVES2 (AS2), a protein that inhibits post-transcriptional gene silencing (PTGS) and positively regulates mRNA decapping and degradation, interacts with the NSP from CaLCuV (Ye et al., 2015). NSP functions as a suppressor of host defense by promoting AS2 expression and triggering AS2's nuclear export, which activates DCP2 decapping activity and suppresses PTGS. As a result, NSP enhances AS2's proviral activity, rendering Arabidopsis more susceptible to begomovirus.

Immunoaffinity capture of proteins using GFP-tagged NSPs (from AbMV) identified unique peptides from Arabidopsis Ras-GAP SH3 domain-binding protein (G3BP) homologs as potential NSP-interacting proteins (Krapp et al., 2017). The host's defense against animal viruses has been linked to the stress granule (SG)-localized protein G3BP (Tourrière et al., 2003; Lloyd, 2012). It plays a crucial role in SG formation. The mRNA-protein aggregates of SGs are formed in response to reduced translation during biotic and abiotic stresses (Buchan & Parker, 2009). Viral proteins bind to G3BP through the FGDF motif and prevent SG assembly (McInerney, 2015; Panas et al., 2015). Further evidence of AbMV-NSP binding to AtG3BP in planta comes from the observation that mutant NSP with an impaired FGDF-like motif fails to co-localize with AtG3BP in SGs (Krapp et al., 2017). Although the biological significance of the G3BP-NSP interaction has not been fully explored, the notion that NSP binds to G3BP to inhibit SG production is analogous to well-established studies on animal virus-host relationships (Lloyd, 2012; Panas et al., 2015).

10.4 IMMUNE CENTER AGAINST BEGOMOVIRUS

A comprehensive understanding of the interactions between plants and pathogens has been revealed through genome-inclusive analyses of plant defense mechanisms and pathogen contamination approaches. Pathogen-effector interactions and plant defense proteins converge to form a subset of highly interconnected host proteins (Mukhtar et al., 2011). NSP from begomoviruses, which facilitates intra- and intercellular transport, may serve as a potential target for identifying components of essential cellular processes, as it interacts with host factors in various organelles and acts as a focal point for host protein-protein interactions. This chapter summarizes the proteins from different subcellular compartments, including nuclear acetylase AtNSI (Carvalho & Lazarowitz, 2004), nuclear histone H3 (Zhou et al., 2011), cytosolic GTPase NIG (Carvalho et al., 2008a), nuclear cytoplasmic AS2 (Ye et al., 2015), and cytoplasmic G3 (Florentino et al., 2006). The NSP-interacting immunological center hosts both proviral proteins such as AtNSI, H3, NIG, NsAK, and AS2, and antiviral proteins such as NIK, G3BP, and AtWWP1. The proviral elements, like AtNSI and H3, may facilitate or maintain NSP binding to vDNA, or they may regulate the dynamic intracellular transport of NSP-vDNA, similar to NIG. The fact that plant cells have developed a defense mechanism to counteract NIG proviral activity demonstrates the significance of NIG-NSP interaction during infection (Calil et al., 2018). Viral contamination triggers the binding of the NB-forming protein AtWWP1, which facilitates the transport of NIG from the cytoplasm to the nucleus, where it becomes confined to AtWWP1 NBs and disrupts cytoplasmic NIG transport function (Calil et al., 2018). However, viral infection may nullify the antiviral effects of AtWWP1-NB. As a defense mechanism, vDNA binds to AtWWP1, reducing the number of AtWWP1 NBs or causing their destruction, thereby restoring the cytosolic localization of NIG (Calil et al., 2018). Consequently, begomoviruses have evolved vDNA-based virulence strategies to surpass nuclear-derived host defense systems.

Similar to this, host immunity is suppressed when NSP binds to defense proteins, including NIK1 and AS2. In the initial instance, upon contamination, NSP prevents NIK1 kinase from transducing an antiviral signal that would otherwise shield plants against begomoviruses (Brustolini et al., 2015; Machado et al., 2015, 2017). In the latter case, NSP triggers nuclear translocation of the innate PTGS suppressor AS2 to the cytosol, where it interacts with DCP2 to enhance decapping activity, decrease siRNA accumulation, and attenuate RNA silencing (Ye et al., 2015).

NSP-interacting immunological hubs connect antimicrobial immune responses through interactions between NIK1 and the flagellin receptor FLS2 and its co-receptor BAK1 (Li et al., 2019). In the absence of bacterial or viral infection, NIK1 binds FLS2 and BAK1 to prevent the initiation of autoimmune responses. Since basal levels of NIK1 determine the strength of the immune response, NIK1 can control how her FLS2-BAK1 complex develops (Li et al., 2019). Upon bacterial infection, PRR FLS2 and its co-receptor BAK1 form active immune complexes and phosphorylate the bound NIK1 at the critical Thr-474 to activate antiviral immunity. Therefore, NIK1-dependent bacterial infection can increase resistance to plant Begomovirus.

10.5 CONCLUSION

In order for viruses to successfully infect a plant systemically, their genomes must travel between and throughout the plant's cells. Therefore, the development of innovative antiviral methods depends on the discovery of host components intricate in viral mobility that interact either directly or indirectly with virus-encoded motion proteins. The nuclear export machinery may be used by NSP to speed up the movement of vDNA from the nucleus to the cytoplasm. Even though there has been significant progress in the previous 10 years in identifying NSP partners, the characterization of a putative NSP-networking immunological core is still far from comprehensive. For example, NIG is known to support intracellular trafficking of NSP-DNA complexes on the cytoplasmic side, but the transport activity and underlying mechanisms of NIG function are poorly understood (Carvalho et al., 2008a, b). The results that NIG interacts with NSP *in vivo* and *in vitro*, can redirect the NSP-DNA complex from the nucleus to the cytoplasm, and function as a proviral factor during begomovirus infection, suggest that NIG can bind to vDNA-NSP. It shows that the trade is involved (Carvalho et al., 2008a, b). These results suggest that NIG may be a susceptibility gene for begomovirus resistance engineering. The role of NIG in begomovirus infection has not been conclusively demonstrated, and no cytosolic NIG-interacting proteins have been discovered that provide intracellular pathways for the complex trafficking of vDNA NSPs to the plasmodesmata.

Studies on protein-protein interactions have also been used to describe the role of NSP in dampening the host defense. In this aspect, NSP induces the production and movement of the endogenous inhibitor of PTGS, AS2, which in turn inversely regulates viral muzzling (Ye et al., 2015). NSP also inhibits the action of the immunoreceptor-like kinases NIK1 and BAK1, which adversely affect innate immune responses (Fontes et al., 2004; Sakamoto et al., 2012). Although there is strong indication that NSP hinders phosphorylation-mediated activation of NIK1 to stop the receptor from signaling (Calil & Fontes, 2017; Gouveia et al., 2017), there is currently less support for the idea that NSP adversely regulates PTI. NSP has been shown to interact with BAK1, the co-receptor for PTI, but its ability to reduce BAK1 function has not been studied. As a result, attempts to exploit the innate immune response as a target for begomovirus resistance in plants have only just begun.

KEYWORDS

- **begomovirus**
- **DNA virus**
- **host range transmission**
- **host-virus relationship**
- **immune center**
- **NSP-interacting proteins**

REFERENCES

Bist, P., Kim, S. S., Pulloor, N. K., McCaffrey, K., Nair, S. K., Liu, Y., et al., (2017). ArfGAP domain-containing protein 2 (ADAP2) integrates upstream and downstream modules of RIG-I signaling and facilitates type I interferon production. *Mol. Cell. Biol., 37*, e00537–36.

Brustolini, O. J. B., Machado, J. P. B., Condori-Apfata, J. A., Coco, D., Deguchi, M., Loriato, V. A. P., et al., (2015). Sustained NIK-mediated antiviral signaling confers broad-spectrum tolerance to *begomoviruses* in cultivated plants. *Plant Biotechnol. J., 13*, 1300–1311. doi: 10.1111/pbi.12349.

Buchan, J. R., & Parker, R., (2009). Review eukaryotic stress granules: The ins and outs of translation. *Mol. Cell, 36*, 932–941. doi: 10.1016/j.molcel.2009.11.020.

Calil, I. P., & Fontes, E. P. B., (2017). Plant immunity against viruses: Antiviral immune receptors in focus. *Ann. Bot., 119*, 711–723. doi: 10.1093/aob/mcw200.

Calil, I. P., Quadros, I. P. S., Araújo, T. C., Duarte, C. E. M., Gouveia-Mageste, B. C., Silva, J. C. F., et al., (2018). A WW domain-containing protein forms immune nuclear bodies against *begomoviruses*. *Mol. Plant., 11*, 1449–1465. doi: 10.1016/j.molp.2018.09.009.

Carvalho, C. M., Fontenelle, M. R., Florentino, L. H., Santos, A. A., Zerbini, F. M., & Fontes, E. P. B., (2008a). A novel nucleocytoplasmic traffic GTPase identified as a functional target of the bipartite geminivirus nuclear shuttle protein. *Plant J., 55*, 869–880. doi: 10.1111/j.1365-313X.2008. 03556.x.

Carvalho, C. M., Machado, J. P. B., Zerbini, F. M., & Fontes, E. P. B., (2008b). NSP-interacting GTPase: A cytosolic protein as cofactor for nuclear shuttle proteins. *Plant Signal. Behav., 3*, 752–754. doi: 10.4161/psb.3.9.6641.

Carvalho, C. M., Santos, A. A., Pires, S. R., Rocha, C. S., Saraiva, D. I., Machado, J. P. B., et al., (2008c). Regulated nuclear trafficking of rpL10A mediated by NIK1 represents a defense strategy of plant cells against virus. *PLoS Pathog., 4*, e1000247. doi: 10.1371/journal.ppat.1000247.

Carvalho, M. F., & Lazarowitz, S. G., (2004). Interaction of the movement protein NSP and the *Arabidopsis* acetyltransferase AtNSI is necessary infection and pathogenicity. *J. Virol., 78*, 11161–11171. doi: 10.1128/JVI.78.20.11161.

Carvalho, M. F., Turgeon, R., & Lazarowitz, S. G., (2006). The geminivirus nuclear shuttle protein NSP inhibits the activity of AtNSI, a vascular-expressed *Arabidopsis* acetyltransferase regulated with the sink-to-source transition. *Plant Physiol., 140*, 1317–1330. doi: 10.1104/pp.105.075556.

Chavrier, P., & Goud, B., (1999). The role of ARF and Rab GTPases in membrane transport. *Curr. Opin. Cell Biol., 11*, 466–475. doi: 10.1016/S0955-0674(99)80067-2.

Chinchilla, D., Zipfel, C., Robatzek, S., Kemmerling, B., Nürnberger, T., Jones, J. D. G., et al., (2007). A flagellin-induced complex of the receptor FLS2 and BAK1 initiates plant defense. *Nature, 448*, 497–500. doi: 10.1038/nature05999.

Decroocq, V., Salvador, B., Sicard, O., Glasa, M., Cosson, P., Svanella-Dumas, L., et al., (2009). The determinant of potyvirus ability to overcome the RTM resistance of *Arabidopsis thaliana* maps to the N-terminal region of the coat protein. *Mol. Plant Microb. Interact., 22*, 1302–1311. doi: 10.1094/mpmi-22-10-1302.

Diamos, A. G., Crawford, J. M., & Mason, H. S., (2019). Fine-tuning expression of begomoviral movement and nuclear shuttle proteins confers cell-to-cell movement to mastreviral replicons in *Nicotiana benthamiana* leaves. *J. Gen. Virol., 100*, 1038–1051. doi: 10.1099/jgv.0.001275.

Florentino, L. H., Santos, A. A., Fontenelle, M. R., Pinheiro, G. L., Zerbini, F. M., Baracat-Pereira, M. C., et al., (2006). A PERK-like receptor kinase interacts with the geminivirus nuclear shuttle protein and potentiates viral infection. *J. Virol., 80*, 6648–6656. doi: 10.1128/jvi.00173-06.

Fontes, E. P. B., Santos, A. A., Luz, D. F., Waclawovsky, A. J., & Chory, J., (2004). The geminivirus nuclear shuttle protein is a virulence factor that suppresses transmembrane receptor kinase activity. *Genes Dev., 18*, 2545–2556. doi: 10.1101/gad.1245904.

Frischmuth, S., Wege, C., Hülser, D., & Jeske, H., (2007). The movement protein BC1 promotes redirection of the nuclear shuttle protein BV1 of *Abutilon mosaic* geminivirus to the plasma membrane in fission yeast. *Protoplasma, 230*, 117–123. doi: 10.1007/s00709-006-0223-x.

Gafni, Y., & Epel, B. L., (2002). The role of host and viral proteins in intra- and inter-cellular trafficking of geminiviruses. *Physiol. Mol. Plant Pathol., 60*, 231–241. doi: 10.1006/pmpp.2002.0402.

Garrido-Ramirez, E. R., Sudarshana, M. R., Lucas, W. J., & Gilbertson, R. L., (2000). *Bean dwarf mosaic virus* BV1 protein is a determinant of the hypersensitive response and avirulence in *Phaseolus vulgaris*. *Mol. Plant Microb. Interact., 13*, 1184–1194. doi: 10.1094/MPMI.2000.13.11.1184.

Gouveia, B. C., Calil, I. P., Machado, J. P. B., Santos, A. A., & Fontes, E. P. B., (2017). Immune receptors and co-receptors in antiviral innate immunity in plants. *Front. Microbiol., 7*, 2139. doi: 10.3389/fmicb.2016.02139.

Hanley-Bowdoin, L., Bejarano, E. R., Robertson, D., & Mansoor, S., (2013). Geminiviruses: Masters at redirecting and reprogramming plant processes. *Nat. Rev. Microbiol., 11*, 777–788. doi: 10.1038/nrmicro3117.

Hehnle, S., Wege, C., & Jeske, H., (2004). Interaction of DNA with the movement proteins of geminiviruses revisited. *J. Virol., 78*, 7698–7706. doi: 10.1128/jvi.78.14.7698-7706.2004.

Jones, J. D. G., & Dangl, J. L., (2006). The plant immune system. *Nature, 444*, 323–329. doi: 10.1038/nature05286.

Kleinow, T., Tanwir, F., Kocher, C., Krenz, B., Wege, C., & Jeske, H., (2009). Expression dynamics and ultrastructural localization of epitope-tagged *Abutilon mosaic virus* nuclear shuttle and movement proteins in *Nicotiana benthamiana* cells. *Virology, 391*, 212–220. doi: 10.1016/j.virol.2009.06.042.

Kong, J., Wei, M., Li, G., Lei, R., Qiu, Y., Wang, C., et al., (2018). The *Cucumber mosaic virus* movement protein suppresses PAMP-triggered immune responses in *Arabidopsis* and tobacco. *Biochem. Biophys. Res. Comm., 498*, 395–401. doi: 10.1016/j.bbrc.2018.01.072.

Krapp, S., Greiner, E., Amin, B., Sonnewald, U., & Krenz, B., (2017). The stress granule component G3BP is a novel interaction partner for the nuclear shuttle proteins of the nanovirus *Pea necrotic yellow dwarf virus* and geminivirus *Abutilon mosaic virus*. *Virus Res., 227*, 6–14. doi: 10.1016/j.virusres.2016.09.021.

Krichevsky, A., Kozlovsky, S. V., Gafni, Y., & Citovsky, V., (2006). Nuclear import and export of plant virus proteins and genomes. *Mol. Plant Pathol., 7*, 131–146. doi: 10.1111/j.1364-3703.2006.00321.x.

Kumar, R. V., (2019). Plant antiviral immunity against geminiviruses and viral counter-defense for survival. *Front. Microbiol., 10*, 1460. doi: 10.3389/fmicb.2019.01460.

Lazarowitz, S. G., & Beachy, R. N., (1999). Viral movement proteins as probes for intracellular and intercellular trafficking in plants. *Plant Cell, 11*, 535–548. doi: 10.1105/tpc.11.4.535.

Li, B., Ferreira, M. A., Huang, M., Camargos, L. F., Yu, X., Teixeira, R. M., et al., (2019). The receptor-like kinase NIK1 targets FLS2/BAK1 immune complex and inversely modulates antiviral and antibacterial immunity. *Nat. Commun., 10*, 4996. doi: 10.1038/s41467-019-12847-6.

Li, F., Yang, X., Bisaro, D. M., & Zhou, X., (2018). The βC1 protein of geminivirus-betasatellite complexes: A target and repressor of host defenses. *Mol. Plant, 11*, 1424–1426. doi: 10.1016/j.molp.2018.10.007.

Lloyd, R. E., (2012). How do viruses interact with stress-associated RNA granules? *PLoS Pathog., 8*, e1002741. doi: 10.1371/journal.ppat.1002741.

Machado, J. P. B., Brustolini, O. J. B., Mendes, G. C., Santos, A. A., & Fontes, E. P. B., (2015). NIK1, a host factor specialized in antiviral defense or a novel general regulator of plant immunity? *Bioessays, 37*, 1236–1242. doi: 10.1002/bies.201500066.

Machado, J. P. B., Calil, I. P., Santos, A. A., & Fontes, E. P. B., (2017). Translational control in plant antiviral immunity. *Genet. Mol. Biol., 40*, 292–304. doi: 10.1590/1678-4685-gmb-2016-0092.

Mariano, A. C., Andrade, M. O., Santos, A. A., Carolino, S. M. B., Oliveira, M. L., Baracat-Pereira, M. C., et al., (2004). Identification of a novel receptor-like protein kinase that interacts with a geminivirus nuclear shuttle protein. *Virology, 318*, 24–31. doi: 10.1016/j.virol.2003.09.038.

McGarry, R. C., Barron, Y. D., Carvalho, M. F., Hill, J. E., Gold, D., Cheung, E., et al., (2003). A novel *Arabidopsis* acetyltransferase interacts with the geminivirus movement protein NSP. *Plant Cell, 15*, 1605–1618. doi: 10.1105/tpc.012120.

McInerney, G. M., (2015). FGDF motif regulation of stress granule formation. *DNA Cell Biol., 34*, 1–4. doi: 10.1089/dna.2015.295.7.

Mukhtar, M. S., Carvunis, A. R., Dreze, M., Epple, P., Steinbrenner, J., Moore, J., et al., (2011). Independently evolved virulence effectors converge onto hubs in a plant immune system network. *Science, 333*, 596–601. doi: 10.1126/science.1203659.

Nicaise, V., & Candresse, T., (2017). *Plum pox virus* capsid protein suppresses plant pathogen-associated molecular pattern (PAMP)-triggered immunity. *Mol. Plant Pathol., 18*, 878–886. doi: 10.1111/mpp.12447.

Noueiry, A. O., Lucas, W. J., & Gilbertson, R. L., (1994). Two proteins of a plant DNA virus coordinate nuclear and plasmodesmal transport. *Cell, 76*, 925–932. doi: 10.1016/0092-8674(94)90366-2.

Panas, M. D., Schulte, T., Thaa, B., Sandalova, T., Kedersha, N., Achour, A., et al., (2015). Viral and cellular proteins containing FGDF motifs bind G3BP to block stress granule formation. *PLoS Pathog., 11*, e004659. doi: 10.1371/journal.ppat.1004659.

Pascal, E., Sanderfoot, A. A., Ward, B. M., Medville, R., Turgeon, R., & Lazarowitz, S. G., (1994). The geminivirus BR1 movement protein binds single-stranded DNA and localizes the cell nucleus. *Plant Cell, 6*, 995–1006. doi: 10.1105/tpc.6.7.995.

Rocha, C. S., Santos, A. A., Machado, J. P. B., & Fontes, E. P. B., (2008). The ribosomal protein L10/QM-like protein is a component of the NIK-mediated antiviral signaling. *Virology, 380*, 165–169. doi: 10.1016/j.virol.2008.08.005.

Rojas, M. R., Macedo, M. A., Maliano, M. R., Soto-Aguilar, M., Souza, J. O., Briddon, R. W., et al., (2018). World management of geminiviruses. *Annu. Rev. Phytopathol., 56*, 637–677. doi: 10.1146/annurev-phyto-080615-100327.

Rojas, M. R., Noueiry, A. O., Lucas, W. J., & Gilbertson, R. L., (1998). Bean dwarf mosaic geminivirus movement proteins recognize DNA in a form- and size-specific manner. *Cell, 95*, 105–113. doi: 10.1016/S0092-8674(00)81786-9.

Sabe, H., Onodera, Y., Mazaki, Y., & Hashimoto, S., (2006). ArfGAP family proteins in cell adhesion, migration, and tumor invasion. *Curr. Opin. Cell Biol., 18*, 558–564. doi: 10.1016/j.ceb.2006.08.002.

Sakamoto, T., Deguchi, M., Brustolini, O. J. B., Santos, A. A., Silva, F. F., & Fontes, E. P. B., (2012). The tomato RLK superfamily: Phylogeny and functional predictions about the role of the LRRII-RLK subfamily in antiviral defense. *BMC Plant Biol., 12*, 229. doi: 10.1186/1471-2229-12-229.

Sánchez-Velar, N., Udofia, E. B., Yu, Z., & Zapp, M. L., (2004). hRIP, a cellular cofactor for rev function, promotes release of HIV RNAs from the perinuclear region. *Genes Dev., 18*, 23–34. doi: 10.1101/gad.1149704.

Sanderfoot, A. A., & Lazarowitz, S. G., (1995). Cooperation in viral movement: The geminivirus BL1 movement protein interacts with BR1 and redirects it from the nucleus to the cell periphery. *Plant Cell, 7*, 1185–1194. doi: 10.1105/tpc.7.8.1185.

Sanderfoot, A. A., & Lazarowitz, S. G., (1996). Getting it together in plant virus movement: Cooperative interactions between bipartite geminivirus movement proteins. *Trends Cell Biol., 6*, 353–358. doi: 10.1016/0962-8924(96)10031-3.

Sanderfoot, A. A., Ingham, D. J., & Lazarowitz, S. G., (1996). A viral movement protein as a nuclear shuttle. The geminivirus *BR1* movement protein contains domains essential for interaction with BL1 and nuclear localization. *Plant Physiol., 110*, 23–33. doi: 10.1104/pp.110.1.23.

Santos, A. A., Carvalho, C. M., Florentino, L. H., Ramos, H. J. O., & Fontes, E. P. B., (2009). Conserved threonine residues within the A-loop of the receptor NIK differentially regulate the kinase function required for antiviral signaling. *PLoS One, 4*, e5781. doi: 10.1371/journal.pone.0005781.

Santos, A. A., Lopes, K. V. G., Apfata, J. A. C., & Fontes, E. P. B., (2010). NSP-interacting kinase, NIK: A transducer of plant defense signaling. *J. Exp. Bot., 61*, 3839–3845. doi: 10.1093/jxb/erq219.

Smakowska-Luzan, E., Mott, G. A., Parys, K., Stegmann, M., Howton, T. C., Layeghifard, M., et al., (2018). An extracellular network of *Arabidopsis leucine*-rich repeat receptor kinases. *Nature, 553*, 342–346. doi: 10.1038/nature25184.

Teixeira, R. M., Ferreira, M. A., Raimundo, G. A. S., Loriato, V. A. P., Reis, P. A. B., & Fontes, E. P. B., (2019). Virus perception at the cell surface: Revisiting the roles of receptor-like kinases as viral pattern recognition receptors. *Mol. Plant Pathol., 20*, 1196–1202. doi: 10.1111/mpp.12816.

Tourrière, H., Chebli, K., Zekri, L., Courselaud, B., Blanchard, J. M., Bertrand, E., et al., (2003). The RasGAP-associated endoribonuclease G3BP assembles stress granules. *J. Cell Biol., 160*, 823–831. doi: 10.1083/jcb.200212128.

Turner, C. E., West, K. A., & Brown, M. C., (2001). Paxillin-ARF GAP signaling and the cytoskeleton. *Curr. Opin. Cell Biol., 13*, 593–599. doi: 10.1016/S0955-0674(00)00256-8.

Ye, J., Yang, J., Sun, Y., Zhao, P., Gao, S., Jung, C., et al., (2015). Geminivirus activates ASYMMETRIC LEAVES 2 to accelerate cytoplasmic DCP2-mediated mRNA turnover and weakens RNA silencing in *Arabidopsis*. *PLoS Pathog., 11*, e1005196. doi: 10.1371/journal.ppat.1005196.

Zerbini, F. M., Briddon, R. W., Idris, A., Martin, D. P., Moriones, E., Navas-Castillo, J., et al., (2017). ICTV virus taxonomy profile: Geminiviridae. *J. Gen. Virol., 98*, 131–133. doi: 10.1099/jgv.0.000738.

Zhang, S. C., Wege, C., & Jeske, H., (2001). Movement proteins (BC1 andBV1) of *Abutilon mosaic* geminivirus are cotransported in and between cells of sink but not of source leaves as detected by green fluorescent protein tagging. *Virology, 290*, 249–260. doi: 10.1006/viro.2001.1185.

Zhou, Y., Rojas, M. R., Park, M. R., Seo, Y. S., Lucas, W. J., & Gilbertson, R. L., (2011). Histone H3 interacts and colocalizes with the nuclear shuttle protein and the movement protein of a geminivirus. *J. Virol., 85*, 11821–11832. doi: 10.1128/jvi.00082-11.

Zipfel, C., Kunze, G., Chinchilla, D., Caniard, A., Jones, J. D. G., Boller, T., et al., (2006). Perception of the bacterial PAMP EF-Tu by the receptor EFR restricts agrobacterium-mediated transformation. *Cell, 125*, 749–760. doi: 10.1016/j.cell.2006.03.037.

Zorzatto, C., Machado, J. P. B., Lopes, K. V. G., Nascimento, K. J. T., Pereira, W. A., Brustolini, O. J. B., et al., (2015). NIK1-mediated translation suppression functions as a plant antiviral immunity mechanism. *Nature, 520*, 679–682. doi: 10.1038/nature14171.

Zvereva, A. S., Golyaev, V., Turco, S., Gubaeva, E. G., Rajeswaran, R., Schepetilnikov, M. V., et al., (2016). Viral protein suppresses oxidative burst and salicylic acid-dependent-autophagy and facilitates bacterial growth on virus-infected plants. *New Phytol., 211*, 1020–1034. doi: 10.1111/nph.13967

CHAPTER 11

Plant Viruses: An Inquisitive Journey from Old to New World

SHAILJA CHAUHAN,[1] APARNA CHODON,[1] P. LEKSHMIPRIYA,[1] GOPAL PANDI,[1] and JEBASINGH TENNYSON[2]

[1]*Department of Plant Biotechnology, School of Biotechnology, Madurai Kamaraj University, Madurai, Tamil Nadu, India*

[2]*Department of Plant Sciences, School of Biological Sciences, Madurai Kamaraj University, Madurai, Tamil Nadu, India*

ABSTRACT

Plant viruses present a significant challenge to global crop production. In recent decades, there have been outbreaks caused by new viruses that have crossed over from reservoir species or by new variants of existing viruses with new pathogenic properties. The study of the origin and evolution of plant viruses has been both fascinating and challenging. However, the lack of a fossil record and the high mutation rate, estimated to be 10^{-2} to 10^{-4} changes per nucleotide per year, make it difficult to gather conclusive evidence for research. Plant virus evolution is driven by mutation, recombination, and reassortment, with gene duplication and imprinting also contributing to divergence. When a virus infects its host, it undergoes quasispecies development, leading to the emergence of new strains of the same virus. Plant RNA viruses have replication mechanisms that are more prone to errors compared to DNA viruses, resulting in a higher likelihood of mutations that produce quasispecies. Virus emergence is typically associated with genetic variability in the virus, population increases, changes in agricultural practices, and the distribution of insect vectors. These mechanisms vary depending on the specific virus and the environment. In order to unravel the evolutionary routes of viruses and effectively control plant viruses, it is essential to

New and Emerging Plant Viruses: The Threat to Food Security.
Alireza Golnaraghi and Rajarshi Kumar Gaur (Eds.)

comprehend the mechanisms of virus evolution and the development of new strains. Genetic diversity occurs not only through recombination between genomic components, but also through the exchange of genomic components, and even by acquiring DNA sequences from viruses of other genera. The increasing complexity of the inter-dependence between the viruses and their satellites indicates that close interactions have developed over time between these different molecules, thereby impacting their co-evolution. This chapter provides complete details about the evolution of various groups of plant viruses, along with satellite entities.

11.1 INTRODUCTION

Viruses, the most abundant life forms with the most diverse gene repertoire, have played a key role in driving host cellular evolution. They are widely distributed everywhere, in every living being. The origin of plant viruses is mysterious, and several theories have been proposed regarding their origin. The resemblances between some plant and animal viruses point to a common lineage, based on genome organization and amino acid sequences. Plant viruses take advantage of several mechanisms to create genetic diversity between and within species. Frequent recombination and reassortment act as evolutionary driving forces, as do other mechanisms such as overprinting and gene duplication. The most important practical effect of virus evolution is its ability to expand host range and niches, resulting in the emergence of new viruses as well as epidemics. Emerging viruses can be new, that is, previously unknown, or frequently known viruses that have become more evident due to changes in the habitat or the generation of a new variant, allowing the virus to expand into new niches. This chapter highlights different theories of plant virus origin, various evolution mechanisms, and the evolution of major groups of plant RNA and DNA viruses.

11.2 VIRUSES EMERGENCE AND EVOLUTION

Viruses are always regarded as perplexing organisms because their entities appear to exist somewhere between the inert and living worlds (Benson, Bamford, Bamford, & Burnett, 2004). Viruses do not leave a physical 'fossil record;' they survive for a short time and then decay quickly, leaving no sign of their existence. However, many viral genomes can integrate into cellular genomes. Paleobiology, or the genomic "fossil record," aids in understanding the long-term evolutionary history of virus-host interactions (Katzourakis, 2013).

Viruses have played a significant role in evolution by exerting pressure on their hosts and altering the entire environment (Rohwer & Thurber, 2009).

They have influenced DNA evolution and DNA replication mechanisms (Forterre, 2002), cellular envelopes (Jalasvuori & Bamford, 2008), pathogenicity (Brüssow, 2007), alternative genetic codes (Shackelton & Holmes, 2008), and the evolution of all three domains of life (Shackelton & Holmes, 2008).

Understanding the evolution of three types of viral genes is vital: (i) genes that have biological homologs; (ii) virus-specific genes with an open reading frame (ORF); and (iii) viral signature genes showing homology with distant relatives. Homologs exist in cellular organisms (Koonin, Senkevich, & Dolja, 2006). Various theories have been proposed for the origin of viruses, but none of them are complete and all have gaps that are difficult to explain. Various hypotheses on virus origin are highlighted in subsections (Figure 11.1).

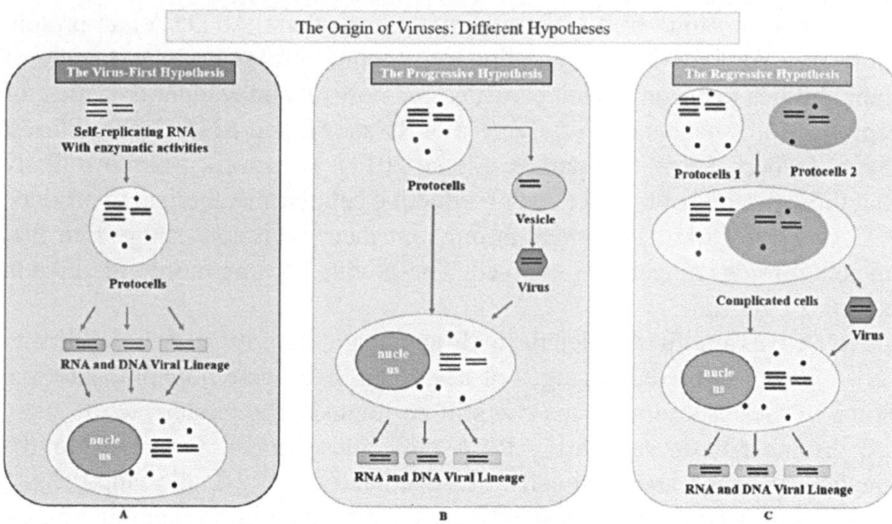

FIGURE 11.1 Three main hypotheses on the origin of viruses are depicted. (A) The virus-first hypothesis suggests that at the beginning of life, there were no cellular forms, only the first RNA molecules with enzymatic activities and the ability to self-replicate, also known as selfish genetic elements. (B) The progressive or escape hypothesis proposes that viruses are derived from cellular RNA or DNA fragments, such as plasmids, retroviruses, and transposons. It suggests that a vesicle, a smaller cell-like entity, engulfed self-replicating RNA and a coat during asymmetrical cell fission. (C) According to the regressive hypothesis, primordial cells lost their ability to encode RNA and evolved into viruses. Their cellular elements evolved over time, but they retained their genetic material and certain elements required for replication. Arrows indicate the course of evolution. Black dots represent ribosomes in protocells, while cells with nuclei represent eukaryotic cells.

Source: Adapted from Diener (1989); Flores, Gago-Zachert, Serra, Sanjuán, & Elena (2014); Claverie (2006); Poole, Jeffares, & Penny (1998); Forterre (2005); and Fililée, Forterre, Sen-Lin, & Laurent (2002).

11.2.1 THE VIRUS-FIRST HYPOTHESIS

The Virus-First Hypothesis provides a whole new perspective on the origin of viruses. According to this hypothesis, viruses predate cells and have evolved from them. According to d'Herelle, viruses are the ancestors of cells (D'Herelle, 1922). Viroids from the *Avsunviroidae* family have a hammerhead ribozyme structure and can cleave RNA into monomeric forms, suggesting that they may be descendants of the oldest biomolecules on Earth (Diener, 1989; Flores, Gago-Zachert, Serra, Sanjuán, & Elena, 2014). Viroids are potential contenders in the RNA world. They possess various unique characteristics, including a small size due to error-prone replication, a high G + C concentration, and a lack of protein-coding ability, which is compatible with a ribosome-free environment, among others. As DNA and protein molecules emerged in the evolutionary timeline, protoviroids lost some of their abilities to become plant parasites and now rely on cellular enzymes for reproduction, such as RNA polymerase, RNase H, and RNA ligase (Flores, Gago-Zachert, Serra, Sanjuán, & Elena, 2014). However, it is more likely that these protoviroids depended on effective cellular metabolism to produce ATP and other ribonucleotides throughout their evolution, suggesting that ancient viroids and cells co-evolved. These units became more ordered and complex over time.

RNA was the first molecule with enzymatic activity and the ability to self-replicate in the early stages of life when there were no cellular forms. Some of these subviral forms can still be found in the modern world; they are the most basic replicating RNA molecules known. Enzymes for the creation of membranes and cell walls eventually evolved, allowing cells to form. Viruses could have evolved before bacteria, archaea, and eukaryotes (Prangishvili, Forterre, & Garrett, 2006).

11.2.1.1 LUCAS CONCEPTS OF VIRUS

At the beginning of life, when no cellular forms existed, there were RNA molecules with enzymatic activities and the ability to self-replicate. Some of these subviral forms can still be found in the modern world; they are known as viroids, which are the smallest and most basic replicating RNA molecules (ranging from 250 to 400 nucleotides) (Kovalskaya & Hammond, 2014).

Furthermore, it is critical to distinguish between viruses that evolved before ancient cells and viruses that evolved before contemporary cells, the offspring of LUCA (last universal common ancestor), while considering concepts of virus origin. The occurrence of homologous capsids and homologous packing ATPases among various viruses infecting all domains of life supports the notion of ancient virus origin. The most well-known example is capsid protein (CP), which is the only protein found in most viruses but not in cellular organisms (Forterre, 2006; Khayat et al., 2005; Bamford, Grimes, & Stuart, 2005).

The advent of these carriers to carry and protect genetic material was one of the criteria for evolution. Hence, the emergence of protein as capsid was a significant evolutionary milestone. A new classification of all living species was proposed a few years ago based on ribosome encoding organism and capsid encoding organism (Raoult & Forterre, 2008).

11.2.2 THE PROGRESSIVE OR ESCAPE HYPOTHESIS

The progressive hypothesis proposes that viruses originated from cellular RNA or DNA fragments that escaped from cells, such as plasmids and transposons. When RNA or DNA fragments acquired a protein sheath, they became self-contained organisms capable of efficiently infecting their ancestral cells (Claverie, 2006). RNA chromosomes spread randomly from cell to cell in ancient times (Poole, Jeffares, & Penny, 1998). It is possible that a vesicle, a smaller cell-like object, developed during asymmetrical cell fission, engulfing a self-replicating RNA and a coat-encoding RNA sequence. The translation machinery was not transported to the newly formed vesicle, leading to the emergence of an ancient RNA virus (Forterre, 2006). In the early stages of virus development, viruses are more likely to have been partners in mutual co-evolution rather than evolving from proto-cells (Hendrix, Lawrence, Hatfull, & Casjens, 2000).

The virus-first and escape theories are combined in this paradigm to create a more complicated explanation. On the other hand, demonstrating how nucleic acids liberated from cells began to code for coat proteins would be challenging. It is easy to picture plasmids evolving late from dsDNA viruses (rather than the other way around) and losing genes coding for coat proteins. It would be difficult to prove how viruses developed from plasmids and efficiently encode capsid (Forterre, 2005, 2006). Moreover, viruses are similar to plasmids in that they do not encode cellular homologs for proteins related to replication machinery (Lipps, Röther, Hart, & Krauss, 2003). This would point to plasmids and viruses sharing a shared evolutionary pathway. Furthermore, viruses derived from cells should show sequence homology with hosts, but bacteriophage T4 proteins show more homology to eukaryotes than to bacterial homologs (Miller et al., 2003). In general, viral proteins are devoid of their cellular homologs (Prangishvili, Forterre, & Garrett, 2006). Although the LUCA hypothesis depicted virus origin to some extent, the Escape hypothesis shows a valid argument that protein sequence homology cannot be the only criteria to explain virus origin (Forterre, 2006).

11.2.3 THE REGRESSIVE HYPOTHESIS

According to the concept of reduction or degeneracy (cell-first), viruses are believed to have originated from small primordial cells (not necessarily primitive) that gradually lost their cellular components over time. However, they retained their genetic material and certain replication-related elements. For a long time, it was assumed that there were no transitional forms between cells and viruses, as parasites from all three domains of life maintained their cellular characteristics, including ribosomes and the ability to produce ATP. This notion changed with the discovery of NCLDVs (nucleocytoplasmic large DNA viruses). Mimivirus, an NCLDV, is larger than most viruses (La Scola et al., 2003), resembling a typical brick-shaped pox virus with dimensions of 200 nm wide and 300 nm long. Mimivirus is twice the size of a pox virus, with a total diameter of 750 nm (Xiao et al., 2005). NCLDVs not only possess extensive DNA, but they also exhibit greater complexity compared to other viruses and rely less on their host for replication. They contain various viral enzymes and components that enable them to produce mRNA in the cytoplasm of their hosts.

Because NCLDV has such a complicated structure, some experts believe that viruses are descendants of more complex free-living organisms. According to this theory, free-living organisms first evolved in a symbiotic relationship. Over time, as one organism increasingly relied on other species and lost crucial genes necessary for independent life, the symbiotic relationship became parasitic. Eventually, it evolved into an absolute parasite, unable to reproduce on its own. The presence of a wide range of putative translation genes associated with Mimivirus, which could be remnants of a once comprehensive translation system, supports this idea. Interestingly, Mimivirus is not so different from parasitic bacteria like *Rickettsia prowazekii* (Raoult et al., 2004).

This is why the reduction theory is easily disproven. However, in a universe of RNA cells, it is much easier to conceive that this reduction led to the birth of a virus, as the initial cell form was much simpler compared to current ones. It is possible that an RNA cell, living as a parasitic endosymbiont within another RNA cell, lost its protein synthesis and energy production machinery, relying instead on the host (Forterre, 2005). The presence of virus signature genes could indicate that they originated from virocells, or that these sequences were acquired from long-dead cells. For instance, bacteriophage and Adenovirus replicate their DNA using a similar unusual protein-priming method and encode a distinct form of DNA polymerase from the polymerase B subfamily. It can initiate replication using such a DNA template and has no counterparts in currently existing cells (Fililée, Forterre, Sen-Lin, & Laurent, 2002). The DNA polymerase gene appears to be a viral signature gene in disguise (Koonin, Senkevich, & Dolja, 2006).

11.3 MECHANISMS OF VIRUS EVOLUTION

The status of viruses as individual living entities or obligate parasites is still a subject of open debate. Viruses maintain tight interactions with the host to complete their infectious life cycle. Considering them as arbiters of life was no longer valid when evidence for infections by other smaller viruses and satellites came to light (La Scola et al., 2008). Many common characteristics in the genome structure and function of plant and animal viruses suggest the possibility of a common origin and evolution mechanisms to some extent, allowing for a common classification. However, the differential constraints resulting from host specificity also play a crucial role in shaping the features of virus genomes. The major driving forces in plant virus evolution are essentially the same forces that facilitate virus evolution, including mutation, recombination, reassortment, etc. (Roossinck, 1997). Many plant RNA viruses have undergone modular evolution and exhibit related replication proteins and movement proteins (MPs). Overprinting to generate *de novo* genes is also a method that has favored evolution. Acquiring host genes that provide an advantage in replication and movement within the host species has also contributed to the extension of viral diversity. The diversity at the nucleotide level within a single species of virus is more extensive than almost any other form of life on Earth. This level of divergence enables viruses to adapt to various new niches. The extent of this variation also depends on the host range, where groups with a broader host range in nature have a higher mutation frequency. While phylogenetic studies in the field of animal virology are extensive, such studies on plant viruses are more limited. However, it is important to adopt the appropriate methodology for these studies, as different methods can yield different results.

11.3.1 MUTATIONS

The lack of proofreading activity of viral polymerases results in high mutation rates, particularly in RNA viruses, which can be almost 10^5 times higher than that of the host. For plant viruses, this ranges from 10^{-6} to 10^{-4} mutations per nucleotide per replication cycle, which is relatively low compared to animal viruses (Drake & Holland, 1999; Malpica et al., 2002; Roossinck, 2003; Duffy, Shackelton, & Holmes, 2008; Elena et al., 2008; Sanjuan, Agudelo-Romero, & Elena, 2009; Sanjuan, Nebot, Chirico, Mansky, & Belshaw, 2010). The importance of base-pairing before polymerization limits the error rates of cellular DNA-dependent DNA polymerases to approximately 10^{-9}. Caulimoviruses, which belong to the group of plant pararetroviruses and have DNA as their genome but replicate *via* an RNA intermediate, are found to have error rates in the range of 4–6 $\times 10^{-4}$ changes per nucleotide within a single plant (Pennington & Melcher, 1993). For ssDNA viruses, including geminiviruses, although they utilize the host replication machinery, mutation rates are often close to those of RNA viruses (Duffy & Holmes, 2008, 2009). These rates are near the error threshold of the replication cycle, beyond which the integrity of the genome is compromised (Holmes, 2003). This increases the potential of the viruses to adapt to changing environments and, in turn, facilitates global spread. From studies conducted on a few plant viruses, most of these mutations are found to be deleterious (Sanjuan, Nebot, Chirico, Mansky, & Belshaw, 2010; Carrasco, De La Iglesia, & Elena, 2007). The rate of mutation and the size of the genome are inversely proportional in almost all cases (Sanjuan, Nebot, Chirico, Mansky, & Belshaw, 2010).

Hotspots for variation have been found in some plant RNA viruses, which are speculated to be the result of hyper mutation induced by polymerases, similar to the observations in the case of human immunodeficiency virus reverse transcriptase (Ji, Hoffmann, & Loeb, 1994). Polymerase pausing has been considered as a possible reason for the presence of a strong secondary structure upstream of the hyper variable region of CMV satellite RNA (Palukaitis & Roossinck, 1995).

Apart from errors arising from misincorporation during replication, RNA-dependent RNA polymerases (RdRps) undergo replication slippage, resulting in short repeats that lead to variations in the genome. Although there is no direct evidence for the existence of such short repeats in some genes of Potyviruses, it suggests the same (Hancock, Chaleeprom, Dale, & Gibbs, 1995). RNA editing is another mechanism for inducing mutations. While this mechanism and all appropriate enzymes are present in plant viruses, there is no report yet describing RNA editing in plant viruses (Scott, 1995).

11.3.2 RECOMBINATION

Since the time of discovery, recombination has been considered a key mechanism for preserving genetic variations in all populations. Recombination plays an important role in facilitating genetic exchanges and preventing the accumulation of deleterious mutations, thereby favoring the selection of improved fitness. This is explained by Muller's ratchet, which states that in the absence of recombination, asexual populations of finite size experience an accumulation of irreversible deleterious mutations, leading to a loss of fitness in a ratchet-like manner with each successive transfer (Muller, 1964; Moury, Desbiez, Jacquemond, & Lecoq, 2006). Although Muller's ratchet has not been tested in plant viruses, it is also connected to the natural evolution of plant viruses.

Recombination is observed in both DNA and RNA viruses. The rate of intra- or interspecific recombination in plant viruses under natural conditions has long been considered to be very low, until phylogenetic evidence came to light. RNA-RNA recombination is one of the major factors responsible for the emergence of novel, often vigorous viral species. Certain types of RNA recombination are known to induce non-template nucleotides. Three major types of RNA recombination generate diversity in viral populations: homologous recombination between two almost identical RNAs, either precisely or imprecisely, or non-homologous recombination between two RNAs with a shorter region of complementarity (Nagy & Bujarski, 1996; Simon & Nagy, 1996). The precision of homologous recombination depends on the sequence. Studies on defective interfering RNAs or the repair of deleted viral genes have provided evidence for the exclusive role of recombination in the evolution of RNA viruses, including luteovirus, cucumovirus, bromovirus, nepovirus, etc. The potential of homologous recombination to allow the regeneration of a functional genome from deleteriously mutated genomes generates better-fit variants.

Acquisition of host genes by horizontal gene transfer has been observed in certain groups of viruses, including *Closteroviridae,* which are known to have the longest RNA genome (Dolja, Kreuze, & Valkonen, 2006). A homolog of heat shock protein 70 (Hsp70h) is present in all members of this family. Phylogenetic evidence supports the acquisition of this gene in the early stages of evolution. In plants, Hsp70h functions as an ATP-dependent chaperone, while in the case of herpes simplex virus, it is involved in apoptosis modulation. Unlike its plant and viral counterparts, closterovirus Hsp70h is involved in virion assembly and cell-to-cell movement (Gober, Wales, & Aurelian, 2005). A cellular Maf/HAM1-like pyrophosphatase, which reduces mutagenesis by impeding non-canonical nucleoside triphosphates not found in any virus groups, has been reported in two members of the *Potyviridae* (Mbanzibwa, Tian, Mukasa, & Valkonen, 2009). Homologs of AlkB proteins have been reported in some RNA viruses, including members of *Flexiviridae, Closteroviridae*, and *Potyviridae* (Dolja, Kreuze, & Valkonen, 2006; Susaimuthu, Tzanetakis, Gergerich, & Martin, 2008). However, these proteins exhibit more similarities with their bacterial counterparts than with plants (Van et al., 2008). This study focuses on the possibility of interkingdom horizontal gene transfer, potentially originating from plant pathogenic or symbiotic bacteria. It is suggested that this transfer may have occurred horizontally between viruses through recombination. Most of these viruses are known to infect perennial and woody hosts, highlighting the importance of these proteins in maintaining genome stability during prolonged periods of infection. Additionally, the integration of non-infectious viral genome segments into the host genome has been reported in members of *Caulimoviridae* and endogenous pararetroviruses (EPRVs) (Bertsch et al., 2009; Harper, Hull, Lockhart, & Olszewski, 2002; Staginnus & Richert-Poggeler, 2006; Geering, Scharaschkin, & Teycheney, 2010). Partial sequences of certain begomoviruses have also been discovered in *Nicotiana* species (Ashby et al., 1997; Bejarano, Khashoggi, Witty, & Lichtenstein, 1996). Furthermore, related sequences of *cardamom mosaic virus* (CdMV) have been observed in cardamom plants, showing similarity to the 3′-region of NIb (RdRP) which contains non-functional motifs. This suggests the integration of viral sequences in *Potyviridae* as well (Jebasingh, Backiyarani, Manohari, & Usha, 2011).

11.3.3 REASSORTMENT

Sequence analysis and hybridization studies have revealed reassortment in some groups of plant viruses. It has been more often reported in mixed infection populations, and this is a not-so-common factor contributing to evolution. *Bean calico mosaic virus* is known to infect only the bean family, while *Melon chlorotic leaf curl virus* can infect a broader range of hosts including *Cucurbitaceae, Fabaceae,* and *Solanaceae.* However, the interaction of DNA-A or DNA-B of these virus groups results in the infection of cucurbits by *Bean calico mosaic virus*. Thus, reassortment contributes to evolution by widening the host range of the viruses (Idris, Mills-Lujan, Martin, & Brown, 2008).

11.3.4 GENE DUPLICATION

Gene duplication followed by rapid divergence is often found in organisms with larger genomes. This event facilitates the development of a novel or modified function. This has been reported in plant comoviruses, where three genes coding CPs are believed to have arisen through gene duplication (Rossman & Johnson, 1989). Another group of closteroviruses has undergone a duplication event of their coat protein gene, which has further diverged (Agranovsky, Lesemann, Maiss, Hull, & Atebekov, 1995). The encoded protein localizes in the unique tail-like region at one end of the virion.

11.3.5 HOST RANGE

It has been observed that the broader the host range of a virus group, the greater the degree of diversity within the viral population. Studies on tobamoviruses could not reveal any significant variation in the genome, while the satellite tobacco mosaic virus (TMV) showed a high degree of diversity, even though it gets replicated by the helper tobamovirus replicase (Fraile et al., 1997). This is justified by the broad host range of TMV, both naturally and experimentally, while tobamoviruses have a much narrower host range. *Cucumber mosaic cucumovirus* and its satellite RNAs possess a broader host range and thus a high degree of variation (Palukaitis & Roossinck, 1995). The higher mutation allows adaptation to more hosts. Groups like lentiviruses have a narrower range of both host and vector, and there is no evidence for recombination or reassortment in them.

11.4 EVOLUTION OF PLANT RNA VIRUSES

Ribose nucleic acids were developed to accommodate accessible peptides in the age of nucleic acid life within a niche of "supramolecular aggregates" (or

SMAs) (Woese, 2002). Alternatively, RNA molecules could have evolved independently of host proto-cells as their "parasites," inhabiting a common environment. From the nucleoprotein world, RNA viruses emerged first, followed by retroelements, and then DNA viruses (Koonin, Senkevich, & Dolja, 2006). Retroviruses and hepatitis B viruses are vestiges of the RNA world, while ancient RNA viruses, specifically (+) ss RNA, are relics of an RNA-to-DNA transition in the evolutionary history of life. This hypothesis is supported by the discovery of reverse transcriptase and the existence of tRNA-like structures (TLSs), which are involved in virus replication (the link between replication and translation) (Forterre, 2006). At the 3′-termini of the genomes of various plant positive-strand RNA viruses, transfer RNA-like structures (TLSs) that mimic tRNAs with three natural aminoacylation identities (valine, histidine, and tyrosine) have been identified (Dreher, 2009). Indeed, tRNA-like motifs may have been inherited from RNA replication signals that were adapted to assist with translation. TLSs are utilized by plant RNA viruses to regulate translation initiation and viral RNA replication (Piñeiro, & Martinez-Salas, 2012). 'Modular evolution' includes the reassortment of functional modules of coding and regulatory sequences from preexisting viral or cellular sources, possibly through RNA recombination, which has played a significant role in RNA virus evolution (Table 11.1).

11.4.1 EVOLUTION OF MAJOR GROUPS OF PLANT RNA VIRUSES

11.4.1.1 SINGLE-STRANDED (+) SENSE RNA VIRUSES

11.4.1.1.1 Bromoviridae

Bromoviridae have spherical or bacilliform virions with three (+) ssRNA segments as the genome. There are six genera reported under this group: bromovirus, cucumovirus, ilavirus, oleavirus, anulavirus, and alfamovirus. Recombination and reassortment between genome segments of different viruses upon coinfection in the same hosts have played a major role in the evolution of this group (Condoňer & Elena, 2008). Recombination studies on each of the three genomic compartments showed crossovers on RNA3 that might have driven the divergence of the prototype genera cucumovirus and bromovirus from the common ancestor. Rearrangements, deletions, and insertions observed in the 5′-UTR of *cucumber mosaic virus* (CMV) RNA3 indicate the possibility of the radiation of the three cucumovirus subgroups (Roossinck, Zhang, & Hellwald, 1999).

TABLE 11.1 Genome Details of the Major Plant Virus Families

Virus	Genetic Material	Genome Size	Mode of Translation	Proteins Encoded
Bromoviridae	Trisegmented, linear (+) sense ssRNA	8 kb	mRNA to protein	1a (Mtr, Hel), 2a (RdRP), 2b, 3a, CP
Closteroviridae	Mono-, bi-, or tripartite (+) sense ssRNA	13–19.3 kb	mRNA to protein	CP, minor CP (CPm), Mtr, Hel, RdRP, HSP70h
Luteoviridae	Single, linear (+) sense ssRNA	5.6–6 kb	mRNA to protein	CP, RdRP, Hel, VPg (viral protein genome-linked)
Potyviridae	Mono- or bipartite (+) sense ssRNA	8.2–11.5 kb	mRNA to protein	P1 (protein 1 protease), HC-Pro (helper component-protease), P3 (protein 3), 6K1 (6-kDa peptide 1), CI (cylindrical inclusion protein), 6K2 (6-kDa peptide 2), VPg, NIa-Pro (nuclear inclusion a-protease), NIb (nuclear inclusion b), CP, PIPO (pretty interesting Potyviridae ORF), PISPO (pretty interesting sweet potato potyvirus ORF).
Rhabdoviridae	Unsegmented or bisegmented (–) ssRNA	10.8–16.1 kb	Anti-genomic RNA to mRNA to protein.	L (RdRP), G (immune response), N, P (cofactor of RdRP), M
Tobamoviridae	Monopartite, linear ssRNA	6.3–6.5 kb	mRNA to protein	RdRP, CP, MP
Partitiviridae	Bisegmented dsRNA	3–4.8 kb	Monocistronic (+) sense mRNA to protein.	RdRP, CP
Reoviridae	9–12 segmented linear dsRNA	18–29 kb	Monocistronic (+) sense mRNA to protein.	RdRP, nucleotide phosphohydrolase, guanylyltransferase, transmethylase, helicase, pyrophosphatase
Geminiviridae	Mono- or bipartite, circular ssDNA	2.5–5.2 kb	dsDNA to mRNA to protein	C1 (Rep), C2 (Ren), C3, C4, V1, V2 (CP), NSP, MP
Nanoviridae	Multipartite (6 or 8), circular ssDNA	Each segment 0.9–1.1 kb	dsDNA to mRNA to protein	M-Rep, CP, Clink, MP, NSP, U1, U2, U3, U4
Caulimoviridae	Circular dsDNA with discontinuities on both strands.	7.1–9.8 kb	Pregenomic RNA to protein.	MP, CP, VAP (virion associated protein), AP (aspartic protease), RT tethered with RNase H1 enzyme.

11.4.1.1.2 *Closteroviridae*

Closteroviruses possess the largest (+) ssRNA genome with a modular organization. These modules include the core or replication module (Methyltransferase or MTR, Helicase or HEL, RNA dependent RNA polymerase or POL), which aid in RNA replication. The accessory module (papain protease or L-Pro) and the cell-to-cell movement module (p6, HSP70h, p60, CP, and modified CP) are separated from the core by an intergenic spacer. The 3′-ORF encodes p23, the RNA silencing suppressor (Dolja, Kreuze, & Valkonen, 2006). The last common ancestor of the family is proposed to have radiated 15–20 million years ago, probably in the late Jurassic period. Rearrangement and recombination of other acquired modules are speculated to have led to the divergence of the groups.

The initial members of the alphavirus superfamily probably had a gene complex encoding MTR, HEL, POL, and CP (Dolja, Karasev, & Koonin, 1994). Phylogenetic analysis reveals the descent of the monophyletic lineage from this alphavirus superfamily, which over time acquired the L-Pro gene and duplicated CP to produce p60. Further duplication events on CP led to modifications on the gene that helped the different genera of closteroviruses adapt to new insect vectors. Additionally, the genomic RNA of the probable ancestral sample virus was split, paving the way for the emergence of Crinivirus as well as the adoption of the p21 silencing suppressor into the Closterovirus ancestor. Later on, each of the groups acquired different genes like p13, p23, etc. The evolution of these members into different lineages is supposed to have been influenced by selection pressure against recombination, gene flow, amino acid changes, etc. (Martín et al., 2009).

11.4.1.1.3 *Luteoviridae*

Luteoviruses have a (+) sense ssRNA genome with two basic modules: 5′-replication proteins and 3′-proteins, including the CP. Phylogenetic studies have established gene transfer between and within these modules in different viruses. Currently, the members of this group are classified into three genera: luteovirus, polerovirus, and enamovirus. The evolutionary diversification of this family is assumed to have started 4,000 years ago, with the divergence of luteovirus and polerovirus occurring 1,500 years ago, and the emergence of distinguished members of luteovirus origin within the last 500 years. There is a high similarity in the p3 genes and ORFs 3, 4, and 5 of both luteovirus and polerovirus, while the 5′-ORFs are different. This similarity suggests the possibility of coinfection by the two parental viruses, and recombination occurring between the viruses due to strand switching at the RNA start sites during RNA replication. Further recombination between the sgRNA start site on the dianthus-like viruses and the homologous polerovirus facilitates the creation of a group with dianthovirus polymerase and polerovirus CP. It is considered that further recombination at the sgRNA site downstream of ORF 5 and between the 5′-sequence of sobemovirus and the 3′-region of luteovirus, followed by attenuation, generated the complete genome of luteovirus and poliovirus, respectively (Miller, Koev, & Mohan, 1997). Three types of recombination events are observed in luteovirus: recombination within a gene, recombination of large parts of the genome within a genus, and recombination between large parts of the genome between genera. The speciation events of this genus mostly occurred within the same host species and the same geographical area, which is called sympatric.

11.4.1.1.4 *Potyviridae*

Potyviridae are the largest known plant virus family. They infect angiosperms in all parts of the world, under all climatic conditions. They infect a wide variety of important crop plants. Unlike tobamoviruses, different species of each group infect plants of different families. The largest lineage of *Bean calico mosaic virus* members is found to infect legumes, cucurbits, aroids, orchids, etc. They have an approximately 10 kb long (+) sense ssRNA genome. They are mostly transmitted by aphids. The filamentous virions are also transmitted through seeds and vegetative propagules.

Most members possess a coat protein gene which has a variable, often repetitive region at the N-terminal. This is assumed to have arisen due to replicase slippage in a discontinuous manner (Ward, Weiller, Shukla, & Gibbs, 1995). The core and the C-terminal of the gene show no such patterns and are presumed to have evolved by undergoing homologous recombination or point mutations. Phylogenetic studies indicate that potyviruses diverged in recent times with rapid speciation in the initial stages and then produced in different lineages. However, the almost identical distance of each of these groups from the initial radiation point indicates that all these branchings happened at almost the same rate (Gibbs, Ohshima, & Gibbs, 2007). *Hubei poty-like virus* 1 (HuPLV1), an interspecies recombinant of *sugarcane mosaic virus* (SCMV) and *bean yellow mosaic virus* (BYMV) assumed to have emerged during their coinfection on *Canna* sp. Studies on recombination breakpoints and the number of recombination events revealed that CP is more likely to undergo recombination followed by the 3′- and 5′-UTRs (Sankaranarayan, Palani, & Tennyson, 2018). For *Wheat streak mosaic titrimovirus* which is speculated to have entered North America from Europe using seeds, the mean pairwise sequence difference of the coherent CP genes from isolates collected from N. America is 2.9% (Stenger, Seifers, & French, 2002; Jones, Coutts, Mackie, & Dwyer, 2005; Dwyer, Gibbs, Gibbs, & Jones, 2007). Considering the similar rate for all the potyviruses showed that they diverge 10,000 times faster than tobamoviruses. The initial radiation can be traced back almost 10,000 years.

11.4.1.2 SINGLE-STRANDED (−) SENSE RNA VIRUSES

11.4.1.2.1 Rhabdoviridae

Genus *Rhabdoviridae* is a sizable and ecologically varied group of viruses that infect both fauna and flora. Among them, there are numerous diseases important to agriculture, fisheries, and public health (Dietzgen et al., 2011). The mode of infection rhabdoviruses use is *via* insects and plants. Using high throughput host genome sequencing, scientists have determined the incorporation of rhabdovirus-like components into the genomes of several insects and plants (Ballinger, Bruenn, & Taylor, 2012; Fort et al., 2012; Chiba et al., 2011; Katzourakis & Gifford, 2010; Li et al., 2015), suggesting a strong evolutionary relationship between the virus and its host.

Lettuce necrotic yellows virus (LNYV) is a member of the *Rhabdoviridae* family and causes severe diseases in lettuce. LNYV is indigenous to Australia and New Zealand, with infrequent reports in Europe. *Hyperomyzus lactucae* transmits LNYV in a propagative way (Boakye & Randles, 1974; Stubbs & Grogan, 1963). On the other hand, *H. carduellimus* (Theob) has a limited geographic distribution, and transmission by *H. carduellimus* appears to be less common (Randles & Carver, 1971). The origins of LNYV are unknown, and it appears to be restricted to Australia and New Zealand. LNYV vector *H. lactucae* is found all over the world, but the virus distribution appears to be limited. Some scientists have suggested that LNYV did not originate in this area (Stubbs, 1963), although Randles and Carver speculated that native plant species could have been the source. In support of this hypothesis, Australian native *Actites megalocarpus, Reichardia tingatana, S. hydrophilus Boulos,* and the New Zealand native *S. kirkii* have been identified as hosts (Randles & Carver, 1971; Francki, Randles, & Dietzgen, 1989).

LNYV contains a negative-sense, single-stranded RNA genome of approximately 12.8 kb and replicates in the cytoplasm of infected cells (Dietzgen, Callaghan, Wetzel, & Dale, 2006; Wetzel, Dietzgen, & Dale, 1994). The genomic RNA is associated with the nucleocapsid protein (N), and together with copies of the phosphoprotein (P) and polymerase (L), forms the infectious nucleocapsid core. This core is surrounded by a host-derived lipid bilayer envelope containing glycosylated G protein projections. The G proteins are linked to the nucleocapsid by the matrix (M) protein and are thought to interact with a receptor in the gut of the aphid vector to facilitate cell entry. The order of genes within the LNYV genome is 3′-leader-N-P-4b-M-G-L-5′-trailer (Wetzel, Dietzgen, & Dale, 1994). The 3′-leader and 5′-trailer are complementary, and coding regions are separated by a conserved short intergenic region GNU(C/U)(N)nACU (Dietzgen, Callaghan, Wetzel, & Dale, 2006). There are few genetic diversity investigations of plant-infecting rhabdoviruses available. According to one study, the cytorhabdovirus *strawberry crinkle virus* is divided into two subgroups based on the examination of an L gene fragment (Klerks et al., 2004; Callaghan & Dietzgen, 2005). Nine LNYV isolates collected from lettuce, garlic, bristly ox tongue (*Picris echioides* L.), and sow thistle in Australia between 1985 and 2000 were found to be divided into two subgroups (I and II). Callaghan and Dietzgen reported that no subgroup I isolate had been identified since 1993 based on examining symptom expression on *N. glutinosa* and N gene sequences of LNYV from Australia.

11.4.1.3 TOBAMOVIRIDAE

Tobamoviruses are typified as the "Mother of Viruses." TMV belongs to this family. Many viruses previously considered as different strains of TMV have now been classified into distinct tobamovirus members. This group of ssRNA viruses is neither placed in the (+) sense nor the (–) sense RNA viruses. Members of this group are classified into three categories. Group 1, including TMV, originated from virion assembly from the MP gene. They mostly infect *Solanaceae* members of the asterid lineage (Lartey, Voss, & Melcher, 1996; Qiu et al., 1999; Soltis, Soltis, & Chase, 1999). Group 2 viruses have their virion assembly within the coat protein gene and thus produce particles containing coat protein mRNA. They mainly infect species of the rosid lineage. The divergence of group 1 and 2 tobamoviruses can be traced back to almost 100–115 million years ago, probably during the emergence of the core Eudicot plants (Chaw, Chang, Chen, & Li, 2004). Hence, it can be speculated that Group 1 and 2 tobamoviruses and their hosts have co-evolved. The first member of Group 3 is the *ribgrass mosaic virus* (RMV), which infects *Plantago lanceolata,* a member of the asterid lineage. Therefore, Group 3 is considered a sublineage of asterid-infecting Group 1. However, the majority of viruses classified under Group 3 are found to infect brassica species. This could be explained by the agroeconomic importance of brassica members, while most of the asterids are weed plants, resulting in unidentified viruses infecting them. Group 3 members thus have a broader host range. Phylogenetic analysis of the entire genome, partial sequences, or translated amino acids of all tobamoviruses show a conserved pattern. The only reported exception to this is the *Odontoglossum ringspot virus* (ORSV), which is found to be a recombinant carrying replicase genes similar to brassica-infecting tobamoviruses, while movement and coat proteins are probably derived from *Solanaceae*-infecting members (Lartey, Voss, & Melcher, 1996; Gibbs et al., 1999).

Solanaceae is a tropical plant family with its earliest fossil records from the Cretaceous period (D'Arcy, 1991; Symon, 1991). Currently, it is mainly distributed in the Americas, around the Himalayas in Eurasia, and Australia. Reports suggest that America has a pool of TMV-resistant genes that have likely arisen and spread through hybridization over the past tens of million years. *Nicotiana* species, *Solanum capsicum,* and *Solanum tuberosum*, native to America, show hypersensitivity towards TMV infection. They harbor a large number of virions and exhibit chlorosis and mottling. Currently, some *Nicotiana* species show no or very few symptoms as they have become long-term hosts of the virus due to exposure and selection by TMV (Gibbs, Keese, Gibbs, & Garcia-Arenal, 1999).

11.4.1.4 DOUBLE-STRANDED RNA (DSRNA) VIRUSES

11.4.1.4.1 Partitiviridae

Members of this group possess a dsRNA genome (3–5 kb) that comprises two segments, each with a single ORF coding for CP and replicase, and occasional association with satellite RNAs or defective RNAs (Chiba, Lin, Kondo, Kanematsu, & Suzuki, 2013; Luque, Mata, Suzuki, Ghabrial, & Castón, 2018). They are known to infect plants, fungi, and protozoa. The host organism is not strictly limited. For example, *Penicillium aurantiogriseum partitivirus* 1 replicates in *N. benthamiana* and *N. tabacum* as well as their fungal host (Nerva, Varese, Falk, & Turina, 2017). Their capability to infect multiple groups of hosts with the aid of two genes makes the studies on their evolution interesting. Similar to other segmented viruses, reassortment is a major driving factor in the evolution of partitiviruses. Reassortment is known to occur more frequently in these separately encapsidated segments than in other groups of viruses where all the segments are encapsidated within a single entity. CP undergoes intense selection pressure than RdRp in distinct host species. This can be correlated with the slower evolutionary rate of RdRp compared to CP (Ong et al., 2017). At the same time, the conservation rate of RdRp motifs in different host species was observed to be higher than that of CP motifs (Karel, 2019). Also, those viruses found to have higher identity among their RdRps were found to have a close relation among their CPs.

11.4.1.4.2 *Reoviridae*

Reoviruses are the most diverse family of dsRNA viruses, with a wide range of host organisms including mammals, birds, fish, insects, arachnids, marine protists, fungi, and plants (Brussaard, Noordeloos, Sandaa, Hedal, & Bratbak, 2004; Hillman, Supyani, Kondo, & Suzuki, 2004; Mertens, Attoui, Duncan, & Dermody, 2005). Their genome is composed of 9–12 segments of double-stranded linear RNA, with a strictly conserved terminal sequence at the termini of each dsRNA segment. While all other multiple-segmented dsRNA viruses show conservation in the RdRp with other monopartite dsRNA viruses, thereby suggesting a common origin (Liu et al., 2012), reoviruses rarely show such sequence similarity. Cross-family horizontal gene transfer has been reported in reoviruses infecting fungi. These horizontal gene transfer events are observed to be limited within the species of rotaviruses and phytoreoviruses (Suzuki, Gojobori, & Nakagomi, 1998; Parra, Bok, Martínez, & Gomez, 2004; Phan, Okitsu, Maneekarn, & Ushijima, 2007). Horizontal gene transfer between reoviruses and other dsRNA viruses was reported in the S7 domain-containing phytoreoviruses as well (Liu et al., 2012).

11.5 EVOLUTION OF DNA VIRUSES OF PLANTS

11.5.1 *SINGLE-STRANDED DNA (SSDNA) VIRUSES*

Viruses with single-stranded DNA (ssDNA) pose a significant hazard to all living creatures. Both plants and animals can be infected by these ssDNA viruses. Examples of ssDNA viruses include Circoviruses (Tischer, Mields,

Wolff, Vagt, & Griem, 1986; Cheung, 2006), bidensoviruses (Hayakawa et al., 2000), small circular (smac) viruses (Ng et al., 2015), Anelloviruses (Biagini, De Micco, & De Lamballerie, 2006; Blatter et al., 2018), redondoviruses (Biagini, De Micco, & De Lamballerie, 2006; Abbas et al., 2019; Abbas et al., 2019), circular replication-associated protein (Rep)-encoding (CRESS) DNA viruses (Dayaram et al., 2014; Rosario et al., 2015), and genomoviruses (Krupovic, Ghabrial, Jiang, & Varsani, 2016). These ssDNA viruses can be monopartite or bipartite, with one or two segments, respectively. Nucleic acid segments are encapsidated into a single virion (viral particle) that propagates as a whole in these viruses. Some viruses are multipartite, with two or more segmented genomes packed into distinct virions, each capable of independent propagation (Randles et al., 2000; Sicard, Michalakis, Gutiérrez, & Blanc, 2016). The International Committee on the Taxonomy of Viruses (ICTV) has classified ssDNA plant viruses into two families based on their genomic organization: (i) Geminiviridae (Zerbini et al., 2017); and (ii) Nanoviridae (Randles et al., 2000; Zerbini et al., 2017; Vetten et al., 2012).

11.5.1.1 GEMINIVIRIDAE

The oldest record of symptoms of geminivirus infection dates back to 1,257 years ago (Saunders, Bedford, Yahara, & Stanley, 2003). Studies have described the origin of geminiviruses from a phytoplasma plasmid that acquired a capsid-coding gene from a plant RNA virus, giving rise to the family of geminiviruses (Rojas, Hagen, Lucas, & Gilbertson, 2005). The hypothesis of the algal origin of geminiviral ancestors suggests that geminiviruses have been evolving for at least 450 million years. What makes this hypothesis interesting is how an extrachromosomal algal plasmid evolved into an independent virus with novel genes, including AC2, AC3, AC4, AV1, and AV2, which enable the hijacking of host systems, intra- and intercellular movement, and PTGS silencing suppression.

Geminiviruses infect dicot plants, redirect the host cell cycle and replication machinery, and are expected to have a high fidelity of replication. However, the rapidly expanding host range and considerable variation between strains within and between populations, as observed through sequence analysis, suggest a higher mutation rate. This can be attributed to the lack of post-replication repair. Geminiviral DNA is not methylated, as methylation hinders replication. As a result, the normal mismatch repair mechanism in hosts is not active during the geminiviral replication cycle. Studies on mismatch repair of heteroduplex geminivirus DNA using tobacco protoplasts have revealed a possible mechanism by which mutations are introduced and maintained in geminiviruses. U-G mismatches on DNA due to cytosine deamination are found to be repaired to C-G with more than 95% efficiency, while the T-G mismatches are not. Methylation of one strand results in the preferential removal of the mispaired residues from the methylated strand and not from the unmethylated strand, thus retaining the introduced mismatch (Inamdar et al., 1992; Brough et al., 1992). Studies on *Tomato yellow leaf curl China virus*, *East African cassava mosaic virus*, and *Maize streak virus* showed transition type point mutations at a rate of 10 substitutions per site per year. Mutations resulting from oxidative damage are also exempted from the host repair strategies since the cells in which viral DNA (vDNA) replication occurs shut off the repair machinery somehow or the vDNA is secluded in regions to which the repair machinery does not have access (Singer & Fraenkel-Conrat, 1965, 1966).

Geminiviridae are responsible for many major crop diseases. Members of this group are more prone to undergo interspecific DNA recombination. The release of infectious vDNA from recombinant plasmids, deletion of foreign sequences, the release of defective DNA molecules, reversal of mutations to wild type, etc., are generally found in geminiviruses, supporting both homologous and non-homologous recombination. Recombination is found to occur mainly in the conserved region (CR) on DNA-A, which aids in obtaining virulence for helpers and the emergence of new satellite components. An experiment on the *Sri Lankan cassava mosaic virus* showed recombination of its DNA-A CR with the DNA-B CR of the *Indian cassava mosaic virus* (ICMV), resulting in greater virulence of the viruses due to the added effect of replication and systemic movement (Saunders et al., 2002).

The presence of geminiviral DNA sequences in the nuclear genome of *Nicotiana* species suggests that vDNA has integrated into the host genome through recombination (Bejarano, Khashoggi, Witty, & Lichtenstein, 1996; Murad et al., 2004). Previous studies have shown sequence similarities between the three conserved motifs of rolling circle replication proteins of geminiviruses and plasmids of bacteria, indicating the replication of geminiviruses within *Agrobacterium tumefaciens* and suggesting that geminiviruses may have originated from bacterial replicons (Jeske, Lutgemeier, & Preiss, 2001). This suggests a possible association between geminiviruses and *A. tumefaciens* or related gram-negative bacteria in the early stages, followed by the acquisition of host sequences over time. This may have facilitated their replication within plant cells.

Pseudorecombination between genomic components can lead to the emergence of new variants of bipartite geminiviruses. An example of such rearrangement was observed in ToMoV DNA A and BDMV DNA B, where the common region on DNA B was replaced by that of DNA A. This recombination demonstrated higher pathogenicity (Hou & Gilbertson, 1996).

The evolution of geminiviruses is closely linked to the evolution of their satellite molecules. These intricate levels of interdependence between geminiviruses and their satellites develop over time. Data from various studies on satellite molecules indicate that alpha satellites originated from nanoviruses and eventually integrated into geminiviruses. Recombination of DNA-A CR with betasatellites has been demonstrated in *Ageratum yellow vein virus* (AYVV), *Tomato yellow leaf curl China virus*, and *cotton leaf curl Multan virus* (CLCuMuV) (Tao & Zhou, 2008). These recombinant satellites encode the pathogenicity determinant βC1 but lack the 220 nt long satellite conserved region (SCR). In the case of *tomato leaf curl virus* (ToLCV) from Australia, the SCR is conserved but βC1 is absent. This replacement of SCR or deletion of βC1 reflects the evolutionary relationship between geminiviruses and their satellites (Briddon & Stanley, 2006; Dry, Krake, Rigden, & Rezaian, 1997).

11.5.1.2 NANOVIRIDAE

Nanoviruses are plant viruses with very small virions that include a multipartite (6–8) circular, ssDNA genome of around 1 kb in length, as well as a few satellite molecules, each with a distinct function (Vetten et al., 2012; Briddon et al., 2018; Malathi & Dasgupta, 2019). Nanoviridae members are known for their wide host ranges and cause a wide range of symptoms (Ji, Hoffmann, & Loeb, 1994; Grigoras et al., 2014; Hull, 2014; Gaafar, Timchenko, & Ziebell, 2017; Gaafar, Cordsen, & Ziebell, 2018). Infected cells' nuclei host viral replication *via* ssDNA rolling circle amplification (RCA) (Rosario, Duffy, & Breitbart, 2012; Jeske, 2018). In a field population, Nanovirus mutation frequencies are 7.5×10^{-4} substitutions per nucleotide, and 5.07×10^{-4} substitutions per nucleotide in a laboratory-maintained population generated from it (Grigoras et al., 2010). The *Nanoviridae* family is one of the fastest-evolving ssDNA viruses infecting plants, with a molecular evolution rate almost 1,000 times faster than the next fastest-evolving ssDNA virus infecting plants (Lal et al., 2020).

Nanoviruses have icosahedral and circular geometries with a T = 1 and a diameter of 18–19 nm. Unlike geminiviruses, nanoviruses are multipartite viruses with 8–10 circular ssDNA molecules, each approximately 1 kb in size (Sano, Wada, Hashimoto, Matsumoto, & Kojima, 1998; Gronenborn, 2004). On the other hand, babuviruses have 12 DNAs of approximately 1.2–1.4 kb in size (Halbert & Baker, 2015; Gronenborn et al., 2018). Recently, additional circular ssDNA molecules (1–1.4 kb) encoding Rep protein, known as satellite molecules, have been discovered in nanoviruses, which are classified as nano-alpha satellites (Briddon et al., 2018).

Members of the *Nanoviridae* family are composed of six or eight components. These components are also independently encapsidated into different virions, each with a distinct function (Randles et al., 2000; Saunders, Bedford, Yahara, & Stanley, 2003; Harrison, 1985). DNA R encodes the master replication (M-Rep) initiator protein (Timchenko et al., 2000; Horser, Harding, & Dale, 2001), DNA M encodes the MP, DNA C encodes the cell-cycle-link (clink) protein (Aronson et al., 2000), DNA S encodes the CP (Wanitchakorn, Harding, & Dale, 1997), and DNA N encodes the nuclear shuttle protein (NSP). Despite several attempts to understand the biological roles of U1, U2, and U4, the functions of these satellite molecules in nanoviruses remain unknown.

The multipartite genome organization and the unusual association of several replication-competent components with viruses from the family Nanoviridae suggest that nanovirus evolution is complicated. More intriguingly, nanoviruses have been proposed to have switched hosts from plants to vertebrates and then given rise to circoviruses through a recombination process with a vertebrate-infecting virus (Gibbs & Weiller, 1999). Phylogenetic analysis indicates that recombination and reassortment may be the cause of BBTV evolution (Hu, Fu, Lin, Su, & Yeh, 2007). BBTV is a plant virus with a single-stranded circular DNA genome. It is the cause of banana bunchy top disease. BBTV has been considered one of the 100 "World's Worst" invaders (Lowe, Browne, Boudjelas, & De Poorter, 2000). BBTV infection is phloem-limited and causes dwarfism, leaf atrophy, chlorosis, and vein clearing. The BBTV genome is composed of six essential components known as DNA-R, -U3, -S, -M, -C, and -N (Gronenborn, 2004; Vetten et al., 2005). DNA-R encodes a replication initiation protein (Rep) (Harding, Burns, Hafner, Dietzgen, & Dale, 1993), whereas DNA-U3 is not always present, and its role in BBTV infection is currently unknown. DNA-S encodes a CP (Wanitchakorn, Harding, & Dale, 1997), DNA-M and DNA-N encode potential MPs (Wanitchakorn, Hafner, Harding, & Dale, 2000), and DNA-C encodes a protein that binds to retinoblastoma and may be involved in cell-cycle manipulation (Aronson et al., 2000; Harding, Burns, Hafner, Dietzgen, & Dale, 1993; Wanitchakorn, Hafner, Harding, & Dale, 2000).

The origin of BBTV is unknown. One hypothesis links the origin and evolution of BBTV to the origin and evolution of bananas in the South and Southeast Asian-Australasian region (Perrier et al., 2011; Stainton et al., 2012). According to the evolutionary analysis, DNA-R and -S in BBTV could be alien components resulting from an ancient co-infection and reassortment event during the formation of the current BBTV genome (Hu, Fu, Lin, Su, & Yeh, 2007).

11.5.2 DOUBLE-STRANDED DNA VIRUSES

11.5.2.1 CAULIMOVIRIDAE

CaMV has a double-stranded DNA and belongs to the family *Caulimoviridae* (King, Adams, Carstens, & Lefkowitz, 2012). The CaMV genome consists of an 8,000-nt circular dsDNA molecule with three short single-stranded regions, two on one strand and one on the other (Franck, Guilley, Jonard, Richards, & Hirth, 1980). There are a total of seven ORFs, along with large and small intergenic regions (King, Adams, Carstens, & Lefkowitz, 2012). The long region between ORF VI and ORF I contain the 35S promoter, the RNA polyA signal, and the primer binding site. Between ORFs V and VI, there is a small intergenic region that contains the promoter for 19S. Six viral gene products are encoded by the genome. The cell-to-cell MP P1, the aphid transmission factor P2, the virion-associated aphid transmission factor P3, and the coat protein precursor P4, P5 act as a precursor protein for certain nucleases, polymerases, and proteases, while P6 is a translation transactivator/viral silencing suppressor as well as a key protein in the inclusion body matrix (Shockey, Gardner, Melcher, & Essenberg, 1999). ORF VII does not encode a protein, and its role is unknown (Haas, Bureau, Geldreich, Yot, & Keller, 2002).

Caulimovirus is a pararetrovirus with a dsDNA genome that replicates through RNA intermediates (Temin, 1985). Pararetroviruses lack integrase, and unlike retroviruses, host genome integration is not mandatory for their replication. Evolutionary studies have shown that Caulimoviridae evolved independently of retrotransposons with LTRs. Viruses are occasionally integrated into the host genome, creating endogenous viral elements (EVEs), also known as 'molecular fossils,' which help us understand the course of evolution. Similarly, endogenous plant pararetroviruses (EPRVs) are viral elements integrated into the plant genome, transferring their genome vertically. Nearly all gymnosperms and ferns contain EPRVs in their genomes, indicating that these viruses are widespread and common in gymnosperms and ferns. The Caulimoviridae family evolved from gymnosperms. CD searches have shown that all EPRVs share certain protein domains with angiosperms. Caulimoviridae and gymnosperm EPRVs share common protein architectures. The lack of any integrase-like domains in CD searches, a pattern similar to Caulimoviridae in angiosperms, raises the possibility that gymnosperm EPRV replication may also not require integration (Gong & Guan-Zhu, 2018).

11.6 ORIGIN OF SATELLITE VIRUSES AND NUCLEIC ACIDS

Satellite viruses are similar to another category of viruses, except for their small size and dependence on helper viruses. They are assumed to have originated from independent viruses undergoing degenerative loss of functions. This probably happened during a mixed infection, from which they are expected to have acquired the helper virus (Francki, Zaitlin, & Palukaitis, 1986). The origin of satellite RNAs is found to be from three different lineages: virus, viroids, and recombination between helper virus and host sequences. Satellite RNAs with a viral origin depend on the helper virus replicase and replicate *via* a unit length (–) sense strand, coding for one to three proteins. In the case of satellite RNAs with viroid origin, replication occurs through a viroid-like rolling circle mechanism and does not produce functional mRNA. The third route of satellite RNA origin, through recombination between the 3′-end of the helper virus and the host sequences, is not affected by the selection pressure on the host sequence, thereby enabling the divergence of the viruses (Keese & Symons, 1987).

Satellite DNAs are mostly found in association with monopartite begomoviruses. ToLCV-sat, the first begomovirus betasatellite to be discovered, was isolated from tomato plants infected with the monopartite ToLCV (Dry, Krake, Rigden, & Rezaian, 1997). Also known as DNAβ, they play an important role in the pathogenicity of begomoviruses. Phylogenetically, betasatellites are classified as those from malvaceous hosts and non-malvaceous hosts (Nawaz-Ul-Rehman & Fauquet, 2009). The origin of betasatellites is traced back to pre-existing DNA elements (Mansoor, Zafar, & Briddon, 2006). Recombination is observed to have important roles in betasatellite evolution, as subviral DNA molecules with betasatellite origin are widely found in plants infected with betasatellite-harboring begomoviruses (Briddon & Stanley, 2006; Stanley, 2004). Higher diversity among betasatellites is observed in Southeast Asia, indicating the possibility of their origin from this region (Venkataravanappa, Reddy, Swaranalatha, Jalali, Briddon, & Reddy, 2011).

The first report of alphasatellites in begomoviruses came from a yellow vein disease in *Ageratum conyzoides* (Saunders & Stanley, 1999). Recently, alphasatellites have been found to be associated with bipartite begomoviruses in Venezuela and Brazil (Paprotka, Metzler, & Jeske, 2010; Romay, Chirinos, Geraud-Pouey, & Desbiez, 2010). Alphasatellites are known to have their origin from nanovirus components that got encapsidated by begomovirus coat protein. This fusion molecule is assumed to have formed as an adaptation strategy of nanovirus for transmission by whitefly feeding on plants co-infected with both begomovirus and nanovirus (Briddon et al., 2004).

11.7 CONCLUSION

In most cases, the strategies adopted to develop virus resistance do not succeed for a longer duration in the field. The high rates of variations in the viral populations are sufficient to compromise the selection pressure exerted by the natural environment. The currently employed control strategies provide new selection pressure for the existing viral populations, which over time are overcome by mutations, recombination, reassortment, etc. Human interference and the imbalance in the ecosystem due to various reasons also contribute to the gain of variation in viral populations. This context emphasizes the importance of understanding plant virus evolution. A better knowledge of plant viruses and their mechanisms of evolution allows the adoption of strategies that provide longer-lasting resistance. However, the generation of virus-free plants seems to be mostly impossible.

KEYWORDS

- **double-stranded DNA viruses**
- **escape hypothesis**
- **Nanoviridae**
- **plant viruses**
- **recombination**
- **RNA viruses**
- **virus evolution**

REFERENCES

Abbas, A. A., Taylor, L. J., Dothard, M. I., Leiby, J. S., Fitzgerald, A. S., Khatib, L. A., et al., (2019). *Redondoviridae*, a family of small, circular DNA viruses of the human Oro-respiratory tract associated with periodontitis and critical illness. *Cell Host Microbe., 25*, 719–729.e714.

Agranovsky, A. A., Lesemann, D. E., Maiss, E., Hull, R., & Atebekov, J. G., (1995). "Rattlesnake" structure of a filamentous plant RNA virus built of two capsid proteins. *Proc. Natl. Acad. Sci. USA., 92*, 2470–2473.

Aronson, M. N., Meyer, A. D., Györgyey, J., Katul, L., Vetten, H. J., Gronenborn, B., et al., (2000). Clink, a nanovirus-encoded protein, binds both pRB and SKP1. *J. Virol., 74*, 2967–2972.

Ashby, M. K., Warry, A., Bejarano, E. R., Khashoggi, A., Burrell, M., & Lichtenstein, C. P., (1997). Analysis of multiple copies of geminiviral DNA in the genome of four closely related *Nicotiana* species suggest a unique integration event. *Plant Mol. Biol., 35*, 313–321.

Ballinger, M. J., Bruenn, J. A., & Taylor, D. J., (2012). Phylogeny, integration, and expression of sigma virus-like genes in *Drosophila*. *Mol. Phylogenet. Evol., 65*, 251–258.

Bamford, D. H., Grimes, J. M., & Stuart, D. I., (2005). What does structure tell us about virus evolution? *Curr. Opin. Struct. Biol., 15*, 655–663.

Bejarano, E. R., Khashoggi, A., Witty, M., & Lichtenstein, C., (1996). Integration of multiple repeats of geminiviral DNA into the nuclear genome of tobacco during evolution. *Proc. Natl. Acad. Sci. U. S. A., 93*, 759–764.

Benson, S. D., Bamford, J. K., Bamford, D. H., & Burnett, R. M., (2004). Does common architecture reveal a viral lineage spanning all three domains of life? *Mol. Cell, 3*(16), 673–685.

Bertsch, C., Beuve, M., Dolja, V. V., Wirth, M., Pelsy, F., Herrbach, E., & Lemaire, O., (2009). Retention of the virus-derived sequences in the nuclear genome of grapevine as a potential pathway to virus resistance. *Biol. Direct., 4*, 21.

Biagini, P., De Micco, P., & De Lamballerie, X., (2006). Identification of a third member of the Anellovirus genus ("small anellovirus") in French blood donors. *Arch. Virol., 151*, 405–408.

Blatter, J. A., Sweet, S. C., Conrad, C., Danziger-Isakov, L. A., Faro, A., Goldfarb, S. B., et al., (2018). Anellovirus loads are associated with outcomes in pediatric lung transplantation. *Pediatr*. Transplant 22.

Boakye, D. B., & Randles, J. W., (1974). Epidemiology of *Lettuce necrotic yellows virus* in South Australia. III: Virus transmission parameters and vector feeding behavior in host and non-host plants. *Aust. J. Agric. Res., 25*, 791–802.

Briddon, R. W., & Stanley, J., (2006). Subviral agents associated with plant single-stranded DNA viruses. *Virology, 344*, 198–210.

Briddon, R. W., Bull, S. E., Amin, I., Mansoor, S., Bedford, I. D., et al., (2004). Diversity of DNA 1: A satellite-like molecule associated with monopartite *Begomovirus*-DNA βcomplexes. *Virology, 324*, 462–474.

Briddon, R. W., Martin, D. P., Roumagnac, P., Navas-Castillo, J., Fiallo-Olivé, E., Moriones, E., et al., (2018). *Alphasatellitidae*: A new family with two subfamilies for the classification of geminivirus-and nanovirus-associated alphasatellites. *Arch. Virol., 163*, 2587–2600.

Brough, C. L., Gardiner, W. E., Inamdar, N. M., Zhang, X. Y., Ehrlich, M., & Bisaro, D. M., (1992). DNA methylation inhibits propagation of *Tomato golden mosaic virus* DNA in transfected protoplasts. *Plant Mol. Biol.*, *18*, 703–712.

Brussaard, C. P., Noordeloos, A., Sandaa, R., Hedal, M., & Bratbak, G., (2004). Discovery of a dsRNA virus infecting the marine photosynthethic protist *Micromonas pusilla*. *Virology*, *319*, 280–291.

Brüssow, H., (2007). Bacteria between protists and phages: From antipredation strategies to the evolution of pathogenicity. *Mol Microbiol.*, *65*, 583–589.

Callaghan, B., & Dietzgen, R. G., (2005). Nucleocapsid gene variability reveals two subgroups of *Lettuce necrotic yellows virus*: Brief report. *Arch Virol.*, *150*, 1661–1667.

Carrasco, P., De La Iglesia, F., & Elena, S. F., (2007). Distribution of fitness and virulence effects caused by single-nucleotide substitutions in *Tobacco etch virus*. *J. Virol.*, *81*, 12979–12984.

Chaw, S. M., Chang, C. C., Chen, H. L., & Li, W. H., (2004). Dating the monocot–dicot divergence and the origin of core eudicots using whole chloroplast genomes. *J. Mol. Evol.*, *58*, 424–441.

Cheung, A. K., (2006). Rolling-circle replication of an animal circovirus genome in a theta-replicating bacterial plasmid in *Escherichia coli*. *J. Virol.*, *80*, 8686–8694.

Chiba, S., Kondo, H., Tani, A., Saisho, D., Sakamoto, W., Kanematsu, S., & Suzuki, N., (2011). Widespread endogenization of genome sequences of non-retroviral RNA viruses into plant genomes. *PLoS Pathog.*, *7*, e1002146.

Chiba, S., Lin, Y. H., Kondo, H., Kanematsu, S., & Suzuki, N., (2013). Effect of defective interfering RNA on symptom induction by, and replication of, a novel partitivirus from a phytopathogenic fungus, *Rosellinia necatrix*. *J. Virol.*, *87*, 2330–2341.

Claverie, J. M., (2006). Viruses take center stage in cellular evolution. *Genome Biol.*, *7*, 110.

Condoñer, F. M., & Elena, S. F., (2008). The promiscuous evolutionary history of the family *Bromoviridae*. *J. Gen. Virol.*, *89*, 1739–1747.

D'Arcy, W. G., (1991). The *Solanaceae* since 1976, with a review of its biogeography. In: Hawkes, J. G., Lester, R. N., Nee, M., & Estrada, N., (eds.), *Solanaceae III: Taxonomy, Chemistry, Evolution*. London: Royal Botanic Gardens and Linnaean society of London.

D'Herelle, F., (1922). *The Bacteriophage; Its Role in Immunity*. Williams and Willkins: Baltimore.

Dayaram, A., Galatowitsch, M., Harding, J. S., Argüello-Astorga, G. R., & Varsani, A., (2014). Novel circular DNA viruses identified in *Procordulia grayi* and *Xanthocnemis zealandica* larvae using metagenomic approaches. *Infect. Genet. Evol.*, *22*, 134–141.

Diener, T. O., (1989). Circular RNAs: Relics of precellular evolution? *Proc. Natl. Acad. Sci. U S A.*, *86*, 9370–9374.

Dietzgen, R. G., Calisher, C. H., Kurath, G., Kuzmin, I. V., Rodriguez, L. L., Stone, D. M., Tesh, R. B., et al., (2011). Family *Rhabdoviridae*. In: King, A. M. Q., Adams, M. J., Carstens, E. B., & Lefkowitz, E. J., (eds.), *Virus Taxonomy* (pp. 686–714). Elsevier Academic Press; Oxford. Ninth Report of the International Committee on Taxonomy of Viruses.

Dietzgen, R. G., Callaghan, B., Wetzel, T., & Dale, J. L., (2006). Completion of the genome sequence of *Lettuce necrotic yellows virus*, type species of the genus *Cytorhabdovirus*. *Virus Res.*, *118*, 16–22.

Dolja, V. V., Karasev, A. V., & Koonin, E. V., (1994). Molecular biology and evolution of closteroviruses: Sophisticated build-up of large RNA genomes. *Annu. Rev. Phytopathol.*, *32*, 261–285.

Dolja, V. V., Kreuze, J. F., & Valkonen, J. P. T., (2006). Comparative and functional genomics of closteroviruses. *Virus Res., 117*, 38–51.

Drake, J. W., & Holland, J. J., (1999). Mutation rates among RNA viruses. *Proc. Natl. Acad. Sci. U. S. A., 96*, 13910–13913.

Dreher, T. W., (2009). Role of tRNA-like structures in controlling plant virus replication. *Virus Res., 139*, 217–229.

Dry, I. B., Krake, L. R., Rigden, J. E., & Rezaian, M. A., (1997). A novel subviral agent associated with a geminivirus: The first report of a DNA satellite. *Proc. Natl. Acad. Sci. USA., 94*, 7088–7093.

Duffy, S., & Holmes, E. C., (2008). Phylogenetic evidence for rapid rates of molecular evolution in the single-stranded DNA *Begomovirus Tomato yellow leaf curl virus*. *J. Virol., 82*, 957–965.

Duffy, S., & Holmes, E. C., (2009). Validation of high rates of nucleotide substitution in geminiviruses: Phylogenetic evidence from *East African cassava mosaic viruses*. *J. Gen. Virol., 90*, 1539–1547.

Duffy, S., Shackelton, L. A., & Holmes, E. C., (2008). Rates of evolutionary change in viruses: Patterns and determinants. *Nat. Rev. Genet., 9*, 267–276.

Dwyer, G. I., Gibbs, M. J., Gibbs, A. J., & Jones, R. A. C., (2007). *Wheat streak mosaic virus* in Australia: Relationship to isolates from the Pacific Northwest of the USA and its dispersion via seed transmission. *Plant Dis., 91*, 164–170.

Elena, S. F., Agudelo-Romero, P., Carrasco, P., Codoner, F. M., Martin, S., Torres-Barcelo, C., & Sanjuan, R., (2008). Experimental evolution of plant RNA viruses. *Heredity, 100*, 478–483.

Fililée, J., Forterre, P., Sen-Lin, T., & Laurent, J., (2002). Evolution of DNA polymerase families: Evidences for multiple gene exchange between cellular and viral proteins. *J. Mol. Evol., 54*, 763–773.

Flores, R., Gago-Zachert, S., Serra, P., Sanjuán, R., & Elena, S. F., (2014). Viroids: Survivors from the RNA world? *Annu. Rev. Microbiol., 68*, 395–414.

Fort, P., Albertini, A., Van-Hua, A., Berthomieu, A., Roche, S., Delsuc, F., Pasteur, N., et al., (2012). Fossil rhabdoviral sequences integrated into arthropod genomes: Ontogeny, evolution, and potential functionality. *Mol Biol Evol., 29*, 381–390.

Forterre, P., (2002). The origin of DNA genomes and DNA replication proteins. *Curr. Opin. Microbiol., 5*, 525–532.

Forterre, P., (2005). The two ages of the RNA world, and the transition to the DNA world: A story of viruses and cells. *Biochimie., 87*, 793–803.

Forterre, P., (2006). The origin of viruses and their possible roles in major evolutionary transitions. *Virus Res., 117*, 5–16.

Fraile, A., Escriu, F., Aranda, M. A., Malpica, J. M., Gibbs, A. J., & García-Arenal, F., (1997). A century of *Tobamovirus* evolution in an Australian population of *Nicotiana glauca*. *J. Virol., 71*, 8316–8320.

Franck, A., Guilley, H., Jonard, G., Richards, K., & Hirth, L., (1980). Nucleotide sequence of *Cauliflower mosaic virus* DNA. *Cell, 21*, 285–294.

Francki, R. I. B., Randles, J. W., & Dietzgen, R. G., (1989). *Lettuce necrotic yellows virus. Description of Plant Viruses.* No 343. Association of Applied Biologists, Wellesbourne.

Francki, R. I. B., Zaitlin, M., & Palukaitis, P., (1986). *In vivo* encapsulation of potato spindle tuber viroid by *Velvet tobacco mottle virus* particles. *Virology, 155*, 469–473.

Gaafar, Y., Cordsen, N. G., & Ziebell, H., (2018). Molecular characterization of the first occurrence of *Pea necrotic yellow dwarf virus* in Denmark. *New Dis. Rep.*, *37*, 16–16.

Gaafar, Y., Timchenko, T., & Ziebell, H., (2017). First report of *Pea necrotic yellow dwarf virus* in the Netherlands. *New Dis. Rep.*, *35*, 23.

Geering, A. D. W., Scharaschkin, T., & Teycheney, P. Y., (2010). The classification and nomenclature of endogenous viruses in the family *Caulimoviridae*. *Arch. Virol.*, *155*, 123–131.

Gibbs, A. J., Keese, P. L., Gibbs, M. J., & Garcia-Arenal, F., (1999). Plant virus evolution; past, present, and future. In: Domingo, E., Webster, R., & Holland, J., (eds.), *Origin and Evolution of Viruses* (pp. 263–285). New York: Academic Press.

Gibbs, M. J., & Weiller, G. F., (1999). Evidence that a plant virus switched hosts to infect a vertebrate and then recombined with a vertebrate-infecting virus. *Proc. Natl. Acad. Sci. USA*, *96*, 8022–8027.

Gibbs, M. J., Ohshima, K., & Gibbs, A. J., (2007). *Potyvirus: A Genus for the Holocene*. In preparation.

Gober, M. D., Wales, S. Q., & Aurelian, L., (2005). Herpes simplex virus type 2 encodes a heat shock protein homologue with apoptosis regulatory functions. *Front. Biosci.*, *10*, 2788–2803.

Gong, Z., & Guan-Zhu, H., (2018). Euphyllophyte paleoviruses illuminate hidden diversity and macroevolutionary mode of *Caulimoviridae*. *Journal of Virology*, *92*(10), e02043–17.

Grigoras, I., Del, C. G. A. I., Martin, D. P., Varsani, A., Romero, J., Mammadov, A. C., et al., (2014). Genome diversity and evidence of recombination and reassortment in nanoviruses from Europe. *J. Gen. Virol.*, *95*, 1178–1191.

Grigoras, I., Timchenko, T., Grande-Pérez, A., Katul, L., Vetten, H. J., & Gronenborn, B., (2010). High variability and rapid evolution of a nanovirus. *Journal of Virology*, *84*(18), 9105–9117.

Gronenborn, B., (2004). Nanoviruses: Genome organization and protein function. *Vet. Microbiol.*, *98*, 103–109.

Gronenborn, B., Randles, J. W., Knierim, D., Barrière, Q., Vetten, H. J., Warthmann, N., et al., (2018). Analysis of DNAs associated with coconut foliar decay disease implicates a unique single-stranded DNA virus representing a new taxon. *Sci. Rep.*, *8*, 5698.

Haas, M., Bureau, M., Geldreich, A., Yot, P., & Keller, M., (2002). *Cauliflower mosaic virus*: Still in the news. *Mol. Plant Pathol.*, *3*, 419–429.

Halbert, S. E., & Baker, C. A., (2015). *Banana bunchy top virus* and its vector *Pentalonia nigronervosa* (*Hemiptera: Aphididae*). *Pathol. Circ.*, *47*, 8.

Hancock, J. M., Chaleeprom, W., Dale, J., & Gibbs, A., (1995). Replication slippage in the evolution of potyviruses. *J. Gen. Virol.*, *76*, 3229–3232.

Harding, R. M., Burns, T. M., Hafner, G., Dietzgen, R. G., & Dale, J. L., (1993). Nucleotide sequence of one component of the banana bunchy top virus genome contains a putative replicase gene. *J. Gen. Virol.*, *74*, 323–328.

Harper, G., Hull, R., Lockhart, B., & Olszewski, N. E., (2002). Viral sequences integrated into plant genomes. *Annu. Rev. Phytopathol.*, *40*, 119–136.

Harrison, B. D., (1985). Advances in geminivirus research. *Annu. Rev. Phytopathol.*, *23*, 55–82.

Hayakawa, T., Kojima, K., Nonaka, K., Nakagaki, M., Sahara, K., Asano, S. I., et al., (2000). Analysis of proteins encoded in the bipartite genome of a new type of parvo-like virus isolated from silkworm -structural protein with DNA polymerase motif. *Virus Res., 66*, 101–108.

Hendrix, R. W., Lawrence, J. G., Hatfull, G. F., & Casjens, S., (2000). The origins and ongoing evolution of viruses. *Trends Microbiol., 8*, 504–508.

Hillman, B. I., Supyani, S., Kondo, H., & Suzuki, N., (2004). A reovirus of the fungus *Cryphonectria parasitica* that is infectious as particles and related to the coltivirus genus of animal pathogens. *J. Virol., 78*, 892–898.

Holmes, E. C., (2003). Error thresholds and the constraints to RNA virus evolution. *Trends Microbiol., 11*, 543–546.

Horser, C. L., Harding, R. M., & Dale, J. L., (2001). *Banana bunchy top nanovirus* DNA-1 encodes the 'master' replication initiation protein. *J. Gen. Virol., 82*, 459–464.

Hou, Y. M., & Gilbertson, R. L., (1996). Increased pathogenicity in a pseudorecombinant bipartite geminivirus correlates with intermolecular recombination. *J. Virol., 70*, 5430–5436.

Hu, J. M., Fu, H. C., Lin, C. H., Su, H. J., & Yeh, H. H., (2007). Reassortment and concerted evolution in *Banana bunchy top virus* genomes. *J. Virol., 81*, 1746–1761.

Hull, R., (2014). *Plant Virology*. Cambridge, MA: Academic press.

Idris, A. M., Mills-Lujan, K., Martin, K., & Brown, J. K., (2008). Melon chlorotic leaf curl virus: Characterization and differential reassortment with closest relatives reveal adaptive virulence in the squash leaf curl virus clade and host shifting by the host-restricted bean calico mosaic virus. *J. Virol., 82*, 1959–1967.

Inamdar, N. M., Zhang, X. Y., Brough, C. L., Gardiner, W. E., Bisaro, D. M., & Ehrlich, M., (1992). Transfection of heteroduplexes containing uracil-guanine or thymine guanine mispairs into plant cells. *Plant Mol. Biol., 20*, 123–131.

Jalasvuori, M., & Bamford, J. K., (2008). Structural co-evolution of viruses and cells in the primordial world. *Orig. Life Evol Biosph., 38*, 165–181.

Jebasingh, T., Backiyarani, S., Manohari, C., & Usha, R., (2011). Detection of *Cardamom mosaic virus*-related sequences in plant genomes. *Indian Journal of Biotechnology, 10*, 369–371.

Jeske, H., (2018). Barcoding of plant viruses with circular single-stranded DNA based on rolling circle amplification. *Viruses, 10*, 469.

Jeske, H., Lutgemeier, M., & Preiss, W., (2001). DNA forms indicate rolling circle and recombination-dependent replication of *Abutilon mosaic virus*. *EMBO J., 20*, 6158–6167.

Ji, J., Hoffmann, J. S., & Loeb, L., (1994). Mutagenicity and pausing of HIV reverse transcriptase during HIV plus strand DNA synthesis. *Nucleic Acids Res., 22*, 47–52.

Jones, R. A. C., Coutts, B. A., Mackie, A. E., & Dwyer, G. I., (2005). Seed transmission of *Wheat streak mosaic virus* shown unequivocally in wheat. *Plant Dis., 89*, 1048–1050.

Karel, P., (2019). Evolutionary forces at work in partitiviruses. *Virus Genes., 55*, 563–573.

Katzourakis, A., & Gifford, R. J., (2010). Endogenous viral elements in animal genomes. *PLoS Genet., 6*, e1001191.

Katzourakis, A., (2013). Paleovirology: Inferring viral evolution from host genome sequence data. *Philos. Trans. R Soc. Lond. B Biol. Sci., 368*, 20120493.

Keese, P., & Symons, R. H., (1987). The structure of viroids and virusoids. In: Semancik, J. S., (ed.), *Viroids and Viroid-Like Pathogens* (pp. 1–47). CRC Press, Boca Raton, FL.

Khayat, R., Tang, L., Larson, E. T., Lawrence, C. M., Young, M., & Johnson, J. E., (2005). Structure of an archaeal virus capsid protein reveals a common ancestry to eukaryotic and bacterial viruses. *Proc Natl Acad Sci U S A., 102*, 18944–18949.

King, A. M. Q., Adams, M. J., Carstens, E. B., & Lefkowitz, E. J., (2012). *Virus Taxonomy: Classification and Nomenclature of Viruses: Ninth Report of the International Committee on Taxonomy of Viruses*. Elsevier Academic Press.

Klerks, M. M., Lindner, J. L., Vasˇkova, D., Sˇpak, J., Thompson, J. R., Jelkmann, W., & Schoen, C. D., (2004). Detection and tentative grouping of *Strawberry crinkle virus* isolates. *Eur. J. Plant Pathol.*, *110*, 45–52.

Koonin, E. V., Senkevich, T. G., & Dolja, V. V., (2006). The ancient virus world and evolution of cells. *Biol. Direct.*, *1*, 29.

Kovalskaya, N., & Hammond, R. W., (2014). Molecular biology of viroid-host interactions and disease control strategies. *Plant Sci.*, *228*, 48–60.

Krupovic, M., Ghabrial, S. A., Jiang, D., & Varsani, A., (2016). Genomoviridae: A new family of widespread single-stranded DNA viruses. *Arch. Virol.*, *161*, 2633–2643.

La Scola, B., Audic, S., Robert, C., Jungang, L., De Lamballerie, X., & Drancourt, M., (2003). A giant virus in amoebae. *Science*, *299*, 2033.

La Scola, B., Desnues, C., Pagnier, I., Robert, C., Barrassi, L., Fournous, G., Merchat, M., et al., (2008). The virophage as a unique parasite of a giant *Mimivirus*. *Nature*, *455*, 100–104.

Lal, A., et al., (2020). Nanovirus disease complexes: An emerging threat in the modern era. *Frontiers in Plant Science*, *11*, 558403.

Lartey, R. T., Voss, T. C., & Melcher, U., (1996). Tobamovirus evolution: Gene overlaps, recombination, and taxonomic implications. *Mol. Biol. Evol.*, *13*, 1327–1338.

Li, C. X., Shi, M., Tian, J. H., Lin, X. D., Kang, Y. J., Chen, L. J., Qin, X. C., et al., (2015). Unprecedented genomic diversity of RNA viruses in arthropods reveals the ancestry of negative sense RNA viruses. *eLife.*, *4*, e05378.

Lipps, G., Röther, S., Hart, C., & Krauss, G., (2003). A novel type of replicative enzyme harboring ATPase, primase and DNA polymerase activity. *EMBO J.*, *22*, 2516–2525.

Liu, H., Fu, Y., Xie, J., Cheng, J., Ghabrial, S. A., et al., (2012). Evolutionary genomics of mycovirus-related dsRNA viruses reveals cross-family horizontal gene transfer and evolution of diverse viral lineages. *BMC Evol. Biol.*, *12*, 91.

Lowe, S., Browne, M., Boudjelas, S., & De Poorter, M., (2000). *100 of the World's Worst Invasive Alien Species a Selection from the Global Invasive Species Database*. Published by The Invasive Species Specialist Group (ISSG) a specialist group of the Species Survival Commission (SSC) of the World Conservation Union (IUCN). www.issg.org/booklet.pdf (accessed on 02 March 2024).

Luque, D., Mata, C. P., Suzuki, N., Ghabrial, S. A., & Castón, J. R., (2018). Capsid structure of dsRNA fungal viruses. *Viruses*, *10*(9), 481.

Malathi, V. G., & Dasgupta, I., (2019). Insights into the world of ssDNA viruses. *Virus Disease*, *30*, 1, 2.

Malpica, J. M., Fraile, A., Moreno, I., Obies, C. I., Drake, J. W., & Garcia-Arenal, F., (2002). The rate and character of spontaneous mutation in an RNA virus. *Genetics*, *162*, 1505–1511.

Mandal, B., (2010). Advances in small isometric multicomponent ssDNA viruses infecting plants. *Indian J. Virol.*, *21*, 18–30.

Mansoor, S., Zafar, Y., & Briddon, R. W., (2006). Geminivirus disease complexes: The threat is spreading. *Trends Plant Sci.*, *11*, 209–212.

Martín, S., Sambade, A., Rubio, L., Vives, M. C., Moya, P., et al., (2009). Contribution of recombination and selection to molecular evolution of *Citrus tristeza virus*. *J. Gen. Virol.*, *90*, 1527–1538.

Mbanzibwa, D. R., Tian, Y., Mukasa, S. B., & Valkonen, J. P. T., (2009). Cassava brown streak virus (*Potyviridae*) encodes a putative Maf/HAM1pyrophosphatase implicated in reduction of mutations and a P1 proteinase that suppresses RNA silencing but contains no HC-pro. *J. Virol., 83*, 6934–6940.

Mertens, P. P., Attoui, H., Duncan, R., & Dermody, T., (2005). Reoviridae. In: Fauquet, C. M., Mayo, M. A., Maniloff, J., Desselberger, U., & Ball, L. A., (eds.), *Virus Taxonomy: Eighth Report of the International Committee on taxonomy of Viruses* (pp. 447–454). Elsevier/Academic Press.

Miller, E. S., Kutter, E., Mosig, G., Arisaka, F., Kunisawa, T., & Rüger, W., (2003). Bacteriophage T4 genome. *Microbiol. Mol. Biol. Rev.*, *67*, 86–156.

Miller, W. A., Koev, G., & Mohan, B. R., (1997). Are there risks associated with transgenic resistance to luteoviruses? *Plant Dis*., *81*, 700–710.

Moury, B., Desbiez, C., Jacquemond, M., & Lecoq, H., (2006). Genetic diversity of plant virus populations: Towards hypothesis testing in molecular epidemiology. *Adv. Virus Res*., *67*, 49–87.

Muller, H. J., (1964). The relation of recombination to mutational advance. *Mutat. Res*., *1*, 2–9.

Murad, L., Bielawski, J. P., Matyasek, R., Kovarık, A., Nichols, R. A., et al., (2004). The origin and evolution of geminivirus-related DNA sequences in *Nicotiana*. *Heredity*, *92*, 352–358.

Nagy, P. D., & Bujarski, J. J., (1996). Different mechanisms of homologous and nonhomologous recombination in *Brome mosaic virus*. *Semin. Virol*., *7*, 363–372.

Nawaz-Ul-Rehman, M. S., & Fauquet, C. M., (2009). Evolution of geminiviruses and their satellites. *FEBS Letters*, *583*, 1873–3468.

Nerva, L., Varese, G. C., Falk, B. W., & Turina, M., (2017). Mycoviruses of an endophytic fungus can replicate in plant cells: Evolutionary implications. *Sci Rep*., *7*(1), 1908.

Ng, T. F. F., Zhang, W., Sachsenröder, J., Kondov, N. O., Da Costa, A. C., Vega, E., et al., (2015). A diverse group of small circular ssDNA viral genomes in human and non-human primate stools. *Virus Evol*., *1*, vev017.

Ong, J. W. L., Li, Sivasithamparam, K., Dixon, K. W., Jones, M. G. K., & Wylie, S. J., (2017). The challenges of using high-throughput sequencing to track multiple bipartite mycoviruses of wild orchid-fungus partnerships over consecutive years. *Virology*, *510*, 297–304.

Palukaitis, P., & Roossinck, M. J., (1995). Variation in the hypervariable region of *Cucumber mosaic virus* satellite RNAs is affected by the helper virus and the initial sequence context. *Virology*, *206*, 765–768.

Paprotka, T., Metzler, V., & Jeske, H., (2010). The first DNA 1-like alpha satellites in association with New World *Begomoviruses* in natural infections. *Virology*, *404*, 148–157.

Parra, G. I., Bok, K., Martínez, M., & Gomez, J. A., (2004). Evidence of rotavirus intragenic recombination between two sublineages of the same genotype. *J Gen Virol*., *85*, 1713–1716.

Pennington, R. E., & Melcher, U., (1993). *In planta* deletion of DNA inserts from the large intergenic region of *Cauliflower mosaic virus* DNA. *Virology*, *192*, 188–196.

Perrier, X., De Langhe, E., Donohue, M., Lentfer, C., Vrydaghs, L., Bakry, F., Carreel, F., et al., (2011). Multidisciplinary perspectives on banana (*Musa* spp.) domestication. *Proc. Natl. Acad. Sci. USA*., *108*, 11311–11318.

Phan, T. G., Okitsu, S., Maneekarn, N., & Ushijima, H., (2007). Evidence of intragenic recombination in G1 rotavirus VP7 genes. *J. Virol*., *81*, 10188–10194.

Piñeiro, D., & Martinez-Salas, E., (2012). RNA structural elements of hepatitis C virus controlling viral RNA translation and the implications for viral pathogenesis. *Viruses, 4*, 2233–2250.

Poole, A. M., Jeffares, D. C., & Penny, D., (1998). The path from the RNA world. *J. Mol. Evol., 46*, 1–17.

Prangishvili, D., Forterre, P., & Garrett, R. A., (2006). Viruses of the *Archaea*: A unifying view. *Nature Reviews Microbiology, 4*, 837–848.

Qiu, Y. L., Lee, J., Bernasconi-Quadroni, F., Soltis, D. E., Soltis, P. S., Zanis, M., et al., (1999). The earliest angiosperms: Evidence from mitochondrial, plastid and nuclear genomes. *Nature, Lond., 402*, 404–407.

Randles, J. W., & Carver, M., (1971). Epidemiology of *Lettuce necrotic yellows virus* in South Australia. II: Distribution of virus, host plants, and vectors. *Aust. J. Agric. Res., 22*, 231–237.

Randles, J., Chu, P., Dale, J., Harding, R., Hu, J., Katul, L., et al., (2000). "Nanoviridae." In: Van, R. M. H. V., et al., (eds.), *Virus Taxonomy: Classification and Nomenclature of Viruses*. Seventh Report of the International Committee on Taxonomy of Viruses. London: Elsevier/Academic Press.

Raoult, D., & Forterre, P., (2008). Redefining viruses: Lessons from mimivirus. *Nat. Rev. Microbiol., 6*, 315–319.

Raoult, D., Audic, S., Robert, C., Abergel, C., Renesto, P., Ogata, H., et al., (2004). The 1.2-megabase genome sequence of *Mimivirus*. *Science, 306*, 1344–1350.

Rohwer, F., & Thurber, R. V., (2009). Viruses manipulate the marine environment. *Nature, 459*, 207–212.

Rojas, M. R., Hagen, C., Lucas, W. J., & Gilbertson, R. L., (2005). Exploiting chinks in the plant's armor: Evolution and emergence of geminiviruses. *Annu. Rev. Phytopathol., 43*, 361–394.

Romay, G., Chirinos, D., Geraud-Pouey, F., & Desbiez, C., (2010). Association of an atypical alphasatellite with a bipartite new world *Begomovirus*. *Arch. Virol., 155*, 1843–1847.

Roossinck, M. J., (1997). Mechanisms of plant virus evolution. *Annu Rev Phytpathol., 35*, 191–209.

Roossinck, M. J., Zhang, L., & Hellwald, K. H., (1999). Rearrangements in the 5′ non translated region and phylogenetic analysis of *Cucumber mosaic virus* RNA 3 indicate radial evolution of three subgroups. *J. Virol., 73*, 6752–6758.

Roossinck, M., (2003). Plant RNA virus evolution. *Curr. Opin. Microbiol., 6*, 406–409.

Rosario, K., Duffy, S., & Breitbart, M., (2012). A field guide to eukaryotic circular single-stranded DNA viruses: Insights gained from metagenomics. *Arch. Virol., 157*, 1851–1871.

Rosario, K., Schenck, R. O., Harbeitner, R. C., Lawler, S. N., & Breitbart, M., (2015). Novel circular single-stranded DNA viruses identified in marine invertebrates reveal high sequence diversity and consistent predicted intrinsic disorder patterns within putative structural proteins. *Front. Microbiol., 6*, 696.

Rossman, M. G., & Johnson, J. E., (1989). Icosahedral RNA virus structure. *Annu. Rev. Biochem., 58*, 533–573.

Sanjuan, R., Agudelo-Romero, P., & Elena, S. F., (2009). Upper-limit mutation rate estimation for a plant RNA virus. *Biol. Lett., 5*, 394–396.

Sanjuan, R., Nebot, M. R., Chirico, N., Mansky, L. M., & Belshaw, R., (2010). Viral mutation rates. *J. Virol., 84*, 9733–9738.

Sankaranarayan, R., Palani, S., & Tennyson, J., (2018). Hubei poty-like virus 1 is likely an interspecies recombinant of *Sugarcane mosaic virus* and putative *Bean yellow mosaic virus*. *Archives of Virology.* https://doi.org/10.1007/s00705-018-4047-8.

Sano, Y., Wada, M., Hashimoto, Y., Matsumoto, T., & Kojima, M., (1998). Sequences of ten circular ssDNA components associated with the milk vetch dwarf virus genome. *J. Gen. Virol.*, *79*, 3111–3118.

Saunders, K., & Stanley, J., (1999). A nanovirus-like DNA component associated with yellow vein disease of *Ageratum conyzoides*: Evidence for interfamilial recombination between plant DNA viruses. *Virology*, *264*, 142–152.

Saunders, K., Bedford, I. D., Yahara, T., & Stanley, J., (2003). Aetiology: The earliest recorded plant virus disease. *Nature*, *422*, 831.

Saunders, K., Salim, N., Mali, V. R., Malathi, V. G., Briddon, R., Markham, P. G., & Stanley, J., (2002). Characterization of *Sri Lankan cassava mosaic virus* and Indian *Cassava mosaic virus*: Evidence for acquisition of a DNA B component by a monopartite *begomovirus*. *Virology*, *293*, 63–74.

Scott, J., (1995). A place in the world for RNA editing. *Cell*, *81*, 833–836.

Shackelton, L. A., & Holmes, E. C., (2008). The role of alternative genetic codes in viral evolution and emergence. *J. Theor. Biol.*, *254*, 128–134.

Shockey, M. W., Gardner, C. O. Jr., Melcher, U., & Essenberg, R. C., (1999). Polypeptides associated with inclusion bodies from leaves of turnip infected with *Cauliflower mosaic virus*. *Virology*, *105*, 575–581.

Sicard, A., Michalakis, Y., Gutiérrez, S., & Blanc, S., (2016). The strange lifestyle of multipartite viruses. *PLoS Pathog.*, *12*, e1005819.

Simon, A. E., & Nagy, P. D., (1996). RNA recombination in turnip crinkle virus: Its role in formation of chimeric RNAs, multimers, and in 3'-end repair. *Semin. Virol.*, *7*, 373–379.

Singer, B., & Fraenkel-Conrat, H., (1965). Effects of light in the presence of iron salts on ribonucleic acid and model compounds. *Biochemistry*, *4*, 226–233.

Singer, B., & Fraenkel-Conrat, H., (1966). Dye-catalyzed photoinactivation of *Tobacco mosaic virus* ribonucleic acid. *Biochemistry*, *5*, 2446–2450.

Soltis, P. S., Soltis, D. E., & Chase, M. W., (1999). Angiosperm phylogeny inferred from multiple genes: A research tool for comparative biology. *Nature*, *402*, 402–404.

Staginnus, C., & Richert-Poggeler, K. R., (2006). Endogenous pararetroviruses: Two-faced travelers in the plant genome. *Trends Plant Sci.*, *11*, 485–491.

Stainton, D., Kraberger, S., Walters, M., Wiltshire, E. J., Rosario, K., Halafihi, M., Lolohea, S., et al., (2012). Evidence of inter-component recombination, intra-component recombination and re-assortment in *Banana bunchy top virus*. *J. Gen. Virol.*, *93*, 1103–1119.

Stanley, J., (2004). Subviral DNAs associated with geminivirus disease complexes. *Vet. Microbiol.*, *98*, 121–129.

Stenger, D. C., Seifers, D. L., & French, R., (2002). Patterns of polymorphism in *Wheat streak mosaic virus*: Sequence space explored by a clade of closely related viral genotypes rivals that between the most divergent strains. *Virology*, *3*(2), 58–70.

Stubbs, L. L., & Grogan, R. G., (1963). Necrotic yellows: A newly recognized virus disease of lettuce. *Aust. J. Agric. Res.*, *14*, 439–459.

Susaimuthu, J., Tzanetakis, I. E., Gergerich, R. C., & Martin, R. R., (2008). A member of a new genus in the potyviridae infects Rubus. *Virus Res.*, *131*, 145–151.

Suzuki, Y., Gojobori, T., & Nakagomi, O., (1998). Intragenic recombinations in rotaviruses. *FEBS Lett.*, *427*, 183–187.

Symon, D. E., (1991). Gondwanan elements of the *Solanaceae*. In: Hawkes, J. G., Lester, R. N., Nee, M., & Estrada, N., (eds.), *Solanaceae III: Taxonomy, Chemistry, Evolution*. London: Royal Botanic Gardens and Linnaean Society of London.

Tao, X., & Zhou, X., (2008). Pathogenicity of a naturally occurring recombinant DNA satellite associated with *Tomato yellow leaf curl China virus*. *J. Gen. Virol.*, *89*, 306–311.

Temin, H. M., (1985). Reverse transcription in the eukaryotic genome: Retroviruses, pararetroviruses, retrotransposons, and retrotranscripts. *Mol. Biol. Evol.*, *2*, 455–468.

Timchenko, T., Katul, L., Sano, Y., De Kouchkovsky, F., Vetten, H. J., & Gronenborn, B., (2000). The master rep concept in nanovirus replication: Identification of missing genome components and potential for natural genetic reassortment. *J. Virol.*, *274*, 189–195.

Tischer, I., Mields, W., Wolff, D., Vagt, M., & Griem, W., (1986). Studies on epidemiology and pathogenicity of porcine circovirus. *Arch. Virol.*, *91*, 271–276.

Van, D. B. E., Omelchenko, M. V., Bekkelund, A., Leihne, V., Koonin, E. V., Dolja, V. V., & Falnes, P. O., (2008). Viral AlkB proteins repair RNA damage by oxidative demethylation. *Nucleic Acids Res.*, *36*, 5451–5461.

Venkataravanappa, V., Reddy, C. N., Swaranalatha, P., Jalali, S., Briddon, R. W., & Reddy, M. K., (2011). Diversity and phylogeography of *Begomovirus*-associated βsatellites of Okra in India. *Virol. J.*, *8*, 555.

Vetten, H. J., Chu, P. W. G., Dale, J. L., Harding, R., Hu, J., Katul, L., Kojima, M., et al., (2005). *Nanoviridae*. In: Fauquet, C. M., Mayo, M. A., Maniloff, J., Desselberger, U., & Ball, L. A., (eds.), *Virus Taxonomy: Eighth Report of the International Committee on Taxonomy of Viruses* (pp. 343–352). Academic Press, San Diego/London.

Vetten, H. J., Dale, J. L., Grigoras, I., Gronenborn, B., Harding, R., Randles, J. W., et al., (2012). *Family Nanoviridae in Virus Taxonomy: Ninth Report of the International Committee on Taxonomy of Viruses*. Amsterdam: Elsevier.

Wanitchakorn, R., Hafner, G. J., Harding, R. M., & Dale, J. L., (2000). Functional analysis of proteins encoded by *Banana bunchy top virus* 822 phytopathology DNA-4 to -6. *J. Gen. Virol.*, *81*, 299–306.

Wanitchakorn, R., Harding, R., & Dale, J., (1997). *Banana bunchy top virus* DNA 3 encodes the viral coat protein. *Arch. Virol.*, *142*, 1673–1680.

Ward, C. W., Weiller, G., Shukla, D. D., & Gibbs, A. J., (1995). Molecular systematics of the potyviridae, the largest plant virus family. In: Gibbs, A. J., Calisher, C. H., & Garcia- Arena, F., (eds.), *Molecular Basis of Virus Evolution* (pp. 477–500). Cambridge: Cambridge University Press.

Wetzel, T., Dietzgen, R. G., & Dale, J. L., (1994). Genomic organization of lettuce necrotic yellows rhabdovirus. *Virology*, *200*, 401–412.

Woese, C. R., (2002). On the evolution of cells. *Proc. Natl. Acad. Sci. U S A.*, *99*, 8742–8747.

Xiao, C., Chipman, P. R., Battisti, A. J., Bowman, V. D., Renesto, P., Raoult, D., & Rossmann, M. G., (2005). Cryo-electron microscopy of the giant *Mimivirus*. *Journal of Molecular Biology*, *353*, 493–496.

Zerbini, F. M., Briddon, R. W., Idris, A., Martin, D. P., Moriones, E., Navas-Castillo, J., et al., (2017). ICTV virus taxonomy profile: *Geminiviridae*. *J. Gen. Virol.*, *98*, 131–133.

Index

H

I

P

T

U

V

W